Engineering Analysis

Interactive Methods and Programs with FORTRAN, QuickBASIC, MATLAB, and Mathematica

Y. C. Pao

CRC Press
Boca Raton London New York Washington, D.C.

Acquiring Editor: Cindy Renee Carelli
Project Editor: Albert W. Starkweather, Jr.
Cover design: Dawn Boyd

Library of Congress Cataloging-in-Publication Data

Catalog record is available from the Library of Congress

International Standard Book Number 0-8493-2016-X
Printed in the United States of America 1 2 3 4 5 6 7 8 9 0
Printed on acid-free paper

Erratum

Catalog No. 2016: *Engineering Analysis: Interactive Methods and Programs with FORTRAN, QuickBASIC, MATLAB, and Mathematica* by Y. C. Pao.

Page 185: The first paragraph should be followed by an equation:

MATLAB has a file **quad.m** which can perform Simpson's Rule. To evaluate the area of a semi-circle by application of Simpson's Rule using **quad.m**, we first prepare the integrand function as a m file as follows:

```
function y=integrnd(x)
         y=(1-(x-1).^2).^.5;
```

Files Available from CRC Press

FORTRAN, **QuickBASIC**, **MATLAB**, and Mathematic files, which contain the source and executable programs associated with this book are available from CRC Press' website — http://www.crcpress.com.

Before downloading, prepare two 3.5-inch, high-density disks — one for the files and one for a backup. Also create a temporary directory named <interactive> on your hard drive, which will expedite downloading. To download these files, type: `http://www.crcpress.com/download/2016/` When prompted, enter `2016` under name and `crcpress` under password. Then store the files in the <interactive> folder. If you encounter a problem, call 1-800-CRC-PRES (272-7737). The dowloaded files may be copied to a 3.5-inch disk. The temporary <interactive> folder then may be deleted. Don't forget to make a backup copy of your 3.5-inch disk.

There are four subdirectories **<FORTRAN>**, **<QB>**, **<mFiles>**, and **<Mathtica>** which contain the **FORTRAN** source and executable programs, **QuickBASIC** source and executable programs, **m** files of **MATLAB**, and input and output statements of for the **Mathematica** operations depicted in this textbook, respectively:

1. **<FORTRAN>** has the following files:

```
Bairstow.FOR   CharacEquationFOR   CubeSpln.FOR   DiffTabl.FOR
EditFOR.EXE    EigenVec.FOR        EigenvIt.FOR   ExactFit.FOR
FindRoot.FOR   FOR1.EXE            FOR2.EXE       FORTRAN.LIB
Gauss.FOR      GauJor.FOR          LagrangI.FOR   LeastSq1.FOR
LeastSqG.FOR   LINK.EXE            MatxInvD.FOR   NewRaphG.FOR
NuIntgra.FOR   OdeBvpFD.FOR        OdeBvpRK.FOR   ParabPDE.FOR
Relaxatn.FOR   RungeKut.FOR        Volume.FOR     WavePDE.FOR
```

EDITFOR.EXE is provided for re-editing the *.FOR source programs such as Bairstow.FOR, CubeSpln.FOR, etc. (refer to the **FORTRAN** programs index) to include supplementary subprograms describing the problem which need to be solved interactively. To re-edit, insert the 3.5-inch disk into Drive A and when the a:\ prompt shows, type cd fortran to switch to the **<FORTRAN>** subdirectory. For example, to solve a polynomial by the Bairstow's method one needs to define the polynomial, for which the roots are to be computed. To reedit Bairstow.FOR, the user enters a:\editfor Bairstow.for to add new **FORTRAN** statements or change them. Notice that both upper and lower case characters are acceptable. While creating a new version of Bairstow.FOR, the old version will be saved in Bairstow.BAK.

To create an object file, FOR1 filename such as Bairstow.FOR and FOR2 need to be implemented. A BAISTOW.OBJ will then be generated. For linking with the **FORTRAN** library functions, **FORTRAN**.LIB, one enters, for example, LINK Bairstow to create an executable file Bairstow.EXE. To *run*, the user simply types Bairstow after the prompt A:\ and then answers questions interactively.

2. **<QuickBASIC>** has the following files:

```
Select.BAS      Select.EXE
Bairstow.EXE    BRUN40.EXE          CharacEq.EXE        CubeSpln.EXE
EigenStb.EXE    EigenVec.EXE        EigenVib.EXE        EigenvIt.EXE
ExactFit.EXE    FindRoot.EXE        Gauss.EXE           LagrangI.EXE
LeastSq1.EXE    LeastSqG.EXE        MatxInvD.EXE        NuIntgra.EXE
OdeBvpFD.EXE    OdeBvpRK.EXE        ParabPDE.EXE        QB.EXE
Relaxatn.EXE    RungeKut.EXE        Volume.EXE
Bairstow.QB     CharacEq.QB         CubeSpln.QB         DiffTabl.QB
EigenStb.QB     EigenVec.QB         EigenVib.QB         EigenvIt.QB
ExactFit.QB     FindRoot.QB         GauJor.QB           Gauss.QB
LagrangI.QB     LeastSq1.QB         LeastSqG.QB         MatxAlgb.QB
MatxMtpy.QB     NewRaphG.QB         NuIntgra.QB         OdeBvpFD.QB
OdeBvpRK.QB     ParabPDE.QB         Relaxatn.QB         RungeKut.QB
Volume.QB       WavePDE.QB
```

To commence **QuickBASIC**, when a:\ is prompted on screen, the user enters QB. QB.EXE and BRUN40.EXE therefore are included in **<QB>**. The program **Select** enables user to select the available **QuickBASIC** program in this textbook. After user responds with C:\Select, the screen shows a menu as shown in Figure 1 and user then follow the screen help-messages to run a desired program.

3. **<mFiles>** is a subdirectory associated with **MATLAB** and has the following files:

```
BVPF.m        DerivatF.m    DiffTabl.m    EigenvIt.m
F.m           FindRoot.m    FP.m          Functns.m
FuncZ.m       FuncZnew.m    FunF.m        GauJor.m
integrnd.m    LagrangI.m    LeastSqG.m    NewRaphG.m
ParabPDE.m    Relaxatn.m    Volume.m      Warping.m
WavePDE.m
```

When the 3.5-inch disk containing all of these **m** files is in Drive A, any of these files can be accessed by enclosing the filename inside a pair of parentheses as illustrated in *Section 3.2* where F.m and FP.m are required for FindRoot.m and in *Section 5.2* where an integrand function **integrnd.m** is defined for numerical integration. If all files have been added into **MATLAB** library m files, then no reference to the Drive A is necessary and the pair of parentheses can also be dropped.

4. **<Mathtica>** is a subdirectory associated with **Mathematica** and has the files of:

```
Bairstow.MTK    CubeSpln.MTK    DiffTabl.MTK    EigenVec.MTK
ExactFit.MTK    FindRoot.MTK    FUNCTNS.MTK     EigenvIt.MTK
Gauss.MTK       GauJor.MTK      LagrangI.MTK    LeastSq1.MTK
LeastSqG.MTK    MatxAlgb.MTK    NewRaphG.MTK    NuIntgra.MTK
OdeBvpFD.MTK    OdeBvpRK.MTK    ParabPDE.MTK    Rexalatn.MTK
RungeKut.MTK    Volume.MTK      WavePDE.MTK
```

Any of the above programs can be executed by **Mathematica** via mouse operation. First, by clicking the **File** option and when the pull-down menu appears, select *Open* and then enter the filename such as a:\Mathtica\MatxAlgb.MTK (assuming the 3.5-inch disk containing **<Mathtica>** is in Drive A) and press the **Enter** key. When all lines of this file is displayed on screen, move cursor to any input line such as *In[1]*: A = {{1,2},{3,4}}; MatrixForm[A] and hit the **Enter** key. **Mathematica** will respond by repeating those lines for *Out[1]*. Hence, user can reproduce all of the output lines by sequentially running the input lines [1] through [9]. However, if user first run In[1] and then In[3], **Mathematica** cannot perform the addition of [A] because [B] is not defined. If after having run In[1], user selects In[5], or, In[6], **Mathematica** then has no problem of giving out results.

Program Select - Menu-Driven Selection of Interactive Computer Programs.

	Bairstow	CharacEq	CubeSpln	DiffTabl
EigenODE	EigenVec	EigenvIt	ExactFit	
FindRoot	Gauss			
		LagrangI	LeastSq1	LeastSqG
MatxInvD	NewRaphG	NuIntGra		
			OdeBvpFD	OdeBvpRK
ParabPDE	Relaxatn			
RungeKut	Volume		Wave PDE	

Press <End> key to quit.

Press ←↑↓→ keys to select a desired program. Next, press F1 key to run.

FIGURE 1. The Select screen.

Dedication

This book is dedicated to Prof. E. J. Marmo,
who offered a congenial work-environment for the author
to grow in the computer-aided engineering field.

Preface and Acknowledgments

Writing textbooks on topics in the field of *Computer Aided Engineering* (CAE) indeed has been a very satisfying experience. First, I had the pleasure of being a coauthor with Prof. Thomas C. Smith of the book *Introduction to Digital Computer Plotting* by Gordon & Breach in 1973. The book *Elements of Computer-Aided Design and Manufacturing, CAD/CAM*, was published in 1982 by John Wiley & Sons. The book *A First Course in Finite Element Analysis* published by Allyn & Bacon followed in 1986, and *Engineering Drafting and Solid Modeling with SilverScreen,* published by CRC Press, appeared in 1993.

Having taught the subjects of computer methods for engineering analysis since 1966, I finally have the courage to organize this textbook out of a large volume of classroom notes collected over the past 31 years.

The rapid growth of computer technology is difficult for any one to keep pace, and to make revision of textbooks in the CAE field. However, the computational methods developed by the pioneers, such as Euler, Gauss, Lagrange, Newton, and Runge, continue to serve us incredibly effective. These computational algorithms remain classic, only are now executed with modern computer technology.

As far as the programming languages are concerned, **FORTRAN** has been dominating the scientific fields for many decades. **BASIC** considered by many to be too plain and cumbersome while **C** is considered by others to be too sophisticated; both, however, are gaining popularity and increasingly replacing **FORTRAN** in the computational community. This is particularly true when **QuickBASIC** was introduced by Microsoft.

MATLAB and **Mathematica** developed by the MathWorks, Inc. and Wolfram Research, Inc., respectively both contain a vast collection of files (similar to **FORTRAN**'s library functions) which can perform the often-encountered computational problems. For implementation, the **MATLAB** and **Mathematica** instructions to be interactively entered through keyboard are extremely simple. And, it also provides very easy-to-use graphic output. When students find it too easy to use, they often become uninterested in learning what are the methods involved. This text is prepared with **FORTRAN**, **QuickBASIC**, **MATLAB** and **Mathematica**, and more importantly gives the algorithms involved in the methods. Ample number of sample problems are solved to demonstrate how the developed programs should be interactively applied. Furthermore, the development of the user-generated supplementary files is emphasized so that more supporting subprograms can be added to the **MATLAB** m-files and **Mathematica** toolkits. It is a text for self-study as well as for the need of general references.

Numerous friends, colleagues, and students have assisted in collecting the materials assembled herein, and they have made a great number of constructive suggestions for the betterment of this work. To them, I am most grateful. Especially, I would like to thank my long-time friends Dr. H. C. Wang, formerly with the IBM Thomas Watson Research Laboratory and now with the Industrial Research Institutes in Hsingchu, Taiwan; Dr. Erik L. Ritman of the Mayo Clinic in Rochester, MN, and Leon Hill of the Boeing Company in Seattle, WA, for their help and encouragement throughout my career in the CAE field. Profs. R. T. DeLorm, L. Kersten, C. W. Martin, R. N. McDougal, G. M. Smith, and E. J. Marmo had assisted in acquiring equipment and research funds which made my development in the CAE field possible, I extend my most sincere gratitude to these colleagues at the University of Nebraska–Lincoln. For providing constructive inputs to my published works, I should give credits to Prof. Gary L. Kinzel of the Ohio State University, Prof. Donald R. Riley of the University of Minnesota, Dr. L. C. Chang of the General Motors' EDS Division, Dr. M. Maheshiwari and Mr. Steve Zitek of the Brunswick Corp., my former graduate assistants J. Nikkola, T. A. Huang, K. A. Peterson, Dr. W. T. Kao, Dr. David S. S. Shy, C. M. Lin, R. M. Sedlacek, L. Shi, J. D. Wilson, Dr. A. J. Wang, Dave Breiner, Q. W. Dong, and Michael Newman, and former students Jeff D. Geiger, Tim Carrizales, Krishna Pendyala, S. Ravikoti, and Mark Smith. I should also express my appreciation to the readers of my other four textbooks mentioned above who have frequently contacted me and provided input regarding various topics that they would like to be considered as connected to the field of CAE and numerical problems that they would like to be solved by application of computer. Such input has proven to be invaluable to me in preparation of this text. CRC Press has been a delightful partner in publishing my previous book and again this book. The completion of this book would not be possible without the diligent effort and superb coordination of Cindy Renee Carelli, Suzanne Lassandro, and Albert Starkweather, I wish to express my deepest appreciation to them and to the other CRC editorial members. Last but not least, I thank my wife, Rosaline, for her patience and encouragement.

Y. C. Pao

Contents

1 Matrix Algebra and Solution of Matrix Equations

1.1 INTRODUCTION

Computers are best suited for repetitive calculations and for organizing data into specialized forms. In this chapter, we review the *matrix* and *vector* notation and their manipulations and applications. Vector is a one-dimensional array of numbers and/or characters arranged as a single column. The number of rows is called the *order* of that vector. Matrix is an extension of vector when a set of numbers and/or characters are arranged in rectangular form. If it has M rows and N column, this matrix then is said to be of order M by N. When M = N, then we say this *square* matrix is of order N (or M). It is obvious that vector is a special case of matrix when there is only one column. Consequently, a vector is referred to as a column matrix as opposed to the row matrix which has only one row. Braces are conventionally used to indicate a vector such as {V} and brackets are for a matrix such as [M].

In writing a computer program, DIMENSION or DIM statements are necessary to declare that a certain variable is a vector or a matrix. Such statements instruct the computer to assign multiple memory spaces for keeping the values of that vector or matrix. When we deal with a large number of different entities in a group, it is better to arrange these entities in vector or matrix form and refer to a particular entity by specifying where it is located in that group by pointing to the row (and column) number(s). Such as in the case of having 100 numbers represented by the variable names A, B, …, or by A(1) through A(100), the former requires 100 different characters or combinations of characters and the latter certainly has the advantage of having only one name. The A(1) through A(100) arrangement is to adopt a vector; these numbers can also be arranged in a matrix of 10 rows and 10 columns, or 20 rows and five columns depending on the characteristics of these numbers. In the cases of collecting the engineering data from tests of 20 samples during five different days, then arranging these 100 data into a matrix of 20 rows and five columns will be better than of 10 rows and 10 columns because each column contains the data collected during a particular day.

In the ensuing sections, we shall introduce more definitions related to vector and matrix such as transpose, inverse, and determinant, and discuss their manipulations such as addition, subtraction, and multiplication, leading to the organizing of systems of linear algebraic equations into matrix equations and to the methods of finding their solutions, specifically the Gaussian Elimination method. An apparent application of the matrix equation is the transformation of the coordinate axes by a rotation about any one of the three axes. It leads to the derivation of the three basic transformation matrices and will be elaborated in detail.

Since the interactive operations of modern personal computers are emphasized in this textbook, how a simple three-dimensional brick can be displayed will be discussed. As an extended application of the display monitor, the transformation of coordinate axes will be applied to demonstrate how animation can be designed to simulate the continuous rotation of the three-dimensional brick. In fact, any three-dimensional object could be selected and its motion animated on a display screen.

Programming languages, **FORTRAN**, **QuickBASIC**, **MATLAB**, and **Mathematica** are to be initiated in this chapter and continuously expanded into higher levels of sophistication in the later chapters to guide the readers into building a collection of their own programs while learning the computational methods for solving engineering problems.

1.2 MANIPULATION OF MATRICES

Two matrices [A] and [B] can be added or subtracted if they are of same order, say M by N which means both having M rows and N columns. If the sum and difference matrices are denoted as [S] and [D], respectively, and they are related to [A] and [B] by the formulas [S] = [A] + [B] and [D] = [A]-[B], and if we denote the elements in [A], [B], [D], and [S] as a_{ij}, b_{ij}, d_{ij}, and s_{ij} for i = 1 to M and j = 1 to N, respectively, then the elements in [S] and [D] are to be calculated with the equations:

$$s_{ij} = a_{ij} + b_{ij} \quad (1)$$

and

$$d_{ij} = a_{ij} - b_{ij} \quad (2)$$

Equations 1 and 2 indicate that the element in the ith row and jth column of [S] is the sum of the elements at the same location in [A] and [B], and the one in [D] is to be calculated by subtracting the one in [B] from that in [A] at the same location. To obtain all elements in the sum matrix [S] and the difference matrix [D], the index i runs from 1 to M and the index j runs from 1 to N.

In the case of *vector* addition and subtraction, only one column is involved (N = 1). As an example of addition and subtraction of two vectors, consider the two vectors in a two-dimensional space as shown in Figure 1, one vector $\{V_1\}$ is directed from the origin of the x-y coordinate axes, point O, to the point 1 on the x-axis which has coordinates $(x_1,y_1) = (4,0)$ and the other vector $\{V_2\}$ is directed from the origin O to the point 2 on the y-axis which has coordinates $(x_2,y_2) = (0,3)$. One may want to find the resultant of $\{R\} = \{V_1\} + \{V_2\}$ which is the vector directed from the origin to the point 3 whose coordinates are $(x_3,y_3) = (4,3)$, or, one may want to find the difference vector $\{D\} = \{V_1\} - \{V_2\}$ which is the vector directed from the origin O to the point 4 whose coordinates are $(x_4,y_4) = (4,-3)$. In fact, the vector {D} can be obtained by adding $\{V_1\}$ to the negative image of $\{V_2\}$, namely $\{V_{2-}\}$ which is a vector directed from the origin O to the point 5 whose coordinates are (x_5,y_5). Mathematically, based on Equations 1 and 2, we can have:

$$\{R\} = \{V_1\} + \{V_2\} = \begin{bmatrix} 4 \\ 0 \end{bmatrix} + \begin{bmatrix} 0 \\ 3 \end{bmatrix} = \begin{bmatrix} 4 \\ 3 \end{bmatrix}$$

and

$$\{D\} = \{V_1\} - \{V_2\} = \begin{bmatrix} 4 \\ 0 \end{bmatrix} - \begin{bmatrix} 0 \\ 3 \end{bmatrix} = \begin{bmatrix} 4 \\ -3 \end{bmatrix}$$

When Equation 1 is applied to two arbitrary two-dimensional vectors which unlike $\{V_1\}$, $\{V_2\}$, and $\{V_{2-}\}$ but are not on either one of the coordinate axes, such as {D} and {E} in Figure 1, we then have the sum vector {F} = {D} + {E} which has components of 1 and –2 units along the x- and y-directions, respectively. Notice that O467 forms a parallelogram in Figure 1 and the two vectors {D} and {E} are the two adjacent sides of the parallelogram at O. To find the sum vector {F} of {D} and {E} graphically, we simply draw a diagonal line from O to the opposite vertex of the parallelogram — this is the well-known *Law of Parallelogram.*

It should be evident that to write out a vector which has a large number of rows will take up a lot of space. If this vector can be rotated to become from one column to one row, space saving would then be possible. This process is called transposition as we will be leading to it by first introducing the length of a vector.

For the calculation of the *length* of a two-dimensional or three-dimensional vector, such as $\{V_1\}$ and $\{V_2\}$ in Figure 1, it would be a simple matter because they are oriented along the directions of the coordinate axes. But for the vectors such as {R}

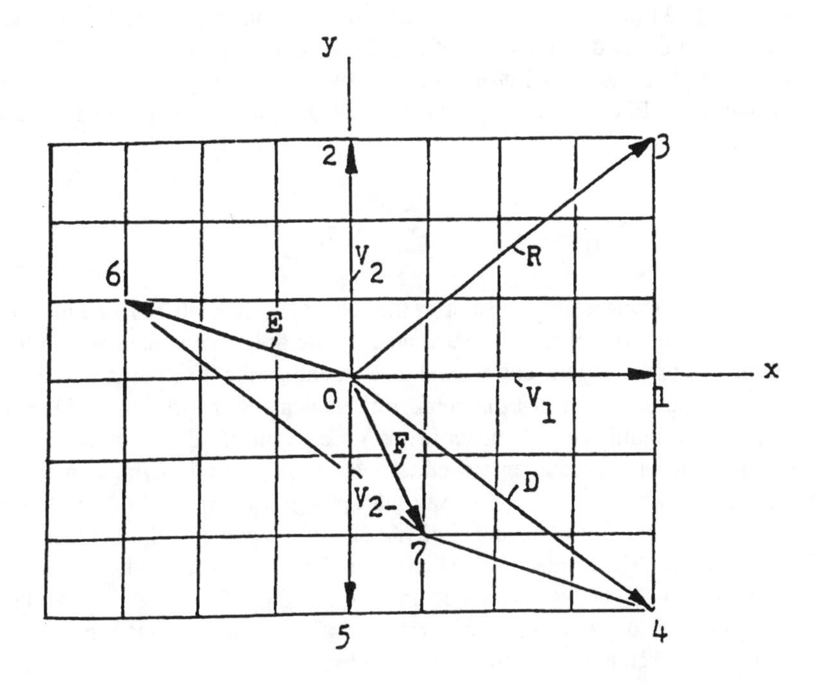

FIGURE 1. Two vectors in a two-dimensional space.

and {D} shown in Figure 1, the calculation of their lengths would need to know the *components* of these vectors in the coordinate axes and then apply the *Pythagorean theorem.* Since the vector {R} has components equal to $r_x = 4$ and $r_y = 3$ units along the x- and y-axis, respectively, its length, here denoted with the symbol | |, is:

$$|\{R\}| = \left[r_x^2 + r_y^2\right]^{0.5} = \left[4^2 + 3^2\right]^{0.5} = 5 \quad (3)$$

To facilitate the calculation of the length of a generalized vector {V} which has N components, denoted as v_1 through v_N, its length is to be calculated with the following formula obtained from extending Equation 3 from two-dimensions to N-dimensions:

$$|\{V\}| = \left[v_1^2 + v_2^2 + \ldots + v_N^2\right]^{0.5} \quad (4)$$

For example, a three-dimensional vector has components $v_1 = v_x = 4$, $v_2 = v_y = 3$, and $v_3 = v_z = 12$, then the length of this vector is $|\{V\}| = [4^2 + 3^2 + 12^2]^{0.5} = 13$. We shall next show that Equation 4 can also be derived through the introduction of the multiplication rule and transposition of matrices.

1.2 MULTIPLICATION OF MATRICES

A matrix [A] of order L (rows) by M (columns) and a matrix [B] of order M by N can be multiplied in the order of [A][B] to produce a new matrix [P] of order L by N. [A][B] is said as [A] *post-multiplied* by [B], or, [B] *pre-multiplied* by [A]. The elements in [P] denoted as p_{ij} for i = 1 to N and j = 1 to M are to be calculated by the formula:

$$p_{ij} = \sum_{k=1}^{M} a_{ik} b_{kj} \quad (5)$$

Equation 5 indicates that the value of the element p_{ij} in the ith row and jth column of the product matrix [P] is to be calculated by multiplying the elements in the ith row of the matrix [A] by the corresponding elements in the jth column of the matrix [B]. It is therefore evident that the number of elements in the ith row of [A] should be equal to the number of elements in the jth column of [B]. In other words, to apply Equation 5 for producing a product matrix [P] by multiplying a matrix [A] on the right by a matrix [B] (or, to say multiplying a matrix [B] on the left by a matrix [A]), the number of columns of [A] should be equal to the number of row of [B]. A matrix [A] of order L by M can therefore be post-multiplied by a matrix [B] of order M by N; but [A] cannot be pre-multiplied by [B] unless L is equal to N!

As a numerical example, consider the case of a square, 3 × 3 matrix post-multiplied by a rectangular matrix of order 3 by 2. Since L = 3, M = 3, and N = 2, the product matrix is thus of order 3 by 2.

$$\begin{bmatrix} 1 & 2 & 3 \\ 4 & 5 & 6 \\ 7 & 8 & 9 \end{bmatrix} \begin{bmatrix} 6 & -3 \\ 5 & -2 \\ 4 & -1 \end{bmatrix} = \begin{bmatrix} 1(6)+2(5)+3(4) & 1(-3)+2(-2)+3(-1) \\ 4(6)+5(5)+6(4) & 4(-3)+5(-2)+6(-1) \\ 7(6)+8(5)+9(4) & 7(-3)+8(-2)+9(-1) \end{bmatrix}$$

$$= \begin{bmatrix} 6+10+12 & -3-4-3 \\ 24+25+24 & -12-10-5 \\ 42+40+32 & -21-16-9 \end{bmatrix} = \begin{bmatrix} 28 & -10 \\ 73 & -27 \\ 114 & -46 \end{bmatrix}$$

More exercises are given in the Problems listed at the end of this chapter for the readers to practice on the matrix multiplications based on Equation 5.

It is of interest to note that the square of the length of a vector {V} which has N components as defined in Equation 4, $|\{V\}|^2$, can be obtained by application of Equation 5 to {V} and its transpose denoted as $\{V\}^T$ which is a row matrix of order 1 by N (one row and N columns). That is:

$$|\{V\}|^2 = \{V\}^T\{V\} = v_1^1 + v_2^2 + \ldots + v_3^2 \tag{6}$$

For a L-by-M matrix having elements e_{ij} where the row index i ranges from 1 to L and the column index j ranges from 1 to M, the transpose of this matrix when its elements are designated as t_{rc} will have a value equal to e_{cr} where the row index r ranges from 1 to M and the column index c ranges from 1 to M because this transpose matrix is of order M by L. As a numerical example, here is a pair of a 3×2 matrix [G] and its 2×3 transpose [H]:

$$\underset{3\times2}{[G]} = \begin{bmatrix} 6 & -3 \\ 5 & -2 \\ 4 & -1 \end{bmatrix} \quad \text{and} \quad \underset{2\times3}{[H]} = [G]^T = \begin{bmatrix} 6 & 5 & 4 \\ -3 & -2 & -1 \end{bmatrix}$$

If the elements of [G] and [H] are designated respectively as g_{ij} and h_{ij}, then $h_{ij} = g_{ji}$. For example, from above, we observe that $h_{12} = g_{21} = 5$, $h_{23} = g_{32} = -1$, and so on. There will be more examples of applications of Equations 5 and 6 in the ensuing sections and chapters.

Having introduced the transpose of a matrix, we can now conveniently revisit the addition of {D} and {E} in Figure 1 in algebraic form as {F} = {D} + {E} = $[4\ -3]^T + [-3\ 1]^T = [4+(-3)\ -3+1]^T = [1\ -2]^T$. The resulting sum vector is indeed correct as it is graphically verified in Figure 1. The saving of space by use of transposes of vectors (row matrices) is not evident in this case because all vectors are two-dimensional; imagine if the vectors are of much higher order.

Another noteworthy application of matrix multiplication and transposition is to reduce a system of linear algebraic equations into a simple, (or, should we say a single) *matrix equation*. For example, if we have three unknowns x, y, and z which are to be solved from the following three linear algebraic equations:

$$\begin{aligned} x+2y+3z&=4 \\ 5x+6y+7z&=8 \\ -2x-37&=9 \end{aligned} \tag{7}$$

Let us introduce two vectors, {V} and {R}, which contain the unknown x, y, and z, and the right-hand-side constants in the above three equations, respectively. That is:

$$\{V\}=[x\ \ y\ \ z]^T=\begin{bmatrix} x \\ y \\ z \end{bmatrix} \quad \text{and} \quad \{R\}=[4\ \ 8\ \ 9]^T=\begin{bmatrix} 4 \\ 8 \\ 9 \end{bmatrix} \tag{8}$$

Then, making use of the multiplication rule of matrices, Equation 5, the system of linear algebraic equations, 7, now can be written simply as:

$$[C]\{V\}=\{R\} \tag{9}$$

where the *coefficient* matrix [C] formed by listing the coefficients of x, y, and z in first equation in the first row and second equation in the second row and so on. That is,

$$[C]=\begin{bmatrix} 1 & 2 & 3 \\ 5 & 6 & 7 \\ -2 & -3 & 0 \end{bmatrix}$$

There will be more applications of matrix multiplication and transposition in the ensuing chapters when we discuss how matrix equations, such as [C]{V} = {R}, can be solved by employing the Gaussian Elimination method, and how ordinary differential equations are approximated by finite differences will lead to the matrix equations. In the abbreviated matrix form, derivation and explanation of computational methods becomes much simpler.

Also, it can be observed from the expressions in Equation 8 how the transposition can be conveniently used to define the two vectors not using the column matrices which take more lines.

FORTRAN Version

Since Equations 1 and 2 require repetitive computation of the elements in the sum matrix [S] and difference matrix [D], machine could certainly help to carry out this laborous task particularly when matrices of very high order are involved. For covering all rows and columns of [S] and [D], looping or application of **DO** statement of the **FORTRAN** programming immediately come to mind. The following program is provided to serve as a first example for generating [S] and [D] of two given matrices [A] and[B]:

```
C     Program MatxAlgb.1 - Matrix addition and subtraction.
      DIMENSION A(3,3),B(3,3),D(3,3),S(3,3)
      Data A/1.,2.,3.,4.,5.,6.,7.,8.,9./,B/1.,2*2.,3*3.,3*4./
      DO 5 I=1,3
      DO 5 J=1,3
      S(I,J)=A(I,J)+B(I,J)
    5 D(I,J)=A(I,J)-B(I,J)
      WRITE (*,*) ' Matrix A'
      WRITE (*,10) ((A(I,J),J=1,3),I=1,3)
   10 FORMAT(3F5.1)
      WRITE (*,*) ' Matrix B'
      WRITE (*,10) ((B(I,J),J=1,3),I=1,3)
      WRITE (*,15) ((S(I,J),J=1,3),(D(I,J),J=1,3),I=1,3)
   15 FORMAT('  Matrix S',7X,'Matrix D'/(6F5.1))
      STOP
      END
```

The resulting display on the screen is:

```
Matrix A
1.0  4.0  7.0
2.0  5.0  8.0
3.0  6.0  9.0
Matrix B
1.0  3.0  4.0
2.0  3.0  4.0
2.0  3.0  4.0
Matrix S             Matrix D
2.0  7.0 11.0         .0  1.0  3.0
4.0  8.0 12.0         .0  2.0  4.0
5.0  9.0 13.0        1.0  3.0  5.0
```

To review **FORTRAN** briefly, we notice that matrices should be declared as variables with two subscripts in a DIMENSION statement. The displayed results of matrices A and B show that the values listed between // in a DATA statment will be filling into the first column and then second column and so on of a matrix. To instruct the computer to take the values provided but to fill them into a matrix row-by-row, a more explicit DATA needs to be given as:

DATA ((A(I,J),J = 1,3),I = 1,3)/1.,4.,7.,2.,5.,8.,3.,6.,9./

When a number needs to be repeated, the * symbol can be conveniently applied in the DATA statement exemplified by those for the matrix [B].

Some sample WRITE and FORMAT statements are also given in the program. The first * inside the parentheses of the WRITE statement when replaced by a number allows a device unit to be specified for saving the message or the values of the variables listed in the statement. * without being replaced means the monitor will be the output unit and consequently the message or the value of the variable(s) will be displayed on screen. The second * inside the parentheses of the WRITE statement if not replaced by a statement number, in which formats for printing the listed variables are specified, means "unformatted" and takes whatever the computer provides. For example, statement number 15 is a FORMAT statement used by the WRITE statement preceding it. There are 18 variables listed in that WRITE statement but only six F5.1 codes are specified. F5.1 requests five column spaces and one digit after the decimal point to be used to print the value of a listed variable. / in a FORMAT statement causes the print/display to begin at the first column of the next line. 6F5.1 is, however, enclosed by the inner pair of parentheses that allows it to be reused and every time it is reused the next six values will be printed or displayed on next line. The use (*,*) in a WRITE statement has the convenience of viewing the results and then making a hardcopy on a connected printer by pressing the *PrtSc* (Print Screen) key.

Interactive Operation

Program **MatxAlgb.1** only allows the two particular matrices having their elements specified in the DATA statement to be added and subtracted. For finding the sum matrix [S] and difference matrix [D] for any two matrices of same order N, we ought to upgrade this program to allow the user to enter from keyboard the order N and then the elements of the two matrices involved. This is *interactive* operation of the program and proper messages should be given to instruct the user what to do which means the program should be *user-friendly*. The program **MatxAlgb.2** listed below is an attempt to achieve that goal:

```
C  Program MatxAlgb.2 - Interactive matrix addition and subtraction.
      DIMENSION A(25,25),B(25,25),D(25,25),S(25,25)
      WRITE (*,*) 'Enter the order of the two matrices, N (<25) :'
      READ (*,*) N
      DO 3 I=1,N
      WRITE (*,2) I
    2 FORMAT(' Enter all elements of [A] in row',I3/
     *       '   then press RETURN/ENTER key!')
    3 READ (*,*) (A(I,J),J=1,N)
      DO 6 I=1,N
      WRITE (*,5) I
    5 FORMAT(' Enter all elements of [B] in row',I3/
     *       '   then press RETURN/ENTER key!')
    6 READ (*,*) (B(I,J),J=1,N)
      DO 7 I=1,N
      DO 7 J=1,N
      S(I,J)=A(I,J)+B(I,J)
    7 D(I,J)=A(I,J)-B(I,J)
      WRITE (*,*)
      WRITE (*,*) 'Matrix A'
      DO 8 I=1,N
    8 WRITE (*,10) I,(A(I,J),J=1,N)
   10 FORMAT(' Row ',I5/(5E15.5))
      WRITE (*,*)
      WRITE (*,*) 'Matrix B'
      DO 12 I=1,N
```

```
12 WRITE (*,10) I,(B(I,J),J=1,N)
   WRITE (*,*)
   WRITE (*,*) 'Matrix S'
   DO 15 I=1,N
15 WRITE (*,10) I,(S(I,J),J=1,N)
   WRITE (*,*)
   WRITE (*,*) 'Matrix D'
   DO 22 I=1,N
22 WRITE (*,10) I,(D(I,J),J=1,N)
   STOP
   END
```

The interactive execution of the problem solved by the previous version **Matxalgb.1** now can proceed as follows:

```
Enter the order of the two matrices, N (<25) :
3
Enter all elements of [A] in row  1
  then press RETURN/ENTER key!
1,4,7
Enter all elements of [A] in row  2
  then press RETURN/ENTER key!
2,5,8
1nter all elements of [A] in row  3
  then press RETURN/ENTER key!
3,6,9
Enter all elements of [B] in row  1
  then press RETURN/ENTER key!
1,3,4
Enter all elements of [B] in row  2
  then press RETURN/ENTER key!
2,3,4
Enter all elements of [B] in row  3
  then press RETURN/ENTER key!
2,3,4


Matrix A
Row      1
     .10000E+01      .40000E+01      .70000E+01
Row      2
     .20000E+01      .50000E+01      .80000E+01

Row      3
     .30000E+01      .60000E+01      .90000E+01
```

```
Matrix B
Row      1
     .10000E+01        .30000E+01        .40000E+01
Row      2
     .20000E+01        .30000E+01        .40000E+01
Row      3
     .20000E+01        .30000E+01        .40000E+01

Matrix S
Row      1
     .20000E+01        .70000E+01        .11000E+02
Row      2
     .40000E+01        .80000E+01        .12000E+02
Row      3
     .50000E+01        .90000E+01        .13000E+02

Matrix D
Row      1
     .00000E+01        .10000E+01        .30000E+01
Row      2
     .00000E+01        .20000E+01        .40000E+01
Row      3
     .10000E+01        .30000E+01        .50000E+01
```

The results are identical to those obtained previously. The READ statement allows the values for the variable(s) to be entered via keyboard. A WRITE statement has no variable listed serves for need of skipping a line to provide better readability of the display. Also the I and E format codes are introduced in the statement 10. Iw where w is an integer in a FORMAT statement requests w columns to be provided for displaying the value of the integer variable listed in the WRITE statement, in which the FORMAT statement is utilized. Ew.d where w and d should both be integer constants requests w columns to be provided for display a real value in the scientific form and carrying d digits after the decimal point. Ew.d format gives more feasibility than Fw.d format because the latter may cause an *error message* of insufficient width if the value to be displayed becomes too large and/or has a negative sign.

More Programming Review

Besides the operation of matrix addition and subtraction, we have also discussed about the transposition and multiplication of matrices. For further review of computer programming, it is opportune to incorporate all these matrix algebraic operations into a single interactive program. In the listing below, three subroutines for matrix addition and subtraction, transposition, and multiplication named as **MatrixSD**, **Transpos**, and **MatxMtpy**, respectively, are created to support a program called **MatxAlgb** (Matrix Algebra).

```
C    Program MatxAlgb - Interactive matrix addition and subtraction,
C                          transposition, and multiplication.
      DIMENSION A(25,25),AT(25,25),B(25,25),P(25,25),SorD(25,25)
      WRITE (*,50)
   50 FORMAT(' Enter -1/1/2/3 for matrix subtraction/addition/',
     *        'transposition/multiplication :')
      READ (*,*) K
      KP2=K+2
      GO TO (100,100,100,200,300) KP2
C     Matrix addition or subtraction.
  100 WRITE (*,103)
  103 FORMAT(' Enter the order of the two matrices [A] & [B],'
     *        ' M & N (both<26) :')
      READ (*,*) M,N
      DO 107 I=1,M
      WRITE (*,105) I
  105 FORMAT(' Enter all elements of [A] in row',I3/
     *        '    then press RETURN/ENTER key!')
  107 READ (*,*) (A(I,J),J=1,N)
      DO 126 I=1,M
      WRITE (*,121) I
  121 FORMAT(' Enter all elements of [B] in row',I3/
     *        '    then press RETURN/ENTER key!')
  126 READ (*,*) (B(I,J),J=1,N)
      CALL MatrixSD(A,B,25,25,M,N,K,SorD)
      IF (K) 131,131,145
  131 WRITE (*,*) 'Matrix [A]-[B]'
  135 DO 138 I=1,M
  138 WRITE (*,141) I,(SorD(I,J),J=1,N)
  141 FORMAT(' Row ',I5/(5E15.5))
      GO TO 500
  145 WRITE (*,*) 'Matrix [A]+[B]'
      GOTO 135
C     Matrix transposition.
  200 WRITE (*,203)
      GO TO 500
C     Matrix multiplication.
  300 WRITE (*,303)
  303 FORMAT(' To find P(L,N)=A(L,M)B(M,N), first enter L, M,'
     *        ' and N (all<26) : ')
      READ (*,*) L,M,N
      DO 307 I=1,L
      WRITE (*,105) I
  307 READ (*,*) (A(I,J),J=1,M)
      DO 326 I=1,M
      WRITE (*,121) I
  326 READ (*,*) (B(I,J),J=1,N)
      CALL MatxMtpy(A,B,25,25,25,L,M,N,P)
  331 WRITE (*,*) 'Matrix [P]'
      DO 338 I=1,L
  338 WRITE (*,141) I,(P(I,J),J=1,N)
  500 STOP
      END

      SUBROUTINE MATXMTPY(A,B,Lmax,Mmax,Nmax,L,M,N,P)
C     Matrix multiplication of P(L,N)=A(L,M)B(M,N).
C     Input arguments: A, B, L, M, and N.
C     Output argument: P
      DIMENSION A(Lmax,Mmax),B(Mmax,Nmax),P(Lmax,Nmax)
      DO 5 I=1,L
      DO 5 J=1,N
      P(I,J)=0.
      DO 5 K=1,M
```

```
    5 P(I,J)=P(I,J)+A(I,K)*B(K,J)
      RETURN
      END

      SUBROUTINE TRANSPOS(A,Mmax,Nmax,M,N,AT)
C     Finds the transpose AT(N,M) for a given matrix A(M,N).
      DIMENSION A(Mmax,Nmax),AT(Nmax,Mmax)
      DO 5 I=1,M
      DO 5 J=1,N
    5 AT(J,I)=A(I,J)
      RETURN
      END

      SUBROUTINE MATRIXSD(A,B,Mmax,Nmax,M,N,K,SORD)
C     Finds the sum matrix SORD(M,N)=A(M,N)+B(M,N) when K=1;
C       or the difference matrix SORD(M,N)=A(M,N)-B(M,N) when K=-1.
      DIMENSION A(Mmax,Nmax),B(Mmax,Nmax),SORD(Mmax,Nmax)
      DO 5 I=1,M
      DO 5 J=1,N
    5 SORD(I,J)=A(I,J)+K*B(I,J)
      RETURN
      END
```

The above program shows that Subroutines are independent units all started with a SUBROUTINE statement which includes a name followed by a pair of parentheses enclosing a number of *arguments*. The Subroutines are called in the main program by specifying which variables or constants should serve as arguments to connect to the subroutines. Some arguments provide input to the subroutine while other arguments transmit out the results determined by the subroutine. These are referred to as *input arguments* and *output arguments*, respectively. In many instances, an argument may serve a dual role for both input and output purposes. To construct as an independent unit, a subprogram which can be in the form of a SUBROUTINE, or **FUNCTION** (to be elaborated later) must have RETURN and END statements.

It should also be remarked that program **MatxAlgb** is arranged to handle any matrix having an order of no higher than 25 by 25. For this restriction and for having the flexibility of handling any matrices of lesser order, the Lmax, Mmax, and Nmax arguments are added in all three subroutines in order not to cause any mismatch of matrix sizes between the main program and the called subroutine when dealing with any L, M, and N values which are interactively entered via keyboard.

Computed GOTO and arithmetic IF statements are also introduced in the program **MatxAlgb**. GOTO (i,j,k,...) C will result in going to (execute) the statement numbered i, j, k, and so on when C has a value equal to 1, 2, 3, and so on, respectively. IF (Expression) a,b,c will result in going to the statement numbered a, b, or c if the value calculated by the expression or a single variable is less than, equal to, or, greater than zero, respectively.

It is important to point out that in describing any derived procedure of numerical computation, *indicial notation* such as Equation 5 should always be preferred to facilitate programming. In that notation, the indices are directly used, or, literally translated into the index variables for the DO loops as can be seen in Subroutine MatxMtpy which is developed according to Equation 5. Subroutine MatrixSD is another example of literally translating Equations 1 and 2. For defining the values of the element in the following *tri-diagonal band matrix:*

$$[C]=\begin{bmatrix} 1 & 2 & 0 & 0 & 0 \\ -3 & 1 & 2 & 0 & 0 \\ 0 & -3 & 1 & 2 & 0 \\ 0 & 0 & -3 & 1 & 2 \\ 0 & 0 & 0 & -3 & 1 \end{bmatrix}$$

we ought not to write 25 separate statements for the 25 elements in this matrix but derive the indicial formulas for i,j = 1 to 5:

$$c_{ij}=0, \quad \text{if } j>i+2, \text{or}, j<i-2$$

$$c_{i,i+1}=2,$$

and

$$c_{i,i-1}=-3$$

Then, the matrix [C] can be generated with the DO loops as follows:

```
      DO 5 I=1,5
         DO 5 J=1,5
            C(I,J)=0.
            IF (J.EQ.I)       C(I,J)=1.
            IF (J.EQ.(I+1)) C(I,J)=2.
            IF (J.EQ.(I-1)) C(I,J)=-3.
    5 CONTINUE
```

The above short program also demonstrates the use of the **CONTINUE** statement for ending the DO loop(s), and the **logical IF** statements. The *true*, or, *false* condition of the expression inside the outer pair of parentheses directs the computer to execute the statement following the parentheses or the next statement immediately below the current IF statement. Reader may want to practice on deriving indicial formulas and then write a short program for calculating the elements of the matrix:

$$[M]=\begin{bmatrix} 1 & 0 & 0 & 0 & 0 & 0 & 0 & 0 \\ 2 & 1 & 0 & 0 & 0 & 0 & 0 & 0 \\ 3 & 2 & 1 & 0 & 0 & 0 & 0 & 0 \\ 4 & 3 & 2 & 1 & 0 & 0 & 0 & 0 \\ 5 & 4 & 3 & 2 & 1 & 0 & 0 & 0 \\ 6 & 5 & 4 & 3 & 2 & 1 & 0 & 0 \\ 7 & 6 & 5 & 4 & 3 & 2 & 1 & 0 \\ 8 & 7 & 6 & 5 & 4 & 3 & 2 & 1 \end{bmatrix} \quad (10)$$

As another example of writing a computer program based on indicial notation, consider the case of calculating e^x based on the infinite series:

$$\begin{aligned} e^x &= 1+\frac{x^1}{1!}+\frac{x^2}{2!}+\frac{x^2}{3!}+\ldots+\frac{x^i}{i!}+\ldots \\ &= \sum_{i=0}^{\infty}\frac{x^i}{i!} \end{aligned} \tag{11}$$

With the understanding that 0! = 1, we have expressed the series as a summation involving the index i which ranges from zero to infinity. A FUNCTION **ExpoFunc** can be developed for calculating e^x based on Equation 11 and taking only a finite number of terms for a partial sum of the series when the contribution of additional term is less than certain percentage of the sum in magnitude, say 0.001%. This FUNCTION may be arranged as:

```
      FUNCTION ExpoFunc(X)
C     Calculates EXP(X)
      ExpoFunc=1.
      NT=1
      UP=X
      FACTO=1.
    5 TERM=UP/FACTO
      IF (ABS(TERM).LT.0.00001*ABS(ExpoFunc)) GOTO 999
      ExpoFunc=ExpoFunc+TERM
      NT=NT+1
      FACTO=FACTO*NT
      UP=UP*X
      GOTO 5
  999 RETURN
      END
```

To further show the advantage of adopting vector and matrix notation, here let us apply FUNCTION **ExpoFunc** to examine the surface $z(x,y) = e^{x+y}$ above the rectangular area $0 \le x \le 2.0$ and $0 \le y \le 1.5$. The following program, ExpTest, will enable us to compare the values of e^{x+y} generated by the FUNCTION **ExpoFunc** and by the function **EXP** available in the **FORTRAN** library (hence called *library function*).

```
C     Program ExpTest - Application of FUNCTION ExpoFunc for z=e**(x+y).
      DIMENSION X(5),Y(4),Z(4,5),ZF(4,5)
      WRITE (*,1)
    1 FORMAT(20X,'Z using ExpoFunc',25X,'Z using EXP'/)
      DO 3 J=1,4
    3 Y(J)=(J-1)*0.5
      WRITE (*,5) ((Y(J),J=1,4),I=1,2)
    5 FORMAT(7X,4(' Y=',F3.1,3X),3X,4(' Y=',F3.1,3X))
      DO 9 I=1,5
      X(I)=(I-1)*0.5
      DO 7 J=1,4
```

```
Z(I,J)=ExpoFunc(X(I))*ExpoFunc(Y(J))
7 ZF(I,J)=EXP(X(I))*EXP(Y(J))
9 WRITE (*,10) X(I),(Z(I,J),J=1,4),(ZF(I,J),J=1,4)
10 FORMAT(' X=',F3.1,4F9.5,3X,4F9.5)
STOP
END
```

The resulting printout is:

	Z using ExpoFunc				Z using EXP			
	Y= .0	Y=0.5	Y=1.0	Y=1.5	Y= .0	Y=0.5	Y=1.0	Y=1.5
X= .0	1.00000	1.64872	2.71825	4.48167	1.00000	1.64872	2.71828	4.48169
X=0.5	1.64872	2.71828	4.48164	7.38902	1.64872	2.71828	4.48169	7.38906
X=1.0	2.71825	4.48164	7.38891	12.18232	2.71828	4.48169	7.38906	12.18249
X=1.5	4.48167	7.38902	12.18232	20.08537	4.48169	7.38906	12.18249	20.08554
X=2.0	7.38900	12.18238	20.08517	33.11504	7.38906	12.18249	20.08554	33.11545

It is apparent that two approaches produce almost identical results, so the 0.001% accuracy appears quite adequate for the x and y ranges studied. Also, arranging the results in vector and matrix forms make the presentation much easy to comprehend.

We have experienced how the summation process for an indicial formula involving a Σ should be programmed. Another operation symbol of importance is Π which is for multiplication of many factors. That is:

$$\prod_{i=1}^{N} a_i = a_1 a_2 \ldots a_N \tag{12}$$

An obvious application of Equation 12 is for the calculation of factorials. For example, $5! = \Pi i$ for i ranges from 1 to 5. As an exercise, we display the values of 1! through 50! with the following program involving a subroutine IFACTO which calculates I! for a specified I value:

```
C   Program FT - Factorial Table
        WRITE (*,*) ' I    I!'
        WRITE (*,*)
        DO 5 I=1,30
        CALL IFACTO(I,RIF)
        WRITE (*,3) I,RIF
      3 FORMAT(I3,E15.7)
        STOP
        END
```

```
      SUBROUTINE IFACTO(I,RIF)
C   Calculates the I factorial and gives answer in real form as
C     RIF.
      RIF=1.
      DO 7 K=1,I
    7 RIF=RIF*K
      RETURN
      END
```

The resulting print out is (listed in three columns for saving space)

I	I!								
1	.1000000E+01	7	.5040000E+04	13	.6227021E+10	19	.1216451E+18	25	.1551121E+26
2	.2000000E+01	8	.4032000E+05	14	.8717829E+11	20	.2432902E+19	26	.4032915E+27
3	.6000000E+01	9	.3628800E+06	15	.1307674E+13	21	.5109094E+20	27	.1088887E+29
4	.2400000E+02	10	.3628800E+07	16	.2092279E+14	22	.1124001E+22	28	.3048884E+30
5	.1200000E+03	11	.3991680E+08	17	.3556874E+15	23	.2585202E+23	29	.8841763E+31
6	.7200000E+03	12	.4790016E+09	18	.6402374E+16	24	.6204485E+24	30	.2653539E+33

Another application of Equation 12 is for calculation of the *binomial coefficients* for a real number r and an integer k defined as:

$$\binom{r}{k} = \frac{r(r-1)(r-2)\ldots(r-k+1)}{k!} = \prod_{i=1}^{k} \frac{r-i+1}{i} \tag{13}$$

We shall have the occasion of applying Equations 12 and 13 when the finite differences and Lagrangian interpolation are discussed.

Sample Applications

Program **MatxAlgb** has been tested interactively, the following are the resulting displays of four test cases:

```
Enter -1/1/2/3 for matrix
subtraction/addition/transposition/multiplication :
-1
Enter the order of the two matrices [A] & [B], M & N
(both<26) :
2,2
Enter all elements of [A] in row  1
  then press RETURN/ENTER key!
1,2
Enter all elements of [A] in row  2
  then press RETURN/ENTER key!
3,4
Enter all elements of [B] in row  1
  then press RETURN/ENTER key!
5,6
```

```
Enter all elements of [B] in row  2
  then press RETURN/ENTER key!
7,8
Matrix [A]-[B]
Row     1
   -.40000E+01    -.40000E+01
Row     2
   -.40000E+01    -.40000E+01
Stop - Program terminated.
Enter -1/1/2/3 for matrix
subtraction/addition/transposition/multiplication :
1
Enter the order of the two matrices [A] & [B], M & N
(both<26) :
2,2
Enter all elements of [A] in row  1
  then press RETURN/ENTER key!
1,2
Enter all elements of [A] in row  2
  then press RETURN/ENTER key!
3,4
Enter all elements of [B] in row  1
  then press RETURN/ENTER key!
5,6
Enter all elements of [b] in row  2
  then press RETURN/ENTER key!
7,8
Matrix [A]+[B]
Row     1
    .60000E+01     .80000E+01
Row     2
    .10000E+02     .12000E+02
Stop - Program terminated.

Enter -1/1/2/3 for matrix
subtraction/addition/transposition/multiplication :
2
Enter the order of the two matrices [A] & [B], M & N
(both<26) :
2,3
Enter all elements of [A] in row  1
  then press RETURN/ENTER key!
1,2,3
Enter all elements of [A] in row  2
  then press RETURN/ENTER key!
4,5,6
Transpose of [A]
Row     1
    .10000E+01     .40000E+01
```

```
Row      2
    .20000E+01      .50000E+01
Row      3
    .30000E+01      .60000E+01
Stop - Program terminated.

Enter -1/1/2/3 for matrix
subtraction/addition/transposition/multiplication :
3
To find P(L.N)=A(L,M)B(M,N), first enter L, M, and N
(all<26) :
2,3,2
Enter all elements of [A] in row  1
  then press RETURN/ENTER key!
1,2,3
Enter all elements of [A] in row  2
  then press RETURN/ENTER key!
4,5,6
Enter all elements of [B] in row  1
  then press RETURN/ENTER key!
1,2
Enter all elements of [B] in row  2
  then press RETURN/ENTER key!
2,3
Enter all elements of [B] in row  3
  then press RETURN/ENTER key!
3,4
Matrix [P]
Row      1
    .14000E+02      .20000E+02
Row      2
    .32000E+02      .47000E+02
Stop - Program terminated.
```

QuickBASIC Version

The **QuickBASIC** language has the advantage over the **FORTRAN** language for making quick changes and then running the revised program without compilation. Furthermore, it offers simple plotting statements. Let us have a **QuickBASIC** version of the program **MatxAlgb** and then discuss its basic features.

```
'   Program MatxAlgb - Interactive matrix addition and subtraction,
'                     transposition, and multiplication.
      DECLARE SUB MatrixSD (A(), B(), M, N, K, SorD())
      DECLARE SUB Transpos (C(), M, N, CT())
      DECLARE SUB MatxMtpy (D(), E(), L, M, N, P())
      PRINT "Enter -1/1/2/3 for matrix subtraction/addition/";
      PRINT "transposition/multiplication :"
      INPUT K
      IF (K = -1) THEN 100
      IF (K = 1) THEN 100
      IF (K = 2) THEN 200
      GOTO 300
```

```
'       Matrix addition or subtraction.
100     PRINT "Enter the order of the two matrices [A] & [B], M & N :"
        INPUT M, N: DIM A(M, N), B(M, N), SorD(M, N)
        FOR I = 1 TO M
            PRINT "Enter all elements of [A] in row"; I
            PRINT "  then press RETURN/ENTER key after entering each number!"
            FOR J = 1 TO N: INPUT ; A(I, J): NEXT J: PRINT : NEXT I
        FOR I = 1 TO M
            PRINT "Enter all elements of [B] in row"; I
            PRINT "  then press RETURN/ENTER key after entering each number!"
            FOR J = 1 TO N: INPUT ; B(I, J): NEXT J: PRINT : NEXT I
        CALL MatrixSD(A(), B(), M, N, K, SorD())
        IF (K < 0) THEN 131 ELSE 145
131     PRINT "Matrix [A]-[B]"
135     FOR I = 1 TO M
            PRINT "Row "; I
            FOR J = 1 TO N: PRINT USING "  #.####^^^^"; SorD(I, J); : NEXT J
            PRINT : NEXT I: GOTO 500
145     PRINT "Matrix [A]+[B]": GOTO 135
'       Matrix transposition.
200     PRINT "Enter the order of the matrix [C], M & N :"
        INPUT M, N: DIM C(M, N), CT(N, M)
        FOR I = 1 TO M
            PRINT "Enter all elements of [C] in row"; I
            PRINT "  then press RETURN/ENTER key after entering each number!"
            FOR J = 1 TO N: INPUT ; C(I, J): NEXT J: PRINT : NEXT I
        CALL Transpos(C(), M, N, CT()): PRINT "Transpose of [C]"
        FOR I = 1 TO N
            PRINT "Row "; I
            FOR J = 1 TO M: PRINT USING "  #.####^^^^"; CT(I, J); : NEXT J
            PRINT : NEXT I: GOTO 500
'       Matrix multiplication.
300     PRINT "To find P(L,N)=D(L,M)E(M,N), first enter L, M, and N :"
        INPUT L, M, N: DIM D(L, M), E(M, N), P(L, N)
        FOR I = 1 TO L
            PRINT "Enter all elements of [D] in row"; I
            PRINT "  then press RETURN/ENTER key after entering each number!"
            FOR J = 1 TO M: INPUT ; D(I, J): NEXT J: PRINT : NEXT I
        FOR I = 1 TO M
            PRINT "Enter all elements of [E] in row"; I
            PRINT "  then press RETURN/ENTER key after entering each number!"
            FOR J = 1 TO N: INPUT ; E(I, J): NEXT J: PRINT : NEXT I
        CALL MatxMtpy(D(), E(), L, M, N, P())
331     PRINT "Matrix [P]"
        FOR I = 1 TO L
            PRINT "Row "; I
            FOR J = 1 TO N: PRINT USING "  #.####^^^^"; P(I, J); : NEXT J
            PRINT : NEXT I
500     END

        SUB MatrixSD (A(), B(), M, N, K, SorD())
'       Finds the sum matrix SORD(M,N)=A(M,N)+B(M,N) when K=1;
'         or the difference matrix SORD(M,N)=A(M,N)-B(M,N) when K=-1.
        FOR I = 1 TO M
            FOR J = 1 TO N
                SorD(I, J) = A(I, J) + K * B(I, J): NEXT J: NEXT I
        END SUB

        SUB MatxMtpy (A(), B(), L, M, N, P())
'       Matrix multiplication of P(L,N)=A(L,M)B(M,N).
'       Input arguments: A, B, L, M, and N.
'       Output argument: P
        FOR I = 1 TO L
            FOR J = 1 TO N
                P(I, J) = 0!
                FOR K = 1 TO M
                    P(I, J) = P(I, J) + A(I, K) * B(K, J)
                    NEXT K: NEXT J: NEXT I
        END SUB
```

```
SUB Transpos (A(), M, N, AT())
Finds the transpose AT(N,M) for a given matrix A(M,N).
FOR I = 1 TO M
    FOR J = 1 TO N
        AT(J, I) = A(I, J): NEXT J: NEXT I
END SUB
```

Notice that the order limit of 25 needed in the **FORTRAN** version is removed in the **QuickBASIC** version which allows the dim statement to be adjustable. ' is replacing C in **FORTRAN** to indicate a comment statement in **QuickBASIC**. READ and WRITE in **FORTRAN** are replaced by INPUT and PRINT in **QuickBASIC**, respectively. The DO loop in **FORTRAN** is replaced by the FOR and NEXT pair in **QuickBASIC**.

Sample Applications

When the four cases previously run by the **FORTRAN** version are executed by the **QuickBASIC** version, the screen prompting messages, the interactively entered data, and the computed results are:

```
Enter -1/1/2/3 for matrix
subtraction/addition/transposition/multiplication :
? -1
Enter the order of the two matrices [A] & [B], M & N :
? 2,2
Enter all elements of [A] row 1
  then press RETURN/ENTER key after entering each
number!
? 1? 2
Enter all elements of [A] row 2
  then press RETURN/ENTER key after entering each
number!
? 3? 4

Enter all elements of [B] row 1
  then press RETURN/ENTER key after entering each
number!
? 5? 6
Enter all elements of [B] row 2
  then press RETURN/ENTER key after entering each
number!
? 7? 8
Matrix [A]-[B]
Row  1
  -.4000E+01  -.4000E+01
Row  2
  -.4000E+01  -.4000E+01

Enter -1/1/2/3 for matrix
subtraction/addition/transposition/multiplication :
```

```
? 1
Enter the order of the two matrices [A] & [B], M & N :
? 2,2
Enter all elements of [A] row 1
  then press RETURN/ENTER key after entering each
number!
? 1? 2
Enter all elements of [A] row 2
  then press RETURN/ENTER key after entering each
number!
? 3? 4
Enter all elements of [B] row 1
  then press RETURN/ENTER key after entering each
number!
? 5? 6
Enter all elements of [B] row 2
  then press RETURN/ENTER key after entering each
number!
? 7? 8
Matrix [A]+[B]
Row  1
   .6000E+01   .8000E+01
Row  2
   .1200E+02   .1200E+02

Enter -1/1/2/3 for matrix
subtraction/addition/transposition/multiplication :
? 2
Enter the order of the matrix [C], M & N :
? 2,3
Enter all elements of [C] row 1
  then press RETURN/ENTER key after entering each
number!
? 1? 2? 3
Enter all elements of [C] row 2
  then press RETURN/ENTER key after entering each
number!
? 4? 5? 6
Transpose of [C]
Row  1
   .1000E+01   .4000E+01
Row  2
   .2000E+01   .5000E+01
Row  1
   .3000E+01   .6000E+01

Enter -1/1/2/3 for matrix
subtraction/addition/transposition/multiplication :
? 3
To find P(L,N)=D(L.M)E(M,N), first enter L, M, and N :
? 2,3,2
```

```
Enter all elements of [D] row 1
  then press RETURN/ENTER key after entering each
number!
? 1? 2? 3
Enter all elements of [D] row 2
  then press RETURN/ENTER key after entering each
number!
? 4? 5? 6
Enter all elements of [E] row 1
  then press RETURN/ENTER key after entering each
number!
? 1? 2
Enter all elements of [E] row 2
  then press RETURN/ENTER key after entering each
number!
? 2? 3
Enter all elements of [E] row 3
  then press RETURN/ENTER key after entering each
number!
? 3? 4
Matrix [P]
Row  1
  0.1400E+02  0.2000E+02
Row  2
  0.3200E+02  0.4700E+02
```

MATLAB Applications

MATLAB developed by the Mathworks, Inc. offers a quick tool for matrix manipulations. To *load* **MATLAB** after it has been set-up and stored in a subdirectory of a hard drive, say C, we first switch to this subdirectory by entering (followed by pressing ENTER)

C:\cd MATLAB

and then switch to its own subdirectory BIN by entering (followed by pressing ENTER)

C:\MATLAB>cd BIN

Next, we type MATLAB to obtain a display of:

C:\MATLAB>BIB>MATLAB

Pressing the ENTER key results in a display of:

>>

which indicates **MATLAB** is ready to begin. Let us rerun the cases of matrix subtraction, addition, transposition, and multiplication previously considered in the **FORTRAN** and **QuickBASIC** versions. First, we enter the matrix [A] in the form of:

>> A = [1,2;3,4]

When the ENTER key is pressed, the displayed result is:

```
A =
  1  2
  3  4
```

Notice that the elements of [A] should be entered row by row. While the rows are separated by ;, in each row elements are separated by comma. After the print out of the above results, >> sign will again appear. To eliminate the unnecessary line space (between A = and the first row 1 2), the statement **format compact** can be entered as follows (the phrase "pressing ENTER key" will be omitted from now on):

```
>> format compact, B = [5,6;7,8]
B =
  5  6
  7  8
```

Notice that *comma* is used to separate the statements. To demonstrate matrix subtraction and addition, we can have:

```
>> A-B
ans =
  –4  –4
  –4  –4
>> A + B
ans =
  6   8
  10  12
```

To apply **MATLAB** for transposition and multiplication of matrices, we can have:

```
>> C = [1,2,3;4,5,6]
C =
  1  2  3
  4  5  6
>> C'
ans =
  1  4
  2  5
  3  6
```

```
>> D = [1,2,3;4,5,6]; E = [1,2;2,3;3,4]; P = D*E

P =

   14  20

   32  47
```

Notice that **MATLAB** uses ' (*single quote*) in place of the superscripted symbol T for transposition. When ; (*semi-colon*) follows a statement such as the D statement, the results will *not* be displayed. As in **FORTRAN** and **QuickBASIC**, * is the multiplication operator as is used in P = D*E, here involving three matrices not three single variables. More examples of **MATLAB** applications including plotting will ensue. To *terminate* the **MATLAB** operation, simply enter *quit* and then the RETURN key.

Mathematica Applications

To *commence* the service of **Mathematica** from **Windows** setup, simply point the mouse to it and double click the left button. The **Input** display bar will appear on screen, applications of **Mathematica** can start by entering commands from keyboard and then press the **Shift** and **Enter** keys. To *terminate* the **Mathematica** application, enter **Quit[]** from keyboard and then press the **Shift** and **Enter** keys.

Mathematica commands, statements, and functions are gradually introduced and applied in increasing degree of difficulty. Its graphic capabilities are also utilized in presentation of the computed results.

For matrix operations, **Mathematica** can compute the sum and difference of two matrices of same order in symbolic forms, such as in the following cases of involving two matrices, A and B, both of order 2 by 2:

In[1]: = A = {{1,2},{3,4}}; MatrixForm[A]

Out[1]//MatrixForm =

```
   1  2

   3  4
```

In[1]: = is shown on screen by **Mathematica** while user types in A = {{1,2},{3,4}}; MatrixForm[A]. Notice that braces are used to enclose the elements in each row of a matrix, the elments in a same row are separated by commas, and the rows are also separated by commas. MatrixForm demands that the matrix be printed in a matrix form. *Out[1]//MatrixForm* = and the rest are response of **Mathematica**.

In[2]: = B = {{5,6},{7,8}}; MatrixForm[B]

Out[2]//MatrixForm =

```
   5  6
   7  8
```

In[3]: = MatrixForm[A + B]

Out[3]//MatrixForm =

6 8

10 12

In[4]: = Dif = A-B; MatrixForm[Dif]

Out[4]//MatrixForm =

–4 –4

–4 –4

In[3] and In[4] illustrate how matrices are to be added and subtracted, respectively. Notice that one can either use A + B directly, or, create a variable Dif to handle the *sum* and *difference* matrices.

Also, **Mathematica** has a function called **Transpose** for transposition of a matrix. Let us reuse the matrix A to demonstrate its application:

In[5]: = AT = Transpose[A]; MatrixForm[AT]

Out[5]//MatrixForm =

1 3

2 4

1.3 SOLUTION OF MATRIX EQUATION

Matrix notation offers the convenience of organizing mathematical expression in an orderly manner and in a form which can be directly followed in coding it into programming languages, particularly in the case of repetitive computation involving the *looping* process. The most notable situation is in the case of solving a system of linear algebraic equation. For example, if we need to determine a linear equation $y = a_1 + a_2x$ which geometrically represents a straight line and it is required to pass through two specified points (x_1,y_1) and (x_2,y_2). To find the values of the coefficients a_1 and a_2 in the y equation, two equations can be obtained by substituting the two given points as:

$$(1)a_1 + (x_1)a_2 = y_1 \tag{1}$$

and

$$(1)a_1 + (x_2)a_2 = y_2 \tag{2}$$

To facilitate programming, it is advantageous to write the above equations in matrix form as:

$$[C]\{A\}=\{Y\} \tag{3}$$

where:

$$[C]=\begin{bmatrix}1 & x_1\\1 & x_2\end{bmatrix},\{A\}=\begin{bmatrix}a_1\\a_2\end{bmatrix}, \text{ and } \{Y\}=\begin{bmatrix}y_1\\y_2\end{bmatrix} \tag{4}$$

The matrix equation 3 in this case is of small order, that is an order of 2. For small systems, Cramer's Rule can be conveniently applied which allows the unknown vector $\{A\}$ to be obtained by the formula:

$$\{A\}=\left[\left|[c_1]\right| \; \left|[c_2]\right|\right]^T/\left|[C]\right| \tag{5}$$

Equation 5 involves the calculation of three determinants, i.e., , $|[C_1]|$, $|[C_2]|$, and $|[C]|$ where $[C_1]$ and $[C_2]$ are matrices derived from the matrix [C] when the first and second columns of [C] are replaced by $\{Y\}$, respectively. If we denote the elements of a general matrix [C] of order 2 by c_{ij} for i,j = 1,2, the determinant of [C] by definition is:

$$\left|[C]\right|=c_{11}c_{22}-c_{12}c_{21} \tag{6}$$

The general definition of the determinant of a matrix [M] of order N and whose elements are denoted as m_{ij} for i,j = 1,2,…,N is to add all possible product of N elements selected one from each row but from different column. There are N! such products and each product carries a positive or negative sign depending on whether even or odd number of exchanges are necessary for rearranging the N subscripts in increasing order. For example, in Equation 6, c_{11} is selected from the first row and first column of [C] and only c_{22} can be selected and multiplied by it while the other possible product is to select c_{12} from the second row and first column of [C] and that leaves only c_{21} from the second row and first column of [C] available as a factor of the second product. In order to arrange the two subscripts in non-decreasing order, one exchange is needed and hence the product $c_{12}c_{21}$ carries a minus sign. We shall explain this sign convention further when a matrix of order 3 is discussed. However, it should be evident here that a matrix whose order is large the task of calculating its determinant would certainly need help from computer. This will be the a topic discussed in Section 1.5.

Let us demonstrate the application of Cramer's Rule by having a numerical case. If the two given points to be passed by the straight line $y = a_1 + a_2x$ are $(x_1,y_1) = (1,2)$ and $(x_2,y_2) = (3,4)$. Then we can have:

$$\|[C]\| = \begin{vmatrix} 1 & x_1 \\ 1 & x_2 \end{vmatrix} = \begin{vmatrix} 1 & 1 \\ 1 & 3 \end{vmatrix} = 1\times3-1\times1=3-1=2$$

$$\|[c_1]\| = \begin{vmatrix} y_1 & x_1 \\ y_2 & x_2 \end{vmatrix} = \begin{vmatrix} 2 & 1 \\ 4 & 3 \end{vmatrix} = 2\times3-1\times4=6-4=2$$

and

$$\|[C_2]\| = \begin{vmatrix} 1 & y_1 \\ 1 & y_2 \end{vmatrix} = \begin{vmatrix} 1 & 2 \\ 1 & 4 \end{vmatrix} = 1\times4-2\times1=4-2=2$$

Consequently, according to Equation 5 we can find the coefficients in the straight-line equation to be:

$$a_1 = \|[C_1]\|/\|[C]\| = 2/2 = 1 \text{ and } a_2 = \|[C_2]\|/\|[C]\| = 2/2 = 1$$

Hence, the line passing through the points (1,2) and (3,4) is $y = a_1 + a_2x = 1 + x$.

Application of Cramer's Rule can be extended for solving three unknowns from three linear algebraic equations. Consider the case of finding a plane which passes three points (x_i,y_i) for i = 1 to 3. The equation of that plane can first be written as $z = a_1 + a_2x + a_3y$. Similar to the derivation of Equation 3, here we substitute the three given points into the z equation and obtain:

$$(1)a_1 + (x_1)a_2 + (y_1)a_3 = z_1 \tag{7}$$

$$(1)a_1 + (x_2)a_2 + (y_2)a_3 = z_2 \tag{8}$$

and

$$(1)a_1 + (x_3)a_2 + (y_3)a_3 = z_3 \tag{9}$$

Again, the above three equations can be written in matrix form as:

$$[C]\{A\} = \{Z\} \tag{10}$$

where the matrix [C] and the vector {A} previously defined in Equation 4 need to be reexpanded and redefined as:

$$[C] = \begin{vmatrix} 1 & x_1 & y_1 \\ 1 & x_2 & y_2 \\ 1 & x_3 & y_3 \end{vmatrix}, \{A\} = \begin{vmatrix} a_1 \\ a_2 \\ a_3 \end{vmatrix}, \text{ and } \{Z\} = \begin{bmatrix} z_1 \\ z_2 \\ z_3 \end{bmatrix} \tag{11}$$

And, the Cramer's Rule for solving Equation 10 can be expressed as:

$$\{A\} = \left[|[C_1]| \ |[C_2]| \ |[C_3]| \right]^T / |[C]| \tag{12}$$

where $[C_i]$ for i = 1 to 3 for matrices formed by replacing the ith column of the matrix [C] by the vector {Z}, respectively. Now, we need the calculation of the determinant of matrices of order 3. If we denote the element in a matrix [M] as m_{ij} for i,j = 1 to 3, the determinant of [M] can be calculated as:

$$\begin{aligned} |[M]| = {} & m_{11}m_{22}m_{33} - m_{11}m_{23}m_{32} + m_{12}m_{23}m_{31} \\ & - m_{12}m_{21}m_{33} + m_{13}m_{21}m_{32} - m_{13}m_{22}m_{31} \end{aligned} \tag{13}$$

To give a numerical example, let us consider a plane passing the three points, $(x_1,y_1,z_1) = (1,2,3)$, $(x_2,y_2,z_2) = (-1,0,1)$, and $(x_3,y_3,z_3) = (-4,-2,0)$. We can then have:

$$|[C]| = \begin{vmatrix} 1 & x_1 & y_1 \\ 1 & x_2 & y_2 \\ 1 & x_3 & y_3 \end{vmatrix} = \begin{vmatrix} 1 & 1 & 2 \\ 1 & -1 & 0 \\ 1 & -4 & -2 \end{vmatrix} = -2$$

$$|[C_1]| = \begin{vmatrix} z_1 & x_1 & y_1 \\ z_2 & x_2 & y_2 \\ z_3 & x_3 & y_3 \end{vmatrix} = \begin{vmatrix} 3 & 1 & 2 \\ 1 & -1 & 0 \\ 0 & -4 & -2 \end{vmatrix} = 0$$

$$|[C_2]| = \begin{vmatrix} 1 & z_1 & y_1 \\ 1 & z_2 & y_2 \\ 1 & z_3 & y_3 \end{vmatrix} = \begin{vmatrix} 1 & 3 & 2 \\ 1 & 1 & 0 \\ 1 & 0 & -2 \end{vmatrix} = 2$$

and

$$|[C_3]| = \begin{vmatrix} 1 & x_1 & z_1 \\ 1 & x_2 & z_2 \\ 1 & x & z \end{vmatrix} = \begin{vmatrix} 1 & 1 & 3 \\ 1 & -1 & 1 \\ 1 & -4 & -2 \end{vmatrix} = -4$$

According to Equation 13, we find $a_1 = |[C_1]|/|[C]| = 0/(-2) = 0$, $a_2 = |[C_2]|/|[C]| = 2/(-2) = -1$, and $a_3 = |[C_3]|/|[C]| = -4/(-2) = 2$. Thus, the required plane equation is $z = a_1 + a_2x + a_3y = -x + 2y$.

QuickBASIC Version of the program CramerR

A computer program called **CramerR** has been developed as a reviewing exercise in programming to solve a matrix equation of order 3 by application of Cramer

Rule and the definition of determinant of a 3 by 3 square matrix according to Equations 12 and 13, respectively. First, a subroutine called **Determ3** is created explicitly following Equation 13 as listed below:

```
SUB Determ3(A(),D)
D=A(1,1)*(A(2,2)*A(3,3)-A(2,3)*A(3,2))
  +A(2,1)*(A(3,2)*A(1,3)-A(1,2)*A(3,3))
  +A(3,1)*(A(1,2)*A(2,3)-A(1,3)*A(2,2))
END SUB
```

To interactively enter the elements of the coefficient matrix [C] and also the elements of the right-hand-side vector {Z} in Equation 12 and to solve for {A}, the program **CramerR** can be arranged as:

```
' PROGRAM CramerR - solves a matrix equation C(3,3)X(3)=V(3) by Cramer Rule.
'
                    DECLARE SUB DETERM3(C(),D)
                 SCREEN 2: CLEAR : CLS : KEY OFF
 PRINT "Program CramerR - solves the matrix equation C(3,3)X(3)=V(3)"
 PRINT "                  by Cramer Rule."
 DIM C(3,3),CT(3,3),DC(3),V(3)
' Input
 PRINT : PRINT "Input the elements of the coefficient matrix [C], row by row"
 PRINT "  and press <Enter> key after entering a number :"
 FOR I=1 TO 3
     FOR J=1 TO 3
         INPUT; C(I,J)
         NEXT J
     PRINT
     NEXT I
 PRINT
 PRINT "Input the elements of the constant vector {V} :";
 FOR I = 1 TO 3
     INPUT ; V(I)
     NEXT I
' Solve for {X}
 CALL DETERM3(C(),DC)
 FOR I=1 TO 3
     FOR IR=1 TO 3
         FOR JC=1 TO 3
             CT(IR.JC)=C(IR,JC)
             IF (JC.EQ.I) CT(IR,JC)=Z(IR)
             NEXT JC:
         NEXT IR
     CALL DETERM3(CT(),DC(I))
     X(I)=DC(I)/DC
     NEXT I
 PRINT : PRINT "The solution vector has elements : ";
 FOR I = 1 TO 3: PRINT X(I);: NEXT I: END
```

1.4 PROGRAM GAUSS

Program **Gauss** is designed for solving N unknowns from N simultaneous, linear algebraic equations by the Gaussian Elimination method. In matrix notation, the problem can be described as to solve a vector {X} from the matrix equation:

$$[C]\{X\} = \{V\} \tag{1}$$

where [C] is an NxN coefficient matrix and {V} is a Nx1 constant vector, and both are prescribed. For example, let us consider the following system:

$$9x_1 + x_2 + x_3 = 10 \tag{2}$$

$$3x_1 + 6x_2 + x_3 = 14 \tag{3}$$

$$2x_1 + 2x_2 + 3x_3 = 3 \tag{4}$$

If the above three equations are expressed in matrix form as Equation 1, then:

$$[C] = \begin{bmatrix} 9 & 1 & 1 \\ 3 & 6 & 1 \\ 2 & 2 & 3 \end{bmatrix}, \qquad \{V\} = \begin{bmatrix} 10 \\ 14 \\ 3 \end{bmatrix}, \tag{5,6}$$

and

$$\{X\} = \begin{bmatrix} x_1 \\ x_2 \\ x_3 \end{bmatrix} = [x_1 \ x_2 \ x_3]^T \tag{7}$$

where T designates the transpose of a matrix.

GAUSSIAN ELIMINATION METHOD

A systematic procedure named after Gauss can be employed for solving x_1, x_2, and x_3 from the above equations. It consists of first dividing Equation 28 by the leading coefficient, 9, to obtain:

$$x_1 + \frac{1}{9}x_2 + \frac{1}{9}x_3 = \frac{10}{9} \tag{8}$$

This step is called *normalization* of the first equation of the system (1). The next step is to eliminate x_1 term from the other (in this case, two) equations. To do that, we multiply Equation 8 by the coefficients associated with x_1 in Equations 3 and 4, respectively, to obtain:

$$3x_1 + \frac{1}{3}x_2 + \frac{1}{3}x_3 = \frac{10}{3} \tag{9}$$

and

$$2x_1 + \frac{2}{9}x_2 + \frac{2}{9}x_3 = \frac{20}{9} \tag{10}$$

If we subtract Equation 9 from Equation 3, and subtract Equation 10 from Equation 4, the x_1 terms are eliminated. The resulting equations are, respectively:

$$\frac{17}{3}x_2 + \frac{2}{3}x_3 = \frac{32}{3} \tag{11}$$

and

$$\frac{16}{9}x_2 + \frac{25}{9}x_3 = \frac{7}{9} \tag{12}$$

This completes the first *elimination* step. The next normalization is applied to Equation 11, and then the x_2 term is to be eliminated from Equation 12. The resulting equations are:

$$x_2 + \frac{2}{17}x_3 = \frac{32}{17} \tag{13}$$

and

$$\frac{393}{153}x_3 = -\frac{393}{153} \tag{14}$$

The last normalization of Equation 14 then gives:

$$x_3 = -1 \tag{15}$$

Equations 8, 13, and 15 can be organized in matrix form as:

$$\{V\} = \begin{bmatrix} 1 & 1/9 & 1/9 \\ 0 & 1 & 2/17 \\ 0 & 0 & 1 \end{bmatrix} \begin{bmatrix} x_1 \\ x_2 \\ x_3 \end{bmatrix} = \begin{bmatrix} 10/9 \\ 32/17 \\ -1 \end{bmatrix} \tag{16}$$

The coefficient matrix is now a so-called *upper triangular matrix* since all elements below the main diagonal are equal to zero.

As x_3 is already obtained in Equation 15, the other two unknowns, x_2 and x_3, can be obtained by a sequential *backward-substitution* process. First, Equation 13 can be used to obtain:

$$x_2 = \frac{32}{17} - \frac{2}{17}x_3 = \frac{32}{17} - \frac{2}{17}(-1) = \frac{32+2}{17} = 2$$

Once, both x_2 and x_3 have been calculated, x_1 can be obtained from Equation 8 as:

$$x_1 = \frac{10}{9} - \frac{1}{9}x_2 - \frac{1}{9}x_3 = \frac{10}{9} - \frac{1}{9}(2) - \frac{1}{9}(-1) = \frac{10-2+1}{9} = 1$$

To derive a general algorithm for the Gaussian elimination method, let us denote the elements in [C], {X}, and {V} as $c_{i,j}$, x_i, and v_i, respectively. Then the normalization of the first equation can be expressed as:

$$\left(c_{1,j}\right)_{new} = \left(c_{1,j}\right)_{old} / \left(c_{1,1}\right)_{old} \tag{17}$$

and

$$\left(v_1\right)_{new} = \left(v_1\right)_{old} / \left(c_{1,1}\right)_{old} \tag{18}$$

Equation 17 is to be used for calculating the new coefficient associated with x_j in the first, normalized equation. So, j should be ranged from 2 to N which is the number of unknowns (equal to 3 in the sample case). The subscripts old and new are added to indicate the values before and after normalization, respectively. Such designation is particularly helpful if no separate storage in computer are assigned for [C] for the values of its elements. Notice that $(c_{1,1})_{new} = 1$ is not calculated. Preserving this diagonal element enables the determinant of [C] to be computed. (See the topic on matrix inversion and determinant.)

The formulas for the elimination of x_1 terms from the second equation are:

$$\left(c_{2,j}\right)_{new} = \left(c_{2,j}\right)_{old} - \left(c_{2,1}\right)_{old}\left(c_{1,j}\right)_{old} \tag{19}$$

for j = 2,3,…,N (there is no need to include j = 1) and

$$\left(v_2\right)_{new} = \left(v_2\right)_{old} - \left(c_{2,1}\right)_{old}\left(v_1\right)_{old} \tag{20}$$

By changing the subscript 2 in Equations 19 and 20, x_1 term in the third equation can be eliminated. In other words, the general formulas for elimination of x_1 terms from all equation other than the first equation are, for k = 2,3,…,N

$$\left(c_{k,j}\right)_{new} = \left(c_{k,j}\right)_{old} - \left(c_{k,1}\right)_{old}\left(c_{1,j}\right)_{old} \tag{21}$$

for j = 2,3,…,N

$$\left(v_k\right)_{new} = \left(v_k\right)_{old} - \left(c_{k,1}\right)_{old}\left(v_1\right)_{old} \tag{22}$$

Instead of normalizing the first equation, we can generalize Equations 17 and 18 for normalization of the ith equation, for i = 1,2,…,N to the expressions:

$$\left(c_{i,j}\right)_{new} = \left(c_{i.j}\right)_{old} / \left(c_{i,i}\right)_{old} \tag{23}$$

for j = i + 1,i + 2,…,N and

$$\left(v_i\right)_{new} = \left(v_i\right)_{old} / \left(c_{i,i}\right)_{old} \tag{24}$$

Note that $(c_{i,i})_{new}$ should be equal to 1 but no need to calculate since it is not involved in later calculation for finding {X}.

Similarly, elimination of x_i term from kth equation for k = i + 1,i + 2,…,N consists of using the general formula:

$$\left(c_{k,j}\right)_{new} = \left(c_{k,j}\right)_{old} - \left(c_{k,i}\right)_{old}\left(c_{i,j}\right)_{old} \tag{25}$$

for j = i + 1,i + 2,…,N and

$$\left(v_k\right)_{new} = \left(v_k\right)_{old} - \left(c_{k,i}\right)_{old}\left(v_i\right)_{old} \tag{26}$$

Backward substitution for finding x_i involves the calculation of:

$$x_i = v_i - \sum_{j=i+1}^{N} c_{i,j} x_j \tag{27}$$

for i = N–1,N–2,…,2,1. Note that x_N is already found equal to v_N after the Nth normalization.

Program **Gauss** listed below in both **QuickBASIC** and **FORTRAN** languages is developed for interactive specification of the number of unknowns, N, and the values of the elements of [C] and {V}. It proceeds to solve for {X} and prints out the computed values. Sample applications of both languages are provided immediately following the listed programs.

A subroutine **Gauss.Sub** is also made available for the frequent need in the other computer programs which require the solution of matrix equations.

QuickBASIC Version

```
' PROGRAM Gauss - solves a matrix equation C(N,N)X(N)=V(N)
'                 by Gaussian Elimination method.
'                 X and V share same storage space.
'
                  SCREEN 2: CLEAR : CLS : KEY OFF
  PRINT "Program Gauss - solves matrix equation C(N,N)X(N)=V(N)"
  PRINT "                by Gaussian-Elimination method."
  PRINT : PRINT "Input the order of matrix equation, N : ";
          INPUT N : DIM C(N, N), V(N)
  PRINT : PRINT "Input the elements of the coefficient matrix, row by row"
  PRINT "  and press <Enter> key after entering a number :"
  FOR I=1 TO N
      FOR J=1 TO N
          INPUT;C(I,J)
          NEXT J
      PRINT
      NEXT I
  PRINT
  PRINT "Input the elements of the constant vector :";
  FOR I = 1 TO N
      INPUT ; V(I)
      NEXT I
  FOR I = 1 TO N                                         ' *** Normalization
      FOR J=I+1 TO N: C(I,J)=C(I,J)/C(I,I): NEXT J
      V(I)=V(I)/C(I,I)  : IF  I = N THEN GOTO 271
      FOR K = I+1 TO N
          IF C(K,I)=0 THEN 265 ELSE V(K)=V(K)-C(K,I)*V(I)     ' *** Elimination
          FOR J=I+1 TO N: C(K,J)=C(K,J)-C(K,I)*C(I,J): NEXT J
265       NEXT K
      NEXT I
271 FOR I=N-1 TO 1 STEP -1
        FOR J=I+1 TO N: V(I)=V(I)-C(I,J)*V(J): NEXT J
        NEXT I
    PRINT : PRINT : PRINT "The solution vector has elements : ";
    FOR I = 1 TO N: PRINT V(I);: NEXT I: PRINT: PRINT: END
```

Sample Application

```
Program Gauss - solves matrix equation C(N,N)X(N)=V(N)
                by Gaussian elimination method.

Input the order of matrix equation, N : 3

Input the elements of the coefficient matrix, row by row
  and press <Enter> key after entering a number :
? 9? 1? 1
? 3? 6? 1
? 2? 2? 3

Input the elements of the constant vector :? 10? 14? 3
The solution vector has elements : 1  2  -1
```

FORTRAN Version

```
C     PROGRAM Gauss - solves matrix equation C(N,N)X(N)=V(N)
C                     by Gaussian elimination method.
C                     X and V share same storage space.
      DIMENSION C(50,50),V(50)
      WRITE (*,*) 'Program Gauss - solves matrix equation C(N,N)X(N)=V(N)'
      WRITE (*,*) '                by Gaussian elimination method.'
      WRITE (*,*) 'Input the order of matrix equation, N : '
      READ  (*,*) N
      WRITE (*,*) 'Input the elements of the coefficient matrix,'
      WRITE (*,*) '  row by row and press <Enter> key after entering'
      WRITE (*,*) '  each row :'
      DO 5 I=1,N
    5 READ (*,*) (C(I,J),J=1,N)
      WRITE (*,*) 'Input the elements of the constant vector :'
      READ (*,*) (V(I),I=1,N)
      CALL GAUSS(C,N,50,V)
      WRITE (*,15) (V(I),I=1,N)
   15 FORMAT(' The vector {V} is'/5E16.5)
      STOP
      END

      SUBROUTINE GAUSS(C,N,M,V)
C
C     SOLVES MATRIX EQUATION C(N,N)*X(N)=V(N) BY GAUSSIAN ELIMINATION.
C       X and V share same storage.  C is dimensioned (M,M) in the calling program.
C
      DIMENSION C(M,M),V(N)
C
      N1=N-1
      DO 25 K=1,N1
         KP1=K+1
C
C     NORMALIZATION
C
         DO 10 J=KP1,N
   10       C(K,J)=C(K,J)/C(K,K)
         V(K)=V(K)/C(K,K)
C
C     ELIMINATION
C
         DO 25 I=KP1,N
            DO 20 J=KP1,N
   20          C(I,J)=C(I,J)-C(I,K)*C(K,J)
   25       V(I)=V(I)-C(I,K)*V(K)
C
C     BACKWARD SUBSTITUTION
C
      V(N)=V(N)/C(N,N)
      DO 35 I=1,N1
      IR=N-I
      IR1=IR+1
      DO 35 J=IR1,N
   35    V(IR)=V(IR)-C(IR,J)*V(J)
      RETURN
      END
```

Sample Application

```
Program Gauss - solves matrix equation C(N,N)X(N)=V(N)
                by Gauss-Jordan elimination method.
Input the order of matrix equation, N :
3
Input the elements of the coefficient matrix,
  row by row and press <Enter> key after entering
  each row :
9,1,1
3,6,1
2,2,3
Input the elements of the constant vector :
10,14,3
The vector {V} is
     .10000E+01      .20000E+01      -.10000E+01
Stop - Program terminated.
```

Gauss-Jordan Method

One slight modification of the elimination step will make the backward substitution steps completely unnecessary. That is, during the elimination of the x_i terms from the linear algebraic equations except the ith one, Equations 25 and 26 should be applied for k equal to 1 through N and excluding $k = i$. For example, the x_3 terms should be eliminated from the first, second, fourth through Nth equations. In this manner, after the Nth normalization, [C] becomes an identity matrix and {V} will have the elements of the required solution {X}. This modified method is called Gauss-Jordan method.

A subroutine called **GauJor** is made available based on the above argument. In this subroutine, a block of statements are also added to include the consideration of the *pivoting* technique which is required if $c_{i,i} = 0$. The normalization steps, Equations 49 and 50, cannot be implemented if $c_{i,i}$ is equal to zero. For such a situation, a search for a nonzero $c_{i,k}$ is necessary for $i = k + 1, k + 2, \ldots, N$. That is, to find in the kth column of [C] and below the kth row a nonzero element. Once this nonzero $c_{i,k}$ is found, then we can then interchange the ith and kth rows of [C] and {V} to allow the normalization steps to be implemented; if no nonzero $c_{i,k}$ can be found then [C] is *singular* because the determinant of [C] is equal to zero! This can be explained by the fact that when $c_{k,k} = 0$ and no pivoting is possible and the determinant D of [C] can be calculated by the formula:

$$D = c_{1,1}c_{2,2}\ldots c_{k,k}\ldots c_{N,N} = \prod_{k=1}^{N} c_{k,k} \tag{28}$$

where Π indicates a product of all listed factors.

A subroutine has been written based on the Gauss-Jordan method and called **GauJor.Sub**. Both **QuickBASIC** and **FORTRAN** versions are made available and they are listed below.

QuickBASIC Version

```
  SUB GauJor (A(), N, C(), D)
' Gauss-Jordan Elimination method for solving [A]{X}={C}.
' {C} exits as {X}.
' Also calculate the determinant D of [A].
        FOR I = 1 TO N
            IF A(I, I) = 0 THEN 220
'
'                       *** Normalization ***
200         FOR J = I + 1 TO N
                A(I, J) = A(I, J) / A(I, I)
                NEXT J
            C(I) = C(I) / A(I, I)
            IF I=N THEN 280 ELSE 250
'
'                          *** Pivoting ***
220         FOR J = I + 1 TO N
                IF A(J, I) = 0 THEN 230
                FOR K = I TO N
                    T = A(I, K)
                    A(I, K) = A(J, K)
                    A(J, K) = T
                    NEXT K
                T = C(I)
                C(I) = C(J)
                C(J) = T
                GOTO 200
230             NEXT J
             PRINT "The coefficient matrix is singular.":
GOTO 300
'
'                         *** Elimination ***
250       FOR K = 1 TO N
              IF K = I THEN 265
              IF A(K, I) = 0 THEN 265
              C(K) = C(K) - A(K, I) * C(I)
              FOR J = I + 1 TO N
                  A(K, J) = A(K, J) - A(K, I) * A(I, J)
                  NEXT J
265           NEXT K
    NEXT I
'
'                      Calculates determinant.
280 D = 1!
    FOR I = 1 TO N
        D = D * A(I, I)
        NEXT I
300 END SUB
```

FORTRAN Version

```
      SUBROUTINE GAUJOR(C,N,M,V,D)
C
C    SOLVES MATRIX EQUATION C(N,N)*X(N)=V(N) BY GAUSSIAN ELIMINATION.
C      CALCULATES DETERMINANT, D, of C.  Pivoting is provided.
C        X and V share same storage.
C        C is dimensioned (M,M) in the calling program.
C
      DIMENSION C(M,M),V(N)
      DO 35 K=1,N
      KP1=K+1
      IF (C(K,K).EQ.0.) GOTO 12
C
C     NORMALIZATION
C
    8    DO 10 J=KP1,N
   10    C(K,J)=C(K,J)/C(K,K)
         V(K)=V(K)/C(K,K)
         IF (I.EQ.N) RETURN
         GOTO 20
C
C     PIVOTING
C
   12    DO 14 I=KP1,N
         IF (C(I,K).NE.0.) GOTO 16
   14    CONTINUE
   15    WRITE (*,*) 'The coefficient matrix is singular!'
         STOP
   16    DO 18 J=KP1,N
         T=C(K,J)
         C(K,J)=C(I,J)
   18    C(I,J)=T
         T=V(K)
         V(K)=V(I)
         V(I)=T
         GOTO 8
C
C     ELIMINATION
C
   20    DO 30 I=1,N
         IF (I.EQ.K) GOTO 30
            DO 25 J=KP1,N
   25       C(I,J)=C(I,J)-C(I,K)*C(K,J)
         V(I)=V(I)-C(I,K)*V(K)
   30    CONTINUE
   35 CONTINUE
C     CALCULATES Determinant
      D=1.
      DO 40 K=1,N
   40 D=D*C(K,K)
      RETURN
      END
```

Sample Applications

The same problem previously solved by the program **Gauss** has been used again but solved by application of subroutine GauJor. The results obtained with the **QuickBASIC** and **FORTRAN** versions are listed, in that order, below:

```
Program GauJor - solves matrix equation A(N,N)X(N)=C(N)
                 by Gauss-Jordan elimination method.

Input the order of matrix equation, N : 3

Input the elements of the coefficient matrix, row by row
  and press <Enter> key after entering a number :
? 9? 1? 1
? 3? 6? 1
? 2? 2? 3

Input the elements of the constant vector :? 10? 14? 3
The solution vector is :
 1  2  -1
Determinant of [A] =  131

Program GauJor - solves matrix equation C(N,N)X(N)=V(N)
                 by Gauss-Jordan elimination method.
Input the order of matrix equation, N :
3
Input the elements of the coefficient matrix,
  row by row and press <Enter> key after entering
  each row :
9,1,1
3,6,1
2,2,3
Input the elements of the constant vector :
10,14,3
The vector {V} is :
     .10000E+01       .20000E+01      -.10000E+01
The determinant of the coefficient matrix =    .13100E+03
Stop - Program terminated.
```

MATLAB Applications

For solving the vector {X} from the matrix equation [C]{X} = {R} when both the coefficient matrix [C] and the right-hand side vector {R} are specified, **MATLAB** simply requires [C] and {R} to be interactively inputted and then uses a statement X = C\R to obtain the solution vector {X} by multiplying the vector {R} on the left of the inverse of [C] or dividing {R} on the left by [C]. More details are discussed in the program **MatxAlgb**. Here, for providing more examples in **MATLAB** applications, a m file called **GauJor.m** is presented below as a companion of the **FORTRAN** and **QuickBASIC** versions:

```
            function [X,D]=GauJor(C,N,R)
% Solves matrix equation [C]{X}={R} of order N by Gauss-Jordan Elimination.
% Also finds the determinant D of [C].
%
ExitFlag=0; D=0;
for I=1:N, ip1=I+1;
    if C(i,i)==0
%                                   Pivoting
       for k=ip1:N
           if C(k,i)~=0
              for j=ip1:N, T=C(k,j); C(k,j)=C(i,j); C(i,j) =T; end
              T=R(k); R(k)=R(I); R(I)=T; break
           end
       end
       ExitFlag=1;
    end
%                                 Normalization
    if ExitFlag==0;
       for j=ip1:N, C(i,j)=C(i,j)/C(i,i); end
       R(I)=R(I)/C(i,i);
%                                  Elimination
       for k=1:N
           if I~=k
              for j=ip1:N, C(k,j)=C(k,j)-C(k,i)*C(i,j); end
              R(k)=R(k)-C(k,i)*R(I);
           end
       end
    end
    if I==N, break
    end
end
if ExitFlag==1;
   error('The coefficient matrix is singular.')
   else D=1; X=R; for I=1:N, D=D*C(i,i); end
end
```

This file **GauJor.m** should then be added into **MATLAB**. As an example of interactive application of this m file, the sample problem used in the **FORTRAN** and **QuickBASIC** versions is again solved by specifying the coefficient matrix [C] and the right hand side vector {R} to obtain the resulting display as follows:

```
>> C=[9,1,1;3,6,1;2,2,3]; R=[10,14,3]'; [X,D]=GauJor(C,3,R)
 X =
      1.0000
      2.0000
      1.0000
 D =
      131
```

The results of the vector {X} and determinant D for the coefficient matrix [C] are same as obtained before.

Mathematica Applications

For solving a system of linear algebraic equations which has been arranged in matrix form as [A]{X} = {R}, **Mathematica**'s function **LinearSolve** can be applied to solve for {X} when the coefficient matrix [A] and the right-hand side vector {R} are both provided. The following is an example of interactive application:

In[1]: = A = {{3,6,14},{6,14,36},{14,36,98}}

Out[1]: =

{{3, 6, 14}, {6, 14, 36}, {14, 36, 98}}

In[2]: = MatrixForm[A]

Out[2]//MatrixForm: =

3 6 14

6 14 36

14 36 98

In[3]: = R = {9,20,48}

Out[3]: =

{9, 20, 48}

In[4]: = LinearSolve[A,R]

Out[4]: =

{–9,13,–3}

Output[2] and *Output[1]* demonstrate the difference in display of matrix [A] when MatrixeForm is requested, or, not requested, respectively. It shows that without requesting of MatrixForm, some screen space saving can be gained. *Output[4]* gives the solution {X} = $[-9\ 13\ -3]^T$ for the matrix equation [A]{X} = {R} where the coefficient matix [A] and vector {R} are provided by *Input[1]* and *Input[3]*, respectively.

1.5 MATRIX INVERSION, DETERMINANT, AND PROGRAM MatxInvD

Given a square matrix [C] of order N, its inverse as $[C]^{-1}$ of the same order is defined by the equation:

$$[C][C]^{-1} = [C]^{-1}[C] = [I] \tag{1}$$

where [I] is an identity matrix having elements equal to one along its main diagonal and equal to zero elsewhere. That is:

$$[I] = \begin{bmatrix} 1 & 0 & . & . & . & . & 0 \\ 0 & 1 & 0 & . & . & . & 0 \\ & . & . & . & . & . & . \\ 0 & 0 & . & . & . & . & 1 \end{bmatrix} \tag{2}$$

To find $[C]^{-1}$, let c_{ij} and d_{ij} be the elements at the ith row and jth column of the matrices [C] and $[C]^{-1}$, respectively. Both i and j range from 1 to N. Furthermore, let $\{D_j\}$ and $\{I_j\}$ be the jth column of the matrices $[C]^{-1}$ and [I], respectively. It is easy to observe that $\{I_j\}$ has elements all equal to zero except the one in the jth row which is equal to unity. Also,

$$\left\{D_j\right\} = \left[d_{1j} d_{2j} \ldots d_{Nj}\right]^T \tag{3}$$

and

$$[C]^{-1} = \left[D_1 D_2 \ldots D_N\right]^T \tag{4}$$

Based on the rules of matrix multiplication, Equation 1 can be interpreted as $[C]\{D_1\} = \{I_1\}$, $[C]\{D_2\} = \{I_2\}$, …, and $[C]\{D_N\} = \{I_N\}$. This indicates that program **Gauss** can be successively employed N times by using the same coefficient matrix [C] and the vectors $\{I_i\}$ to find the vectors $\{D_i\}$ for i = 1,2,…,N. Program **MatxInvD** is developed with this concept by modifying the program **Gauss**. It is listed below along with a sample interactive run.

QuickBASIC Version

```
' * Program MatxInvD - Calculates inverse and determinant of a square matrix
          SCREEN 2: CLS : CLEAR : KEY OFF
PRINT " * Program MatxInvD - Calculates inverse and determinant of a square matrix
*"
PRINT : INPUT "Enter the order of the matrix : ", N : DIM C(N, N), C1(N, N)
PRINT : PRINT "Enter the elements of the matrix row-by-row and press <Enter> key"
PRINT "  after entering each element : ": PRINT
FOR J = 1 TO N
    FOR J1=1 TO N: INPUT ; C(J, J1): NEXT J1: PRINT: NEXT J: GOSUB 195
PRINT : PRINT "Determinant = "; D  : IF D <> 0 THEN 170
    PRINT : PRINT "The matrix is singular!": PRINT : END
170 PRINT : PRINT "The inverse matrix is :": PRINT
    FOR J = 1 TO N
        FOR J1=1 TO N: PRINT USING " ##.###^^^^";C1(J,J1);: NEXT J1
        PRINT : NEXT J: PRINT : END
    '
195 ' * Find inverse C1 of C(N,N) by Gaussian elimination *
    '
FOR T1 = 1 TO N
    FOR T2=1 TO N: C1(T1,T2)=0: NEXT T2: C1(T1,T1)=1: NEXT T1: N1=N-1: D=1
FOR T1 = 1 TO N1
'
' * PIVOTING *
'
    Q1 = T1 + 1: IF C(T1, T1) <> 0 THEN 285
    FOR T2 = Q1 TO N: IF C(T2,T1)=0 THEN 245 ELSE D=D*(-1)^(T1+T2): GOTO 260
245     NEXT T2      : D = 0: RETURN
    '
260 ' * Interchanging rows *
    '
    FOR T4=1 TO N
        T = C (T2, T4):  C(T2, T4) =  C(T1, T4):  C(T1,T4)=T
        T = C1(T2, T4): C1(T2, T4) = C1(T1, T4): C1(T1,T4)=T: NEXT T4
    '
285 ' * Normalization *
    '
    FOR T4= Q1 TO N:  C(T1,T4) =  C(T1,T4) / C(T1,T1): NEXT T4
    FOR T4=  1 TO N: C1(T1,T4) = C1(T1,T4) / C(T1,T1): NEXT T4
```

```
    ' * Elimination *
    '
    FOR T4 = T1 + 1 TO N
        FOR T5 = Q1 TO N: C(T4,T5) = C(T4,T5) - C(T4,T1)* C(T1,T5): NEXT T5
        FOR T7 =  1 TO N: C1(T4,T7)=C1(T4,T7) - C(T4,T1)*C1(T1,T7): NEXT T7
        NEXT T4: NEXT T1
    '
    ' * Backward Substitution *
    '
FOR T6 = 1 TO N: C1(N, T6) = C1(N, T6) / C(N, N)
    FOR T4 = 1 TO N1: T1 = N - T4
        FOR T5 = T1 + 1 TO N: C1(T1,T6) = C1(T1,T6)-C(T1,T5)*C1(T5,T6): NEXT T5
        NEXT T4: D = D * C(T6, T6): NEXT T6: RETURN
```

Sample Application

```
 * Program MatxInvD - Calculates inverse and determinant of a square matrix *

Enter the order of the matrix : 3

Enter the elements of the matrix row-by-row and press <Enter> key
  after entering each element :

? 3? 0? 0
? 0? 4? 0
? 0? 0? 5

Determinant = 60

The inverse matrix is :

  3.333E-01  0.000E+00  0.000E+00
  0.000E+00  2.500E-01  0.000E+00
  0.000E+00  0.000E+00  2.000E-01
```

FORTRAN Version

```
C * Program MatxInvD - Calculates inverse and determinant of a square matrix of order N
      DIMENSION C(50,50),C1(50,50)
      WRITE (*,2)
    2 FORMAT(' * Program MatxInvD - Calculates inverse and '
     *        'determinant of a square matrix *')
      WRITE (*,4)
    4 FORMAT(' Enter the order of the matrix : ')
      READ (*,*) N
      WRITE (*,6)
    6 FORMAT(' Enter the elements of the matrix row-by-row and ',
     *        'press <Enter> key after entering an entire row : ')
      DO 10 I=1,N
   10 READ (*,*) (C(I,J),J=1,N)
      CALL MATXINVD(C,50,N,C1,D)
      WRITE (*,*) 'Determinant = ',D
      IF (D.NE.0.) GO TO 20
      WRITE (*,*) 'The matrix is singular!'
      STOP
   20 WRITE (*,*) 'The inverse matrix is :'
      DO 25 I=1,N
   25 WRITE (*,30) (C1(I,J),J=1,N)
   30 FORMAT(4E20.5)
      END

      SUBROUTINE MATXINVD(C,M,N,C1,D)
C
C * Find inverse C1 of C(N,N) by Gaussian elimination *
C   Both C & C1 are dimensioned M by M in the calling program.
C
```

```
      DIMENSION C(M,M),C1(M,M)
      DO 8 I=1,N
         DO 5 J=1,N
    5         C1(I,J)=0
    8    C1(I,I)=1
      N1=N-1
      D=1
      DO 800 IE=1,N1
C
C * PIVOTING *
C
         I1=IE+1
         IF (C(IE,IE).NE.0) GO TO 285
         DO 245 K=I1,N
            IF (C(K,IE).EQ.0.) GO TO 245
            D=D*(-1)**(IE+K)
            GO TO 260
  245       CONTINUE
         D=0
         RETURN


C
C * Interchanging rows *
C
  260 DO 270 J=1,N
         T=C(K,J)
         C(K,J)=C(IE,J)
         C(IE,J)=T
         T=C1(K,J)
         C1(K,J)=C1(IE,J)
  270    C1(IE,J)=T
C
C * Normalization *
C
  285 DO 300 J=I1,N
  300    C(IE,J)=C(IE,J)/C(IE,IE)
      DO 400 J=1,N
  400    C1(IE,J)=C1(IE,J)/C(IE,IE)
C
C * Elimination *
C
      DO 422 K=IE+1,N
         DO 411 J=I1,N
  411       C(K,J)=C(K,J)-C(K,IE)*C(IE,J)
         DO 416 J=1,N
  416       C1(K,J)=C1(K,J)-C(K,IE)*C1(IE,J)
  422    CONTINUE
  800 CONTINUE
C
C * Backward Substitution *
C
      DO 900 J=1,N
         C1(N,J)=C1(N,J)/C(N,N)
         DO 810 I=1,N1
            IR=N-I
            DO 805 K=IR+1,N
  805          C1(IR,J)=C1(IR,J)-C(IR,K)*C1(K,J)
  810       CONTINUE
  900 D=D*C(J,J)
      RETURN
      END
```

Sample Applications

```
* Program MatxInvD - Calculates inverse and determinant of a square matrix *
Enter the order of the matrix :
3
Enter the elements of the matrix row-by-row and press <Enter> key after entering
 an entire row :
3,0,0
0,4,0
0,0,5
Determinant =          60.0000000
The inverse matrix is :
         .33333E+00          .00000E+00          .00000E+00
         .00000E+00          .25000E+00          .00000E+00
         .00000E+00          .00000E+00          .20000E+00
```

MATLAB Application

MATLAB offers very simple matrix operations. For example, matrix inversion can be implemented as:

```
>> A=[1,2;3,4]
   A =
        1     2
        3     4
>> Ainv=inv(A)
   Ainv =
      -2.0000    1.0000
       1.5000   -0.5000
```

To check if the obtained inversion indeed satisfies the equation $[A]\}[A]^{-1} = [I]$ where [I] is the identity matrix, we enter:

```
>> I=A*Ainv
   I =
       1.0000         0
       0.0000    1.0000
```

Once $[A]^{-1}$ becomes available, we can solve the vector {X} in the matrix equation $[A]\{X\} = \{R\}$ if {R} is prescribed, namely $\{X\} = [A]^{-1}\{R\}$. For example, may enter a {R} vector and find {X} such as:

```
>> R=[13;31]
   R =
         13
         31
>> X=A\R
   X =
         5.0000
         4.0000
```

Mathematica Applications

Mathematica has a function called **Inverse** for inversion of a matrix. Let us reuse the matrix A that we have entered in earlier examples and continue to demonstrate the application of **Inverse**:

In[1]: = A = {{1,2},{3,4}}; MatrixForm[A]

Out[1]//MatrixForm =

1 2

3 4

In[2]: = B = {{5,6},{7,8}}; MatrixForm[B]

Out[2]//MatrixForm =

5 6

7 8

In[3]: = MatrixForm[A + B]

Out[3]//MatrixForm =

6 8

10 12

In[4]: = Dif = A-B; MatrixForm[Dif]

Out[4]//MatrixForm =

–4 –4

–4 –4

In[5]: = AT = Transpose[A]; MatrixForm[AT]

Out[5]//MatrixForm =

1 3

2 4

In[6]: = Ainv = Inverse[A]; MatrixForm[Ainv]

Out[6]//MatrixForm =

–2 1

$\frac{3}{2} - \left(\frac{1}{2}\right)$

To verify whether or not the inverse matrix Ainv obtained in *Output[6]* indeed satisfies the equations $[A][A]^{-1} = [I]$ which is the identity matrix, we apply **Mathematica** for matrix multiplication:

In[7]: = Iden = A.Ainv; MatrixForm[Iden]

Out[7]//MatrixForm =

1 0

0 1

A dot is to separate the two matrices A and Ainv which is to be multiplied in that order. Output[7] proves that the computed matrix, Ainv, is the inverse of A! It should be noted that D and I are two *reserved variables* in **Mathematica** for the determinant of a matrix and the identity matrix. In their places, here Dif and Iden are adopted, respectively. For further testing, we show that $[A][A]^T$ is a symmetric matrix:

In[8]: = S = A.AT; MatrixForm[S]

Out[8]//MatrixForm =

5 11

11 25

And, the unknown vector {X} in the matrix equation [A]{X} = {R} can be solved easily if {R} is given and $[A]^{-1}$ are available:

In[9]: = R = {13,31}; X = Ainv.R

Out[9] = {5, 4}

The solution of $x_1 = 5$ and $x_2 = 4$ do satisfy the equations $x_1 + 2x_2 = 13$ and $3x_1 + 4x_2 = 31$.

Transformation of Coordinate Systems, Rotation, and Animation

Matrix algebra can be effectively applied for transformation of coordinate systems. When the cartesian coordinate system, x-y-z, is rotated by an angle Θ_z about the z-axis to arrive at the system x′-y′-z′ as shown in Figure 2, where z and z′ axes coincide and directed outward normal to the plane of paper, the new coordinates of a typical point P whose coordinates are (x_P, y_P, z_P) can be easily obtained as follows:

$$x'_P = OP\cos(\theta_P - \theta_z) = (OP\cos\theta_P)\cos\theta_z + (OP\sin\theta_P)\sin\theta_z$$

$$= x_P\cos\theta_z + y_p\sin\theta_z$$

$$y'_P = OP\sin(\theta_P - \theta_z) = (OP\sin\theta_P)\cos\theta_z - (OP\cos\theta_P)\sin\theta_z$$

$$= x_p\sin\theta_z + y_p\sin\theta_z$$

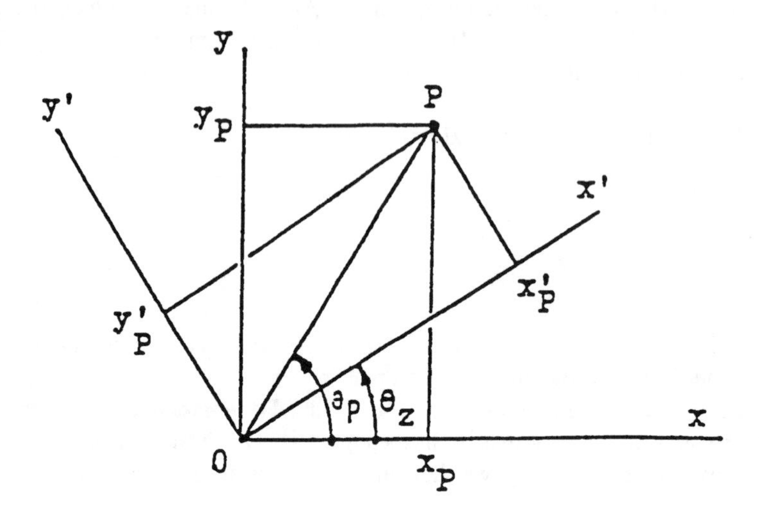

FIGURE 2. The cartesian coordinate system, x-y-z, is rotated by an angle Θ_z about the z-axis to arrive at the system x′-y′-z′.

and

$$z'_P = z_P$$

In matrix notation, we may define $\{P\} = [x_P\ y_P\ z_P]^T$ and $\{P'\} = [x_P'\ y_P'\ z_P']^T$ and write the above equations as $\{P'\} = [T_z]\{P\}$ where the transformation matrix for a rotation of z-axis by Θ_z is:

$$[T_z] = \begin{bmatrix} \cos\theta_z & \sin\theta_z & 0 \\ -\sin\theta_z & \cos\theta_z & 0 \\ 0 & 0 & 1 \end{bmatrix} \tag{5}$$

In a similar manner, it can be shown that the transformation matrices for rotating about the x- and y-axes by angles Θ_x and Θ_y, respectively, are:

$$[T_x] = \begin{bmatrix} 1 & 0 & 0 \\ 0 & \cos\theta_x & \sin\theta_x \\ 0 & -\sin\theta_x & \cos\theta_x \end{bmatrix} \tag{6}$$

and

$$[T_y] = \begin{bmatrix} \cos\theta_y & 0 & -\sin\theta_y \\ 0 & 1 & 0 \\ \sin\theta_y & 0 & \cos\theta_y \end{bmatrix} \tag{7}$$

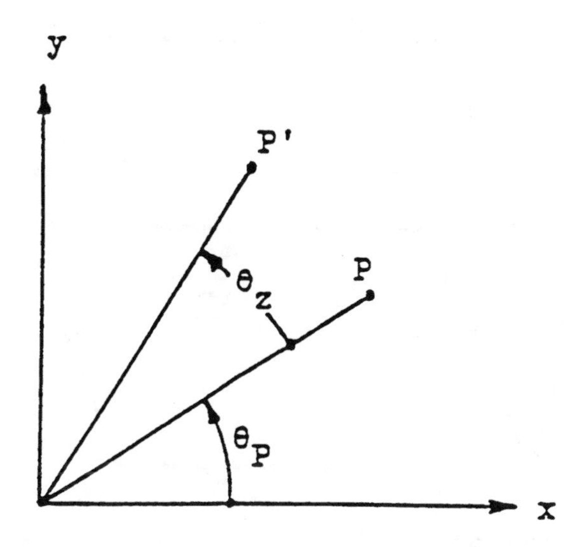

FIGURE 3. Point P whose coordinates are (x_P, y_P, z_P) is rotated to the point P' by a rotation of Θ_z.

It is interesting to note that if a point P whose coordinates are (x_P, y_P, z_P) is rotated to the point P' by a rotation of Θ_z as shown in Figure 3, the coordinates of P' can be easily obtained by the formula $\{P'\} = [R_z]\{P\}$ where $[R_z] = [T_z]^T$. If the rotation is by an angle Θ_x or Θ_y, then $\{P'\} = [R_x]\{P\}$ or $\{P'\} = [R_y]\{P\}$ where $[R_x] = [T_x]^T$ and $[R_y] = [T_y]^T$.

Having discussed about transformations and rotations of coordinate systems, we are ready to utilize the derived formulas to demonstrate the concept of **animation**. Motion can be simulated by first generating a series of rotated views of a three-dimensional object, and showing them one at a time. By erasing each displayed view and then showing the next one at an adequate speed, a smooth motion of the object is achievable to produce the desired animation. Program **Animate1.m** is developed to demonstrate this concept of animation by using a $4 \times 2 \times 3$ brick and rotating it about the x-axis by an angle of 25° and then rotating about the y-axis as many revolutions as desired. The front side of the block (x-y plane) is marked with a character F, and the right side (y-z plane) is marked with a character R, and the top side (x-z plane) is marked with a character T for helping the viewer to have a better three-dimensional perspective of the rotated brick (Figure 4). The x-rotation prior to y-rotation is needed to tilt the top side of the brick toward the front. The speed of animation is controlled by a parameter Damping. This parameter and the desired number of y-revolutions, Ncycle, are both to be interactively specified by the viewer (Figure 5).

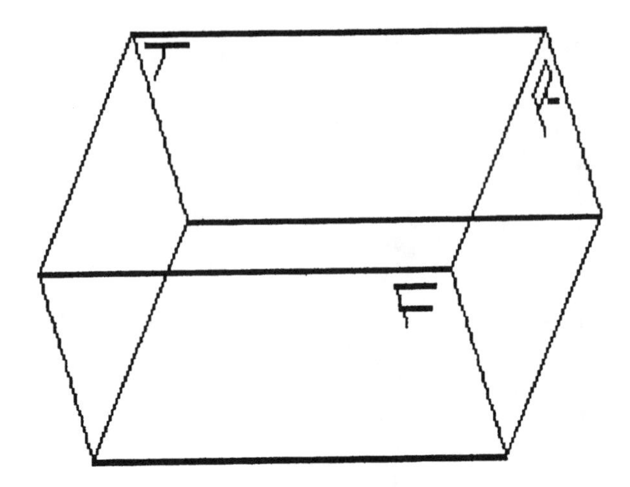

FIGURE 4. The characters F, R, and T help the viewer to have a better three-dimensional perspective of the rotated brick.

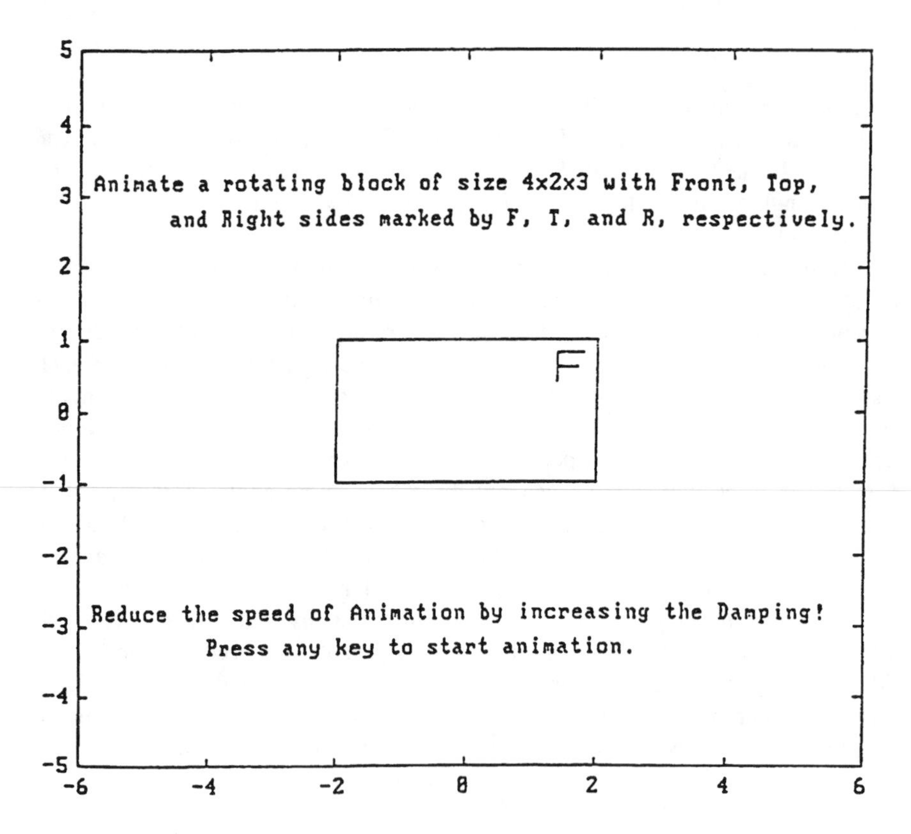

FIGURE 5. The speed of animation is controlled by a parameter Damping. This parameter and the desired number of y-revolutions, Ncycle, are both to be interactively specified by the viewer.

FUNCTION ANIMATE1(NCYCLE,DAMPING)

```
%
% Define the points for drawing block, F, T, and R
  xb=[ -2,  2,   2,  -2, -2, -2,  2,   2,  -2, -2,  2,  2,   2,   2,  -2,  -2];
  yb=[  1,  1,   1,   1,  1, -1, -1,  -1,  -1, -1, -1,  1,   1,  -1,  -1,   1];
  zb=[1.5,1.5,-1.5,-1.5,1.5,1.5,1.5,-1.5,-1.5,1.5,1.5,1.5,-1.5,-1.5,-1.5,-1.5];
  xs=[-6,6,6,-6,-6]; ys=[5,5,-5,-5,5];
  xf=[1.4,1.4,1.8,1.4,1.4,1.7]; yf=[.4,.8,.8,.8,.6,.6];
                              zf=[1.5,1.5,1.5,1.5,1.5,1.5];
  xt=[-1.8,-1.4,-1.6,-1.6]; yt=[1,1,1,1]; zt=[-1.3,-1.3,-1.3,-.9];
  xr=[2,2,2,2,2,2,2]; yr=[.4,.8,.8,.6,.6,.6,.4];
                 zr=[-.9,-.9,-1.3,-1.3,-.9,-1.1,-1.3];
%
% Describe the animation
  plot(xs,ys), hold, plot(xb,yb), plot(xf,yf), plot(xt,yt), plot(xr,yr)
  text(-5.8,3,'Animate a rotating block of size 4x2x3 with Front, Top,')
  text(-4.6,2.5,'and Right sides marked by F, T, and R, respectively.')
  text(-5.8,-3,'Reduce the speed of Animation by increasing the Damping!')
  text(-4,-3.5,'Press any key to start animation.')
  pause
%
% X-rotation
  for kxr=1:5; ax=5*kxr*pi/180; c=cos(ax); s=sin(ax);
      ybn=yb*c-zb*s; zbn=yb*s+zb*c; yfn=yf*c-zf*s; zfn=yf*s+zf*c;
      ytn=yt*c-zt*s; ztn=yt*s+zt*c; yrn=yr*c-zr*s; zrn=yr*s+zr*c;
      clg, plot(xb,ybn), plot(xf,yfn), plot(xt,ytn), plot(xr,yrn)
      for hold=1:Damping; end
      end
%
% Y-rotation
  kend=36*Ncycle;
  for kyr=1:kend; ay=-kyr*pi/18; c=cos(ay); s=sin(ay);
      xbn=xb*c+zbn*s; xfn=xf*c+zfn*s; xtn=xt*c+ztn*s; xrn=xr*c+zrn*s;
      clg, plot(xbn,ybn), plot(xfn,yfn), plot(xtn,ytn), plot(xrn,yrn)
      for hold=1:Damping; end
      end
  end
```

Notice that the coordinates for the corners of the brick are defined in arrays xb, yb, and zb. The coordinates of the points to be connected by linear segments for drawing the characters F, R, ant T are defined in arrays xf, yf, and zf, and xr, yr, and zr, and xt, yt, and zt, respectively.

The equations in deriving $[R_x]$ (= $[T_x]^T$) and $[R_y]$ (= $[T_y]^T$) are applied for x- and y- rotations in the above program. Angle increments of 5 and 10° are arranged for the x- and y-rotations, respectively. The rotated views are plotted using the new coordinates of the points, (xbn,ybn,zbn), (xfn,yfn,zfn), etc. Not all of these new arrays but only those needed in subsequent plot are calculated in this m file.

MATLAB command **clg** is used to erase the graphic window before a new rotated view the brick is displayed. The speed of animation is retarded by the "hold" loops in both x- and y-rotations involving the interactively entered value of the parameter Damping. The **MATLAB** command **pause** enables Figure 4 to be read and requires the viewer to press any key on the keyboard to commence the animation. Notice that a statement begins with a **%** character making that a comment statement, and that **%** can also be utilized for spacing purpose.

The xs and ys arrays allow the graphic window to be scaled by plotting them and then held (by command **hold**) so that all subsequent plots are using the same

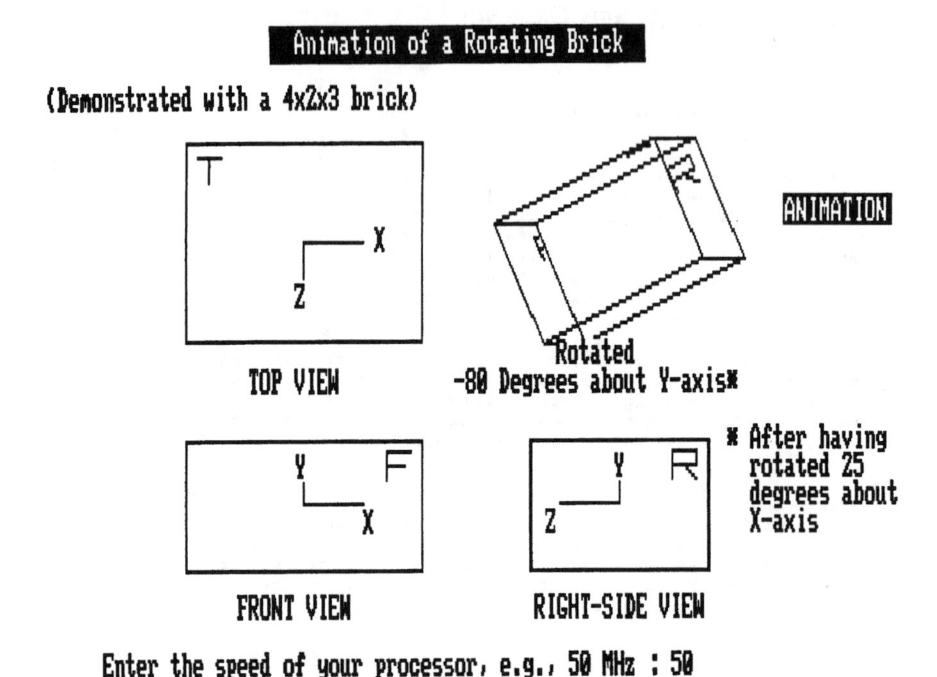

FIGURE 6. Animation of a rotating brick.

scales in both x- and y-directions. The values in xs and ys arrays also control where to properly place the texts in Figure 4 as indicated in the **text** statements.

QuickBASIC Version

A **QuickBASIC** version of the program **Animatel.m** called **Animatel.QB** also is provided. It uses commands GET and PUT to animate the rotation of the 4 × 3 × 2 brick. More features have been added to show the three principal views of the brick and also the rotated view at the northeast corner of screen, as illustrated in Figure 6.

The window-viewport transformation of the rotated brick for displaying on the screen is implemented through the functions FNTX and FNTY. The actual ranges of the x and y measurements of the points used for drawing the brick are described by the values of V1 and V2, and V3 and V4, respectively. These ranges are mapped onto the screen matching the ranges of W1 and W2, and W4 and W3, respectively.

The rotated views of the brick are stored in arrays S1 through S10 using the **GET command**. Animation retrieves these views by application of the **PUT** command. Presently, animation is set for 10 y-swings (Ncycle = 10 in the program **Animatel.m**, arranged in Line 600). The parameter **Damping** described in the program **Animatel.m** here is set equal to 1500 (in Line 695).

```
        * Program Animat1.QB - animates a Rotating Brick *
  DECLARE SUB XRotate (N,AX,XS(),YS(),ZS(),XN(),YN(),ZN())
  DECLARE SUB YRotate (N,AY,XS(),YS(),ZS(),XN(),YN(),ZN())
  CLEAR: CLS: KEY OFF: CV = 3.1416/180: SCREEN 2: W=8: H=8: ASPR=11/25: N=24: NS=1250
  DIM S1(NS),S2(NS),S3(NS),S4(NS),S5(NS),S6(NS),S7(NS),S8(NS),S9(NS),S10(NS),SV(200),
  DIM X(N),X0(N),XN(N),XS(N),Y(N),Y0(N),YN(N),YS(N),Z0(N),ZN(N),ZS(N)
  FOR I = 1 TO N: READ X0(I), Y0(I), Z0(I): NEXT I
  DATA   -2,-1,-1.5,    2,-1,-1.5,    2, 1,-1.5,   -2, 1,-1.5,   -2,-1,1.5,    2,-1,1.5,
2, 1, 1.5
  DATA   -2, 1, 1.5,  1.4,.4, 1.5,  1.4,.6, 1.5, 1.4,.8, 1.5, 1.8,.8,1.5, 1.7,.6,1.5,
-1.8, 1,-1.3
  DATA -1.6, 1,-1.3, -1.4, 1,-1.3, -1.6, 1, -.9,   2,.4, -.9,   2,.6,-.9,   2,.8,-.9,
2,.8,-1.3      DATA    2,.6,-1.3,     2,.6,-1.1,     2,.4,-1.3
  LOCATE 2, 26: PRINT " Animation of a Rotating Brick"
    GET (24.5 * W, H - 1)-(56.5 * W, 2 * H + 1), S1: PUT (24.5 * W, H - 1), S1, PRESET
  LOCATE 4, 5: PRINT "(Demonstrated with a 4x2x3 brick)"
    DEF FNTX (X) = V1 + (X - W1) * TS1 : DEF FNTY (Y) = V4 - (Y - W3) * TS2
    V1 =11*W: V2 = V1 + 31*W: V3 = 10*H: V4 = V3 + 31*W*ASPR:  W1=-3 : W2=3 : W3=-3 : W4=3
    TS1=(V2-V1)/(W2-W1): TS2=(V4-V3)/(W4-W3)  :  LOCATE 22, 22: PRINT "FRONT VIEW"
    FOR I=1 TO N: X(I)=X0(I): Y(I)=Y0(I): NEXT I               'DRAW FRONTAL VIEW, Transform
and plot
    GOSUB 480: GOSUB 490
    PSET (FNTX(1),FNTY(0)): LINE -(FNTX(0),FNTY(0)): LINE -(FNTX(0),FNTY(.5))
  LOCATE 18, 33: PRINT "X": LOCATE 16, 26: PRINT "Y"
    CALL YRotate(N,-90*CV,X0(),Y0(),Z0(),XN(),YN(),ZN())
         V1 = 39 * W: V2 = V1 + 31 * W                                                'DRAW
RIGHT VIEW
  LOCATE 22,  48: PRINT "RIGHT-SIDE VIEW"
    FOR I=1 TO N: X(I)=XN(I): Y(I)=YN(I): NEXT I: GOSUB 480: GOSUB 490        'Transform
and plot
    PSET (FNTX(-1),FNTY(0)) : LINE -(FNTX(0),FNTY(0)): LINE -(FNTX(0),FNTY(.5))
  LOCATE 18, 49 :  PRINT "Z": LOCATE 16, 56: PRINT "Y"
    CALL XRotate(N,90*CV,X0(),Y0(),Z0(),XN(),YN(),ZN())
         V1=11*W: V2=V1+31*W: V3=1.25*H: V4=V3+31*W*ASPR
  LOCATE 14, 23 : PRINT "TOP VIEW"                                                 'DRAW
TOP VIEW
    FOR I=1 TO N: X(I)=XN(I): Y(I)=YN(I): NEXT I: GOSUB 480: GOSUB 490        'Transform
and plot
    PSET (FNTX(1),FNTY(0))  : LINE -(FNTX(0),FNTY(0)): LINE -(FNTX(0),FNTY(-.5))
  LOCATE 9, 34  : PRINT "X" : LOCATE 11, 27: PRINT "Z"
    V1=43*W      : V2=V1+20*W: V3=3*H       : V4=V3+20*W*ASPR                   'ROTATE
ABOUT X-AXIS
    FOR XL=1 TO 5              : TAX=TAX+5    : GET (43*W,3*H)-(68*W,14*H),S1
         TS1 = (V2 - V1) / (W2 - W1): TS2 = (V4-V3)/(W4-W3)
         IF XL = 1 THEN 300 ELSE GOSUB 580: GOTO 305                              'Update
before rotate
300      FOR I = 1 TO N: XS(I) = X0(I): YS(I) = Y0(I): ZS(I) = Z0(I): NEXT I
305      CALL XRotate(N,5*CV,XS(),YS(),ZS(),XN(),YN(),ZN()): PUT (43*W,3*H),S1,XOR
         FOR I =1 TO N: X(I)=XN(I): Y(I)=YN(I): NEXT I
         GOSUB 480: GOSUB 490: GOSUB 695                                         'Transform
and plot
         LOCATE 12,53: PRINT "Rotated" : LOCATE 13,45: PRINT USING "## Degrees about
X-axis";TAX
         NEXT XL      : AY = -10          : TAY=0
    FOR YL=1 TO 9: TAY=TAY+AY: X1=43*W: Y1 =3*H: X2=68*W: Y2=14*H: GOSUB 580   'ROTATE
ABOUT Y-AXIS
         ON YL GOTO 340, 345, 350, 355, 360, 365, 370, 375, 380
340      GET (X1, Y1)-(X2, Y2), S1: GOTO 385
345      GET (X1, Y1)-(X2, Y2), S2: GOTO 385
350      GET (X1, Y1)-(X2, Y2), S3: GOTO 385
355      GET (X1, Y1)-(X2, Y2), S4: GOTO 385
360      GET (X1, Y1)-(X2, Y2), S5: GOTO 385
365      GET (X1, Y1)-(X2, Y2), S6: GOTO 385
370      GET (X1, Y1)-(X2, Y2), S7: GOTO 385
375      GET (X1, Y1)-(X2, Y2), S8: GOTO 385
380      GET (X1, Y1)-(X2, Y2), S9
385      CALL YRotate(N,AY*CV,XS(),YS(),ZS(),XN(),YN(),ZN())
         FOR I = 1 TO N: X(I) = XN(I): Y(I) = YN(I): NEXT I
         ON YL GOTO 395, 400, 405, 410, 415, 420, 425, 430, 435
395      PUT (X1, Y1), S1, XOR: GOTO 440
400      PUT (X1, Y1), S2, XOR: GOTO 440
405      PUT (X1, Y1), S3, XOR: GOTO 440
410      PUT (X1, Y1), S4, XOR: GOTO 440
415      PUT (X1, Y1), S5, XOR: GOTO 440
420      PUT (X1, Y1), S6, XOR: GOTO 440
425      PUT (X1, Y1), S7, XOR: GOTO 440
430      PUT (X1, Y1), S8, XOR: GOTO 440
435      PUT (X1, Y1), S9, XOR
440      GOSUB 480: GOSUB 490                                                    'Transform
and plot
```

```
  LOCATE 12,53: PRINT "Rotated": LOCATE 13,44: PRINT USING "### Degrees about Y-axis*";TAY
  IF YL <> 1 GOTO 470
     LOCATE 16,65: PRINT "* After having": LOCATE 17,67: PRINT "rotated 25"
     LOCATE 18,67: PRINT "degrees about" : LOCATE 19,67: PRINT "X-axis"
470        NEXT YL: GET (X1, Y1)-(X2, Y2), S10
  FOR I=1 TO 500 : NEXT: GOSUB 590: LOCATE 22,1     :           END
480 FOR I=1 TO N : X(I)=FNTX(X(I)): Y(I)=FNTY(Y(I)): NEXT I: RETURN     'Converts window to
viewport
490 PSET  (X(5),Y(5)): LINE -(X(6),Y(6)): LINE -(X(2),Y(2)): LINE -(X(1),Y(1))
'DRAW BRICK
    LINE -(X(5),Y(5)): LINE -(X(8),Y(8)): LINE -(X(7),Y(7)): LINE -(X(3),Y(3)): LINE
-(X(4),Y(4))
    LINE -(X(8),Y(8)): PSET  (X(6),Y(6)): LINE -(X(7),Y(7)): PSET  (X(3),Y(3)): LINE
-(X(2),Y(2))
    PSET  (X(1),Y(1)): LINE -(X(4),Y(4))
    PSET  (X(9),Y(9)): LINE -(X(11),Y(11)): LINE -(X(12),Y(12)): PSET (X(10),Y(10))
'DRAW "F"
                       LINE -(X(13),Y(13))
    PSET (X(14),Y(14)): LINE -(X(16),Y(16)): PSET  (X(17),Y(17)): LINE -(X(15),Y(15))
'DRAW "T"
    PSET (X(18),Y(18)): LINE -(X(20),Y(20)): LINE -(X(21),Y(21)): LINE -(X(22), Y(22))
'DRAW "R"
    LINE -(X(19),Y(19)):PSET  (X(23),Y(23)): LINE -(X(24),Y(24)): RETURN
580 FOR I=1 TO N: XS(I)=XN(I): YS(I)=YN(I)  : ZS(I)=ZN(I): NEXT I: RETURN
'Updating
590 LOCATE 8, 70: PRINT "ANIMATION": GET (68.5*W,7*H-1)-(78.5*W,8*H),SV
'Animation
                              PUT (68.5*W,7*H-1),SV,PRESET
600 FOR I=1 TO 10: PUT (X1,Y1),S10,PSET: GOSUB 695: PUT (X1,Y1),S9,PSET: GOSUB 695
                   PUT (X1,Y1), S8,PSET: GOSUB 695: PUT (X1,Y1),S7,PSET: GOSUB 695
                   PUT (X1,Y1), S6,PSET: GOSUB 695: PUT (X1,Y1),S5,PSET: GOSUB 695
                   PUT (X1,Y1), S4,PSET: GOSUB 695: PUT (X1,Y1),S3,PSET: GOSUB 695
                   PUT (X1,Y1), S2,PSET: GOSUB 695: PUT (X1,Y1),S1,PSET: GOSUB 695
                   PUT (X1,Y1), S2,PSET: GOSUB 695: PUT (X1,Y1),S3,PSET: GOSUB 695
                   PUT (X1,Y1), S4,PSET: GOSUB 695: PUT (X1,Y1),S5,PSET: GOSUB 695
                   PUT (X1,Y1), S6,PSET: GOSUB 695: PUT (X1,Y1),S7,PSET: GOSUB 695
                   PUT (X1,Y1), S8,PSET: GOSUB 695: PUT (X1,Y1),S9,PSET: GOSUB 695: NEXT
I: RETURN
695 FOR PS = 1 TO 1500: NEXT PS: RETURN
 'pause

    SUB XRotate (N,AX,XS(),YS(),ZS(),XN(),YN(),ZN())
'   X-ROTATION
      CS = COS(AX): SN = SIN(AX)
    FOR I=1 TO N: XN(I)=XS(I): YN(I)=YS(I)*CS-ZS(I)*SN: ZN(I)=YS(I)*SN+ZS(I)*CS: NEXT I
    END SUB

    SUB YRotate (N,AY,XS(),YS(),ZS(),XN(),YN(),ZN())
'   Y-ROTATION
      CS = COS(AY): SN = SIN(AY)
    FOR I=1 TO N: YN(I)=YS(I): XN(I)=XS(I)*CS+ZS(I)*SN: ZN(I)=-XS(I)*SN+ZS(I)*CS: NEXT I
    END SUB
```

1.6 PROBLEMS

MATRIX ALGEBRA

1. Calculate the product [A][B][C] by (1) finding [T] = [A][B] and then [T][C], and (2) finding [T] = [B][C] and then [A][T] where:

$$[A]=\begin{bmatrix}1 & 2 & 3\\ 4 & 5 & 6\end{bmatrix} \quad [B]=\begin{bmatrix}6 & 5\\ 4 & 3\\ 2 & 1\end{bmatrix} \quad [C]=\begin{bmatrix}-1 & -2\\ -3 & -4\end{bmatrix}$$

2. Calculate [A][B] of the two matrices given above and then take the transpose of product matrix. Is it equal to the product of $[B]^T[A]^T$?
3. Are $([A][B][C])^T$ and the product $[C]^T[B]^T[A]^T$ identical to each other?

4. Apply the **QuickBASIC** and **FORTRAN** versions of the program **Matx-Algb** to verify the results of Problems 1, 2, and 3.
5. Repeat Problem 4 but use **MATLAB.**
6. Apply the program **MatxInvD** to find $[C]^{-1}$ of the matrix [C] given in Problem 1 and also to $([C]^T)^{-1}$. Is $([C]^{-1})^T$ equal to $([C]^T)^{-1}$?
7. Repeat Problem 6 but use **MATLAB.**
8. For statistical analysis of a set of N given data $X_1, X_2, \ldots, X_N$, it is often necessary to calculate the *mean*, m, and *standard deviation*, 5, by use of the formulas:

$$m = \frac{1}{N}(X_1 + X_2 + \ldots + X_N)$$

and

$$\sigma = \left\{\frac{1}{N}\left[(X_1 - m)^2 + (X_2 - m)^2 + \ldots + (X_N - m)^2\right]\right\}^{0.5}$$

Use indicial notation to express the above two equations and then develop a subroutine meanSD(X,N,RM,SD) for taking the N values of X to compute the real value of mean, RM, and standard deviation, SD.

9. Express the ith term in the following series in indicial notation and then write an interactive program **SinePgrm** allowing input of the x value to calculate sin(x) by terminating the series when additional term contributes less than 0.001% of the partial sum of series in magnitude:

$$\text{Sin } x = \frac{x^1}{1!} - \frac{x^3}{3!} + \frac{x^5}{5!} - \ldots$$

Notice that Sin(x) is an *odd function* so the series contains only terms of odd powers of x and the series carries alternating signs. Compare the result of the program SinePgrm with those obtained by application of the library function Sin available in **FORTRAN** and **QuickBASIC.**

10. Same as Problem 9, but for the cosine series:

$$\text{Cos } x = 1 - \frac{x^2}{2!} + \frac{x^4}{4!} - \frac{x^6}{6!} + \ldots$$

Notice that Cos(x) is an *even function* so the series contains only terms of even powers of x and the series also carries alternating signs.

11. Repeat Problem 4 but use **Mathematica.**
12. Repeat Problem 6 but use **Mathematica.**

GAUSS

1. Run the program **GAUSS** to solve the problem:

$$\begin{bmatrix} 1 & 2 & 3 \\ 4 & 5 & 6 \\ 7 & 8 & 10 \end{bmatrix} \begin{bmatrix} x_1 \\ x_2 \\ x_3 \end{bmatrix} = \begin{bmatrix} 2 \\ 8 \\ 14 \end{bmatrix}$$

2. Run the program **GAUSS** to solve the problem:

$$\begin{bmatrix} 0 & 2 & 3 \\ 4 & 5 & 6 \\ 7 & 8 & 9 \end{bmatrix} \begin{bmatrix} x_1 \\ x_2 \\ x_3 \end{bmatrix} = \begin{bmatrix} -1 \\ 8 \\ 14 \end{bmatrix}$$

   What kind of problem do you encounter? "Divided by zero" is the message! This happens because the coefficient associated with x_1 in the first equation is equal to zero and the normalization in the program **GAUSS** cannot be implemented. In this case, the order of the given equations needs to be interchanged. That is to put the second equation on top or find *below* the first equation an equation which has a coefficient associated with x_1 not equal to zero and is to be interchanged with the first equation. This procedure is called "pivoting." Subroutine **GauJor** has such a feature incorporated, apply it for solving the given matrix equation.

3. Modify the program **GAUSS** by following the Gauss-Jordan elimination procedure and excluding the back-substitution steps. Name this new program **GauJor** and test it by solving the matrix equations given in Problems 1 and 2.
4. Show all details of the normalization, elimination, and backward substitution steps involved in solving the following equations by application of Gaussian Elimination method:

$$4x_1 + 2x_2 - 3x_3 = 8$$

$$5x_1 - 3x_2 + 7x_3 = 26$$

$$-x_1 + 9x_2 - 8x_3 = -10$$

5. Present every normalization and elimination steps involved in solving the following system of linear algebraic equations by the Gaussian Elimination Method:

$$5x_1 - 2x_2 + 2x_3 = 9, \; -2x_1 + 7x_2 - 2x_3 = 9, \text{ and } 2x_1 - 2x_2 + 9x_3 = 41$$

6. Apply the Gauss-Jordan elimination method to solve for x_1, x_2, and x_3 from the following equations:

$$\begin{bmatrix} 0 & 1 & -1 \\ 2 & 9 & 3 \\ 4 & 24 & 7 \end{bmatrix} \begin{bmatrix} x_1 \\ x_2 \\ x_3 \end{bmatrix} = \begin{bmatrix} 1 \\ 1 \\ 1 \end{bmatrix}$$

Show every normalization, elimination, and pivoting (if necessary) steps of your calculation.

7. Solve the matrix equation [A]{X} = {C} by **Gauss-Jordan** method where:

$$\begin{bmatrix} 3 & 2 & 1 \\ 2 & 5 & -1 \\ 4 & 1 & 7 \end{bmatrix} \begin{bmatrix} x_1 \\ x_2 \\ x_3 \end{bmatrix} = \begin{bmatrix} -2 \\ -3 \\ 3 \end{bmatrix}$$

Show every interchange of rows (if you are required to do pivoting before normalization), normalization, and elimination steps by indicating the changes in [A] and {C}.

8. Apply the program **GauJor** to solve Problem 7.
9. Present every normalization and elimination steps involved in solving the following system of linear algebraic equations by the *Gauss-Jordan* Elimination Method:

$$5x_1 - 2x_2 + x_3 = 4$$

$$-2x_1 + 7x_2 - 2x_3 = 9$$

$$x_1 - 2x_2 + 9x_3 = 40$$

10. Apply the program **Gauss** to solve Problem 9 described above.
11. Use **MATLAB** to solve the matrix equation given in Problem 7.
12. Use **MATLAB** to solve the matrix equation given in Problem 9.
13. Use **Mathematica** to solve the matrix equation given in Problem 7.
14. Use **Mathematica** to solve the matrix equation given in Problem 9.

Matrix Inversion

1. Run the program **MatxInvD** for finding the inverse of the matrix:

$$[A] = \begin{bmatrix} 3 & 0 & 2 \\ 0 & 5 & 0 \\ 2 & 0 & 3 \end{bmatrix}$$

2. Write a program Invert3 which inverts a given 3 × 3 matrix [A] by using the cofactor method. A subroutine COFAC should be developed for calculating the cofactor of the element at Ith row and Jth column of [A] in term of the elements of [A] and the user-specified values of I and J. Let the inverse of [A] be designated as [AI] and the determinant of [A] be designated as D. Apply the developed program **Invert3** to generate all elements of [AI] by calling the subroutine **COFAC** and by using D.
3. Write a **QuickBASIC** or **FORTRAN** program **MatxSorD** which will perform the addition and subtraction of two matrices of same order.
4. Write a **QuickBASIC** or **FORTRAN** program **MxTransp** which will perform the transposition of a given matrix.
5. Translate the **FORTRAN** subroutine **MatxMtpy** into a **MATLAB** m file so that by entering the matrices [A] and [B] of order L by M and M by N, respectively, it will produce a product matrix [P] of order L by N.
6. Enter **MATLAB** commands interactively first a square matrix [A] and then calculate its trace.
7. Use **MATLAB** commands to first define the elements in its upper right corner including the diagonal, and then use the symmetric properties to define those in the lower left corner.
8. Convert either **QuickBasic** or FORTRAN version of the program MatxInvD into a MATLAB function file MatxInvD.m with a leading statement function [Cinv,D] = MatxInvD(C,N)
9. Apply the program MatxInvD to invert the matrix:

$$[A] = \begin{bmatrix} 1 & 3 & 4 \\ 5 & 6 & 7 \\ 8 & 9 & 10 \end{bmatrix}$$

   Verify the answer by using Equation 1.
10. Repeat Problem 9 but by **MATLAB** operation.
11. Apply the program **MatxInvD** to invert the matrix:

$$[A] = \begin{bmatrix} -9 & -1 & -2 \\ -3 & -4 & -5 \\ -6 & -7 & -8 \end{bmatrix}$$

   Verify the answer by using Equation 1.
12. Repeat Problem 11 but by **MATLAB** operations.
13. Derive $[R_x]$ and verify that it is indeed equal to $[T_x]^T$. Repeat for $[R_y]$ and $[R_z]$.
14. Apply **MATLAB** to generate a matrix $[R_z]$ for $\theta_z = 45°$ and then to use $[R_z]$ to find the rotated coordinates of a point P whose coordinates before rotation are (1,–2,5).

FIGURE 7. Problem 18.

15. What will be the coordinates for the point P mentioned in Problem 14 if the coordinate axes are rotated counterclockwise about the z-axis by 45°? Use **MATLAB** to find your answer.
16. Apply **MATLAB** to find the location of a point whose coordinates are (1,2,3) after three rotations in succession: (1) about y-axis by 30°, (2) about z-axis by 45° and then (3) about x-axis by –60°.
17. Change m file Animate1.m to animate just the rotation of the front (F) side of the 4 × 2 × 3 brick in the graphic window.
18. Write a **MATLAB** m file for animation of pendulum swing[1] as shown in Figure 7.
19. Write a **MATLAB** m file for animation of a bouncing ball[1] using an equation of $y = 3e^{-0.1x}\sin(2x + 1.5708)$ as shown in Figure 8.
20. Write a **MATLAB** m file for animation of the motion of crank-piston system as shown in Figure 9.
21. Write a **MATLAB** m file to animate the vibrating system of a mass attached to a spring as shown in Figure 10.

* Bouncing-Ball Animation *

* To see how a ball bounces, we need a ball ! Let us create a ball first.

Now, let it bounce !!

Want to see it again, enter Y/N ?

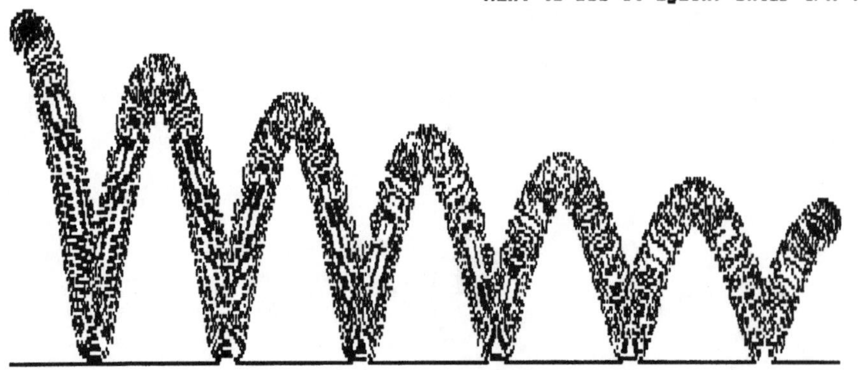

Enter the speed of your processor, e.g., enter 50 for 50 MHz : 50

FIGURE 8. Problem 19.

22. Write a **MATLAB** m file to animate the motion of a cam-follower system as shown in Figure 11.
23. Write a **MATLAB** m file to animate the rotary motion of a wankel cam as shown in Figure 12.
24. Repeat Problem 9 but by **Mathematica** operation.
25. Repeat Problem 11 but by **Mathematica** operation.
26. Repeat Problem 14 but by **Mathematica** operation.
27. Repeat Problem 15 but by **Mathematica** operation.
28. Repeat Problem 16 but by **Mathematica** operation.

Reciprocating Piston Engine

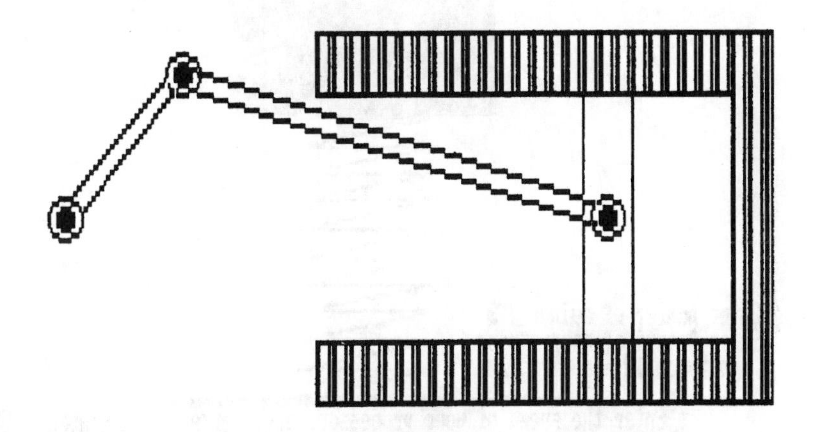

Crank rotates

Connecting Rod converts the motion

Piston slides

Developed by Chris L. Jensen,
enhanced by Dr. Y. C. Pao, May, 1986

FIGURE 9. Problem 20.

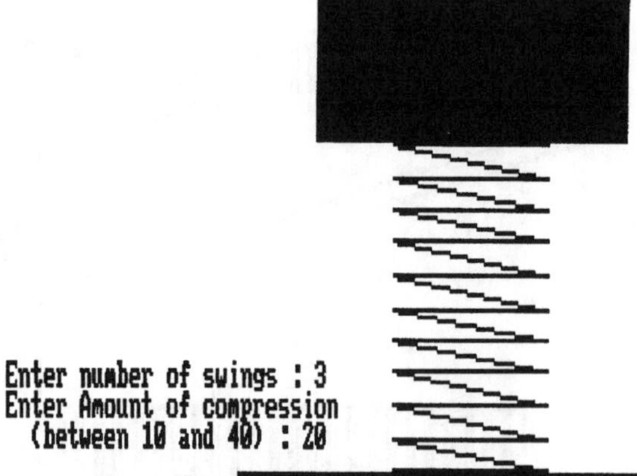

FIGURE 10. Problem 21.

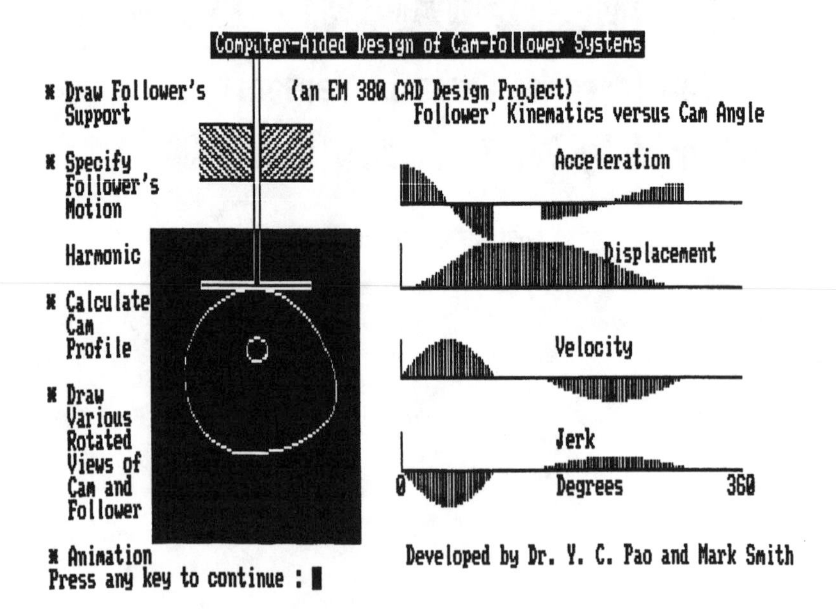

FIGURE 11. Problem 22.

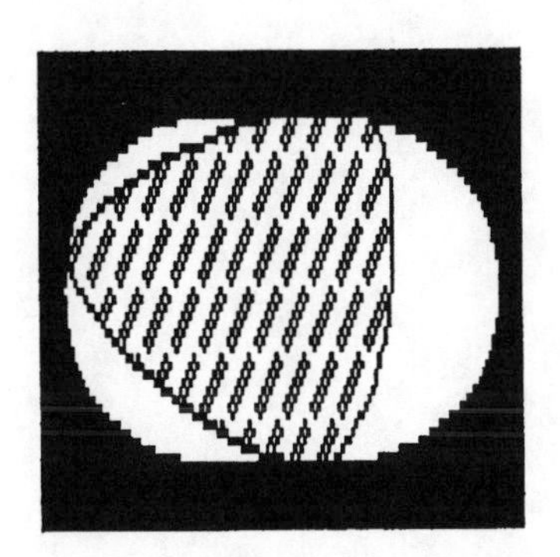

FIGURE 12. Problem 23.

1.7 REFERENCE

1. Y. C. Pao, "On Development of Engineering Animation Software," in *Computers in Engineering*, edited by K. Ishii, ASME Publications, New York, 1994, pp. 851–855.

2 Exact, Least-Squares, and Cubic Spline Curve-Fits

2.1 INTRODUCTION

Engineers conduct experiments and collect data in the laboratories. To make use of the collected data, these data often need to be fitted with some particularly selected curves. For example, one may want to find a parabolic equation $y = c_1 + c_2x + c_3x^2$ which passes three given points (x_i,y_i) for i = 1,2,3. This is a problem of *exact curve-fit*. Or, since knowing in advance that these three points should all fall on a straight line, but the reason that they are not is because of bad calibration of the measuring equipment or because of presence of noises in the testing environment.

In case that we may want express this straight line by the equation $y = c_1 + c_2x$ for the stress and strain data collected for a stretching test of a metal bar in the elastic range, then the question of how to determine the two coefficients c_1 and c_2 is a matter of deciding on which criterion to adopt. The *Least-Squares* method is one of the criteria which is most popularly used. The two cases cited are the consideration of adopting the two and three lowest *polynomial* terms, x^0, x^1, and x^2, and linearly combining them.

If the collected data are supposed to represent a sinusoidal function of time, the curve to be determined may have to be assumed as $x(t) = c_1\sin t + c_2\sin 3t + c_3\sin 5t + c_4\sin 7t$ by linearly combining 4 odd sine terms. This is the case of selecting four particular functions, namely, $f_i(t) = \sin(2i-1)t$ for i = 1,2,3,4., and to determine the coefficients c_{1-4} by application of the least-squares method.

Often some special form of curve needs to be selected to fit a given set of data, the least-squares criterion can still be applied if mathematical transformations can be found to convert the equation describing the curve into linear equations. This is discussed in a section devoted to transformed least-squares curve-fit.

Another commonly applied curve-fit technique is the cubic spline method which allows smooth cubic equations to be derived to ensure continuous slopes and curvatures passing all given points. The mathematics involved in this method will be presented.

In the following sections, we shall discuss the development of the programs **ExactFit**, **LeastSq1**, **LeastSqG**, and **CubeSpln** for the four curve-fit needs mentioned above.

2.2 EXACT CURVE FIT

As another example of solving a matrix equation, let us consider the problem of finding a parabolic equation $y = c_1 + c_2x + c_3x^2$ which passes three given points

(x_i,y_i) for i = 1,2,3. This is a problem of *exact curve-fit*. By simple substitutions of the three points into the parabolic equation, we can obtain:

$$c_1 + c_2 x_i + c_3 x_i^2 = y_i \qquad \text{for } i = 1,2,3 \tag{1}$$

In matrix form, we write these equations as:

$$[A]\{C\} = \{Y\} \tag{2}$$

where $\{C\} = [c_1\ c_2\ c_3]^T$, $\{Y\} = [y_1\ y_2\ y_3]^T$, and [A] is a three-by-three coefficient matrix whose elements if denoted as $a_{i,j}$ are to be calculated using the formula:

$$a_{i,j} = x_i^{j-1} \qquad \text{for } i,j = 1,2,3 \tag{3}$$

It is easy to extend the above argument for the general case of exactly fitting N given points by a N-1st degree polynomial $y = c_1 + c_2 x + \cdots + c_N x^{N-1}$. The only modification needed is to allow the indices i and j in Equations 1 and 3 to be extended to reach a value of N. A program called **ExactFit** has been prepared for this need by utilizing the subroutine **GauJor** to solve for the vector {C} from Equation 1 for the general case of exactly fitting N given points. Listings for both **FORTRAN** and **QuickBASIC** versions along with sample numerical results are presented below.

FORTRAN VERSION

```
C PROGRAM ExactFit - Exact-fit of N given Points with a N-1st degree polynomial.
C                    Setup for N no greater than 50.
      DIMENSION A(50,50),C(50),X(50),Y(50)
      WRITE (*,*) 'Program ExactFit - Exact-fit of N given points by'
      WRITE (*,*) '                   a N-1st degree polynomial.'
      WRITE (*,*) 'Input the number of points, N : '
      READ  (*,*) N
      WRITE (*,*) 'Input the coordinates (X,Y) of the points,'
      WRITE (*,*) '  two numbers each time and press <Enter> key'
      WRITE (*,*) '  after entering.'
      DO 5 I=1,N
      READ (*,*) X(I),Y(I)
      C(I)=Y(I)
      DO 5 J=1,N
    5 A(I,J)=X(I)**(J-1)
      CALL GauJor(A,N,50,C,D)
      WRITE (*,*) 'The coefficients in Y=C(1)+C(2)X+...+C(N)X**(N-1)'
      WRITE (*,*) '  are :'
      WRITE (*,15) (C(I),I=1,N)
   15 FORMAT(5E16.5)
      STOP
      END
```

Sample Applications

```
Program ExactFit - Exact-fit of N given points by
                   a N-1st degree polynomial.
Input the number of points, N :
2
Input the coordinates (X,Y) of the points,
  two numbers each time and press <Enter> key
  after entering.
1,2
2,2
The coefficients in Y=C(1)+C(2)X+...+C(N)X**(N-1)
  are :
     .20000E+01      .00000E+00
Stop - Program terminated.

Program ExactFit - Exact-fit of N given points by
                   a N-1st degree polynomial.
Input the number of points, N :
3
Input the coordinates (X,Y) of the points,
  two numbers each time and press <Enter> key
  after entering.
1,2
1.5,2.5
3,4
The coefficients in Y=C(1)+C(2)X+...+C(N)X**(N-1)
  are :
     .10000E+01      .10000E+00      .00000E+00
Stop - Program terminated.

Program ExactFit - Exact-fit of N given points by
                   a N-1st degree polynomial.
Input the number of points, N :
3
Input the coordinates (X,Y) of the points,
  two numbers each time and press <Enter> key
  after entering.
1,1
2,5
4,4
The coefficients in Y=C(1)+C(2)X+...+C(N)X**(N-1)
  are :
    -.60000E+01      .85000E+01     -.15000E+01
Stop - Program terminated.
```

QuickBASIC Version

```
'*** PROGRAM ExactFit - Exact-fit of N given points by a N-1st degree polynomial.
'                       Setup for N on greater than 50.
     DECLARE SUB GauJor (A(), N, C(), D)
     SCREEN 2: CLEAR : CLS : KEY OFF
PRINT "Program ExactFit - Exact-fit of N given points by a N-1st degree polynomial."
PRINT
PRINT "Input the number of points, N : "; : INPUT ; "", N
                  DIM A(N, N), C(N), X(N), Y(N)
PRINT
PRINT "Input the coordinates of the points, ";
PRINT "two numbers each time"
PRINT "  and press <Enter> key after entering :"
  FOR I = 1 TO N
      INPUT ; "", X(I), Y(I)
      C(I) = Y(I)
      FOR J = 1 TO N
          A(I, J) = X(I) ^ (J - 1)
          NEXT J
      PRINT
      NEXT I
  CALL GauJor(A(), N, C(), D)
PRINT "The coefficients in Y=C(1)+C(2)X+...+C(N)X^(N-1) are : "
       FOR I = 1 TO N
           PRINT USING "    ##.#####^^^^"; C(I);
           NEXT I
       PRINT
       END
```

Sample Application

```
Program ExactFit - Exact-fit of N given points by a N-1st degree polynomial.
Input the number of points, N :
3
Input the coordinates (X,Y) of the points, two numbers each time
  and press <Enter> key after entering.
1,1
2,5
3,3
The coefficients in Y=C(1)+C(2)X+...+C(N)X**(N-1) are :
    -9.0000E+01       1.30000E+01     -3.0000E+10
Stop - Program terminated.
```

MATLAB Application

For handling the exact curve fit of N given points with a N-1st degree polynomial, there is no need to convert either the **FORTRAN** or **QuickBASIC** program **ExactFit**. The sample problems therein can be readily solved by **MATLAB** as follows:

```
>> X=[1,2]'; Y=[2,2]'; A=[ones(2,1),X]; C=A\Y
            C =
                 2
                 0
>> X=[1,1.5,2]'; Y=[2,2.5,4]'; A=[ones(3,1),X,X.^2]; C=A\Y
            C =
                 1
                 1
                 0
```

```
>> X=[1,2,4]'; Y=[1,5,4]'; A=[ones(3,1),X,X.^2]; C=A\Y
            C =
                -6.0000
                 8.5000
                -1.5000
>> X=[1,2,3]'; Y=[1,5,3]'; A=[ones(3,1),X,X.^2]; C=A\Y
            C =
                  9
                 13
                 -3
```

Notice that the coefficient {C} for the curve-fit polynomial is obtained by solving [A]{C} = {Y} where matrix [A] is formed by substituting the X values into the x^i terms for i = 0 to N–1 where N is the number of points provided. **MATLAB** function **ones** has been used to generate the first column of [A] and **MATLAB** matrix operation of C = A\Y which premultiplies {Y} by $[A]^{-1}$ to obtain {C}.

Also, this exact curve-fit problem can be treated as a special case of fitting N given points by a linear combination of N selected functions which in this case happens to be the polynomial terms of x^0 to x^{N-1}, by the least-squares method. A m file called **LeastSqG.m** which is discussed in the program **LeastSqG** can be readily applied to treat such a exact curve-fit problem. Here, we demonstrate how **LeastSqG.m** is used by **MATLAB** interactive operation in solving the sample problems previously presented in the **FORTRAN** and **QuickBASIC** versions of the program **ExactFit**. First, a function must be prepared to describe the ith selected function x^i as follows:

```
function F=FSPoly(i,xv)
            F=xv.^(i-1);
```

The results of four **MATLAB** applications are:

```
>> X=[1,2]; Y=[2,3]; C=LeastSqG('a:FSPoly',X,Y,2,2)
       A =
            2        3
            3        5
       R =
            4
            6
       C =
            2.0000
            0.0000
>> X=[1,1.5,3]; Y=[2,2.5,4]; C=LeastSqG('a:FSPoly',X,Y,3,3)
       A =
            3.0000     5.5000    12.2500
            5.5000    12.2500    31.3750
           12.2500    31.3750    87.0625
```

```
        R =
           8.5000
          17.7500
          43.6250
        C =
           1.0000
           1.0000
           0.0000
>> X=[1,2,4]; Y=[1,5,4]; C=LeastSqG('a:FSPoly',X,Y,3,3)
        A =
             3       7      21
             7      21      73
            21      73     273
        R =
            10
            27
            85
        C =
          -6.0000
           8.5000
          -1.5000
>> X=[1,2,3]; Y=[1,5,3]; C=LeastSqG('a:FSPoly',X,Y,3,3)
        A =
             3       6      14
             6      14      36
            14      36      98
        R =
             9
            20
            48
        C =
          -9.0000
          13.0000
          -3.0000
```

Notice the coefficient vector {C} in the curve-fit polynomial $p(x) = c_1 + c^2x + \ldots + c_Nx^{N-1}$ is solved from the matrix equation [A]{C} = {R} where {A} and {R} are generated using the specified points based on the least squares criterion. The solution of [A]{C} = {R} is simply implemented by **MATLAB** with the statement C = A\R in the file LeastSqG.m.

To verify whether the points have really been fitted exactly, Figure 1 is presented. It is plotted with the following **MATLAB** statements, adding to those already listed above:

```
>> XC=[1:0.1:3]; for I=1:3, Creverse(I)=C(4-I); YC=polyval(Creverse,XC);
>> plot(XC,YC,X,Y,'*'), xlabel('X'), ylabel('Y')
```

Notice that for application of polyval.m, **MATLAB** needs the coefficients of the polynomial arranged in descending order. Since the array C contains the coefficients

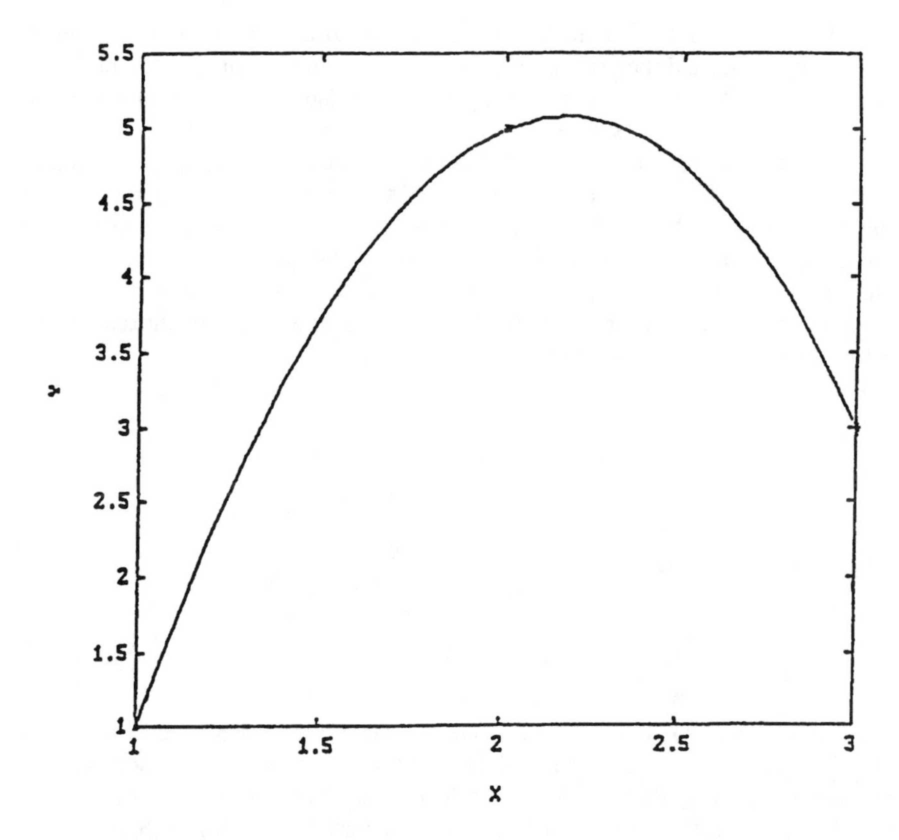

FIGURE 1. The parabolic curve passes through all of the three given points.

in ascending order, a new array called Creverse is thus created for calculation of the curve values for $1 \leq X \leq 3$ and with an increment of 0.1. Figure 1 shows that the parabolic curve passes through all of the three given points.

2.3 PROGRAM LEASTSQ1 — LEAST-SQUARES LINEAR CURVE-FIT

Program **LeastSq1** is designed for curve-fitting of a set of given data by a linear equation based on the least-squares criterion.[2] If only two points are specified, a linear equation which geometrically describes a straight line can be uniquely derived because the line must pass the two specified points. This is the case of *exact fit.* (See programs **Gauss** and **LagrangI** for examples of exact fit.) However, the specified data are often recorded from a certain experiment and due to inaccurate calibration of equipment or due to environmental disturbances such as noise, heat, and so on, these data not necessarily follow an expected behavior which may be described by a type of predetermined equation. Under such a circumstance, a so-called *forced fit* is then required. As a simple example, supposing that we expect the measured set of three data points (X_i,Y_i) for i = 1,2,3 to satisfy a linear law $Y = c_1 + c_2X$. If these three points happen to fall on a straight line, then we have a case of exact fit and the coefficients c_1 and c_2 can be *uniquely* computed. If these three points are not all on a straight line and they still need to be fitted by a linear equation $Y = c_1 + c_2X$, then they are forced to be fitted by a particular straight line based on a selected criterion, permitting errors to exist between the data points and the line.

The least-squares curve-fit for finding a linear equation $Y = c_1 + c_2X$ best representing the set of N given points, denoted as (X_i,Y_i) for i = 1 to N, is to minimize the errors between the straight line and the points to be as small as possible. Since the given points may be above, on, or below the straight line, and these errors may cancel from each other if they are added together, we consider the sum of the squares of these errors. Let us denote y_i to be the value of Y at X_i and S be the sum of the errors denoted as e_i for i = 1 to N, then we can write:

$$S = e_1^2 + e_2^2 + \ldots + e_N^2 = \sum_{i=1}^{N} e_i^2 \tag{1}$$

where for i = 1,2,...,N

$$e_i = y_i - y_i = c_1 + c_2 X_i - Y_i \tag{2}$$

It is obvious that since X_i and Y_i are constants, the sum S of the errors is a function of c_i and c_2. To find a particular set of values of c_1 and c_2 such that S reaches a minimum, we may therefore base on calculus[3] and utilize the conditions $\partial S/\partial c_1 = 0$ and $\partial S/\partial c_2 = 0$. From Equation 1, we have the partial derivatives of S as:

$$\frac{\partial S}{\partial c_1} = 2\left(e_1 \frac{\partial e_1}{\partial c_1} + e_2 \frac{\partial e_2}{\partial c_1} + \ldots + e_N \frac{\partial e_N}{\partial c_1}\right) = 2\sum_{i=1}^{N} e_i \frac{\partial e_i}{\partial c_1}$$

and

$$\frac{\partial S}{\partial c_2} = 2\left(e_1 \frac{\partial e_1}{\partial c_2} + e_2 \frac{\partial e_2}{\partial c_2} + \ldots + e_N \frac{\partial e_N}{\partial c_2}\right) = 2\sum_{i=1}^{N} e_i \frac{\partial e_i}{\partial c_2}$$

From Equation 2, we note that $\partial e_i/\partial c_1 = 1$ and $\partial e_i/\partial c_2 = X_i$. Consequently, the two extremum conditions lead to two algebraic equations for solving the coefficients c_1 and c_2:

$$\left(\sum_{i=1}^{N} 1\right)c_1 + \left(\sum_{i=1}^{N} X_i\right)c_2 = \sum_{i=1}^{N} Y_i \tag{3}$$

and

$$\left(\sum_{i=1}^{N} X_i\right)c_1 + \left(\sum_{i=1}^{N} X_i^2\right)c_2 = \sum_{i=1}^{N} X_i Y_i \tag{4}$$

Program **LeastSq1** provided in both **FORTRAN** and **QuickBASIC** versions is developed using the above two equations. It can be readily applied for calculating the coefficients c_1 and c_2. Two versions are listed and sample applications are presented below.

QuickBASIC Version

```
' * Program LeastSq1 - Linear Least-Squares Curve Fitting *
'
CLEAR : CLS : SCREEN 2: KEY OFF
W = 8 : H = 8
PRINT " * Program LeastSq1 - Fits N given points by a line; Y = mX
+ b * "
PRINT : INPUT "Enter total number of points : ", N: DIM X(N), Y(N):
PRINT
PRINT "Input the X values and press <Enter> key after entering each
number."
       FOR T = 1 TO N
           INPUT ; X(T)
           NEXT T
PRINT
PRINT "Input the Y values and press <Enter> key after entering each
number."
       FOR T = 1 TO N
           INPUT ; Y(T)
           NEXT T
SUMX  = 0
SUMXX = 0
SUMXY = 0
SUMY  = 0
FOR J = 1 TO N
    SUMX  = SUMX  + X(J)
    SUMXX = SUMXX + X(J)  ^ 2
    SUMY  = SUMY  + Y(J)
    SUMXY = SUMXY + X(J)  * Y(J)
    NEXT J
DELTA =  SUMX *  SUMX -    N * SUMXX
    M = (SUMY *  SUMX -    N * SUMXY) / DELTA
    b = (SUMX * SUMXY - SUMY * SUMXX) / DELTA
PRINT
PRINT
PRINT "The results are :"
PRINT : PRINT USING "      m = ##.####^^^^" ; M;
        PRINT USING "  and  b = ##.####^^^^"; b
PRINT
END
```

Sample Application

```
 * Program LeastSq1 - Fits N given points by a line: Y = aX + b *

Enter total number of points : 5

Input the X values and press <Enter> key after entering each number.
? 1? 2? 3? 5? 8
Input the Y values and press <Enter> key after entering each number.
? 2? 5? 8? 11? 24

The results are :
      a =  3.0195E+00  and  b = -1.4740E+00
```

FORTRAN Version

```
C * Program LeastSq1 - Linear Least-Squares Curve Fitting *
C
      DIMENSION X(500),Y(500)
      REAL M
      WRITE (*,2)
    2 FORMAT(' * Program LeastSq1 - Fits N given points by a line;
',
     *        'Y = mX + b * ')
      WRITE (*,4)
    4 FORMAT(' Enter total number of points : ')
      READ (*,*) N
      WRITE (*,6)
    6 FORMAT(' Input the X values : ')
      READ (*,*) (X(I),I=1,N)
      WRITE (*,8)
    8 FORMAT(' Input the Y values : ')
      READ (*,*) (Y(I),I=1,N)
      SUMX =0
      SUMXX=0
      SUMXY=0
      SUMY =0
      DO 10 J=1,N
         SUMX  = SUMX  + X(J)
         SUMXX = SUMXX + X(J) ** 2
         SUMY  = SUMY  + Y(J)
   10    SUMXY = SUMXY + X(J) * Y(J)
      DELTA =  SUMX * SUMX -     N * SUMXX
          M = (SUMY * SUMX -     N * SUMXY) / DELTA
          b = (SUMX * SUMXY - SUMY * SUMXX) / DELTA
      WRITE (*,*) 'The results are :'
      WRITE (*,15) M
   15 FORMAT('      m = ',E12.5)
      WRITE (*,20) b
   20 FORMAT(' and  b = ',E12.5)
      END
```

Sample Application

```
* Program LeastSq1 - Fits N given points by a line; Y = mX + b *
Enter total number of points :
5
Input the X values :
1,2,3,5,8
Input the Y values :
2,5,8,11,24
The results are :
     m =   .30195E+01
and  b = -.14740E+01
```

MATLAB Application

A m file in **MATLAB** called **polyfit.m** can be applied to fit a set of given points (X_i,Y_i) for i = 1 to N by a linear equation $Y = C_1X + C_2$ based on the least-squares criterion. The function **polyfit** has three arguments, the first and second arguments are the X and Y coordinate arrays of the given points, and the third argument specifies to what degree the fitted polynomial is required. For linear fit, the third argument should be set equal to 1. The following shows how the results obtained for the sample problem used in the **FORTRAN** and **QuickBASIC** program **LeastSq1**:

```
>> X = [1,2,3,5,8]; Y = [2,5,8,11,24]; A = polyfit(X,Y,1)

                    C = 3.0195  - 1.4740
```

If the third argument for the function **polyfit** is changed (from 1) to 2, 3, and 4, we also can obtain the least-squares fits of the five given points with a quadratic, cubic, and quartic polynomials, respectively. When the third argument is set equal to 4, we then have the case of exact curve-fit of five points by a fourth-order polynomial. Readers are referred to the program **ExactFit** for more discussions.

Also, it is of interest to know whether one may select an arbitrary set of functions and linearly combine them for least-squares fit, instead of the unbroken set of polynomial terms X^0, X^1, X^2, ..., X^N. Program **LeastSqG** to be presented in the next section will discuss such generalized least-squares curve-fit. But before we do that, let us first look into a situation where program **LeastLQ1** can be applied for a given set of data after some mathematical transformations are employed to modify the data.

Transformed Least-Squares Curve-Fit

There are occasions when we know in advance that a given set of data supposed to fall on a curve described by exponential equations of the type:

$$y = b_1 e^{b_2 x} \tag{5}$$

or

$$y = c_1 x e^{c_2 x} \tag{6}$$

To determine the coefficients b_1 and b_2, or, c_1 and c_2 based on the least-squares criterion, Equation 5 or 6 need to be first transformed into a linear form. To do so, let us first consider Equation 5 and take natural logarithm of both sides. It gives:

$$\ell n\, y = \ell n\, b_1 + \ell n\, e^{b_2 x} = \ell n\, b_1 + b_2 x \tag{7}$$

If we introduce new variable z, and new coefficients a_1 and a_2 such that:

$$z = \ell n\, y, \quad a_1 = \ell n\, b_1, \quad \text{and} \quad a_2 = b_2 \tag{8,9,10}$$

Then Equation 7 becomes:

$$z = a_1 + a_2 x \tag{11}$$

Hence, if we need to determine the coefficients b_1 and b_2 for Equation 2.8 based on N pairs of (x_i,y_i), for i = 1 to N, values and on the least-squares criterion, we simply generate N z values according to Equation 2.11 and then use the N (x_i,z_i) values as input for the program LeastSq1 and expect the program to calculate a_1 and a_2. Equations 9 and 10 suggest that b_2 is to have the value of a_2 while b_1 should be equal to e raised to the a_1 power, or, EXP(a_1) with EXP being the exponential function available in the **FORTRAN** or **QuickBASIC** library.

Equation 6 can be treated in a similar manner by taking logarithms of both sides to obtain:

$$\ell n\, y = \ell n\, c_1 + \ell n\, x + \ell n\, e^{c_2 x} = \ell n\, c_1 + \ell n\, x + c_2 x$$

or

$$\ell n\, y - \ell n\, x = \ell n \frac{y}{x} = \ell n\, c_1 + c_2 x \tag{12}$$

If we introduce new variable z, and new coefficients a_1 and a_2 such that:

$$z = \ell n(y/x), \quad a_1 = \ell n\, c_1, \quad \text{and} \quad a_2 = c_2 \tag{13,14,15}$$

Then, Equation 12 becomes Equation 11 and a_1 and a_2 can be obtained by the program **LeastSq1** using the data set of $(x_i,y_i/x_i)$ values.

As a numerical example, consider the case of a set of nine stress-versus-strain (σ vs. ϵ) data collected from a stretching test of a long bar: (.265,1025), (.4,1400), (.5,1710), (.7,2080), (.95,2425), (1.36,2760), (2.08,3005), (2.45,2850), and

(2.94,2675) where the units for the strains and stresses are in microinch x 10^2 and lb/in^2, respectively. When program LeastSq1 is applied for the modified data of (ϵ,σ/ϵ), the coefficients for the linear fit are $a_1 = 15.288$ and $a_2 = -537.71$. Consequently, according to Equations 13 and 14, and realizing that x and y are now σ and ϵ, respectively, we have arrived at $\sigma = 4.3615 \times 10^6 \epsilon e^{-537.71}$. The derived curve and the given points are plotted in Figure 2 which shows the curve passes the origin as it should.

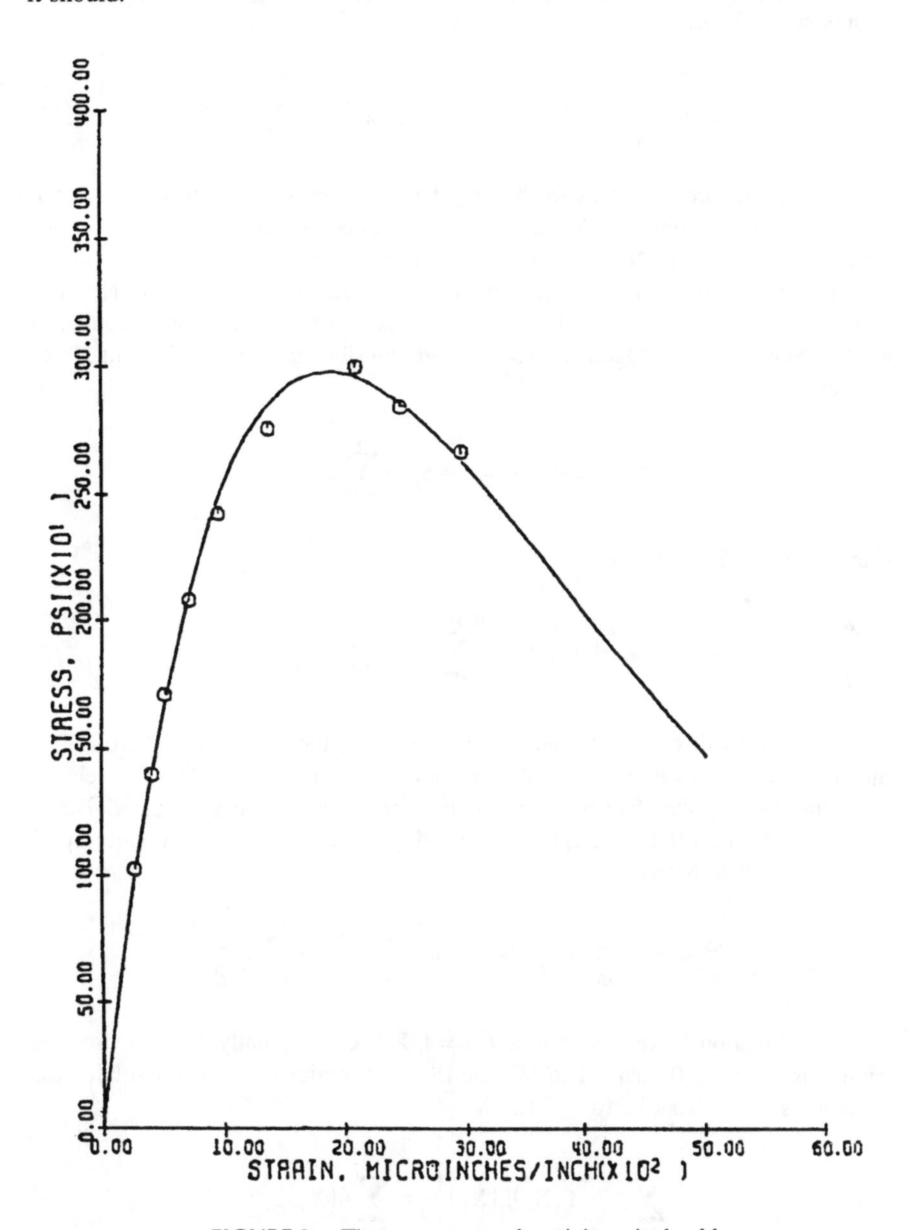

FIGURE 2. The curve passes the origin as it should.

2.4 PROGRAM LEASTSQG — GENERALIZED LEAST-SQUARES CURVEFIT

Program **LeastSqG** is designed for curve-fitting of a set of given data by a linear combination of a set of selected functions based on the least-squares criterion.[2]

Let us consider N points whose coordinates are (X_k,Y_k) for k = 1 to N and let the M selected function be $f_1(X)$ to $f_M(X)$ and the equation determined by the least-squares curve-fit be:

$$Y(X)=a_1f_1(X)+a_2f_2(X)+\ldots+a_Mf_M(X)=\sum_{j=1}^{M}a_jf_j(X) \tag{1}$$

The least-squares curve-fit for finding the coefficients c_{1-M} is to minimize the errors between the computed Y values based on Equation 1 and the given Y values at all X_k's for k = 1 to N. Let us denote y_k to be the value of Y calculated at X_k using Equation 1 and S be the sum of the errors denoted as e_k for k = 1 to N. Since the y_k could either be greater or less than Y_k, these errors e_k's may cancel from each other if they are added together. We therefore consider the sum of the squares of these errors and write:

$$S=e_1^2+e_2^2+\ldots+e_N^2=\sum_{k=1}^{N}e_k^2 \tag{2}$$

where for k = 1,2,…,N

$$e_k=y_k-Y_k=\left[\sum_{j=1}^{M}a_jf_j(X_k)\right]-Y_k \tag{3}$$

It is obvious that since X_k and Y_k are constants, the sum S of the errors is a function of $a_{1\text{ to }M}$. To find a particular set of values of $a_{1\text{ to }M}$ such that S reaches a minimum, we may therefore base on calculus[3] and utilize the conditions $\partial S/\partial a_i = 0$ for i = 1 to M. From Equation 2, we can have the partial derivatives of S with respect to a_i's, for i = 1 to M, as:

$$\frac{\partial S}{\partial a_i}=2\left(e_1\frac{\partial e_1}{\partial a_1}+e_2\frac{\partial e_2}{\partial a_i}+\ldots+e_N\frac{\partial e_N}{\partial a_i}\right)=2\sum_{k=1}^{N}e_k\frac{\partial e_k}{\partial a_i}$$

From Equation 3, we note that $\partial e_k/\partial a_i = f_i(X_k)$. Consequently, the M extremum conditions, $\partial S/\partial a_i = 0$ for i = 1 to M, lead to M algebraic equations for solving the coefficients $a_{1\text{ to }M}$. That is, for i = 1 to M:

$$\sum_{j=1}^{M}\left[\sum_{k=1}^{N}f_i(X_k)f_j(X_k)\right]a_j=\sum_{k=1}^{N}f_i(X_k)Y_k \tag{4}$$

If we express Equation 4 in matrix notation, it has the simple form:

$$[C]\{A\}=\{R\} \tag{5}$$

where [C] is a MxM square coefficient matrix, and {A} and {R} are Mx1 column matrices (vectors). {A} contains the unknown coefficients $a_{1\text{ to }M}$ needed in Equation 1. If we denote the elements in [C] and {R} as c_{ij} and r_i, respectively, Equation 5 indicates that these elements are to be calculated using the formulas, for i = 1,2,...,M:

$$c_{ij}=\sum_{k=1}^{N} f_i(X_k)f_j(X_k) \tag{6}$$

and

$$r_i=\sum_{k=1}^{N} f_i(X_k)Y_k \tag{7}$$

The above derivation appears to be too mathematical; a few examples of actual curve-fit will clarify the procedure involved. As a first example, consider the case of selecting two (M = 2) functions $f_1(X) = 1$ and $f_2(X) = X$ to fit three given points (N = 3), $(X_1,Y_1) = (1,1)$, $(X_2,Y_2) = (2.6,2)$, and $(X_3,Y_3) = (2.8,2)$. Equations 6 and 7 then provide the following:

$$c_{11}=\sum_{k=1}^{3} f_1(X_k)f_1(X_k) = f_1(X_1)f_1(X_1)+f_1(X_2)f_1(X_2)+f_1(X_3)f_1(X_3) = 1x1+1x1+1x1 = 3$$

$$c_{12}=\sum_{k=1}^{3} f_1(X_k)f_2(X_k) = f_1(X_1)f_2(X_1)+f_1(X_2)f_2(X_2)+f_1(X_3)f_2(X_3) = 1x1+1x2.6+1x2.8=6.4$$

$$c_{21}=\sum_{k=1}^{3} f_2(X_k)f_1(X_k) = c_{12} = 6.4$$

$$c_{22}=\sum_{k=1}^{3} f_2(X_k)f_2(X_k) = f_2(X_1)f_2(X_1)+f_2(X_2)f_2(X_2)+f_2(X_3)f_2(X_3) = 1x1+2.6x2.5+2.8x2.8=15.6$$

and

$$r_1=\sum_{k=1}^{3} f_1(X_k)Y_k = f_1(X_1)Y_1+f_1(X_2)Y_2+f_1(X_3)Y_3 = 1x1+1x2+1x2 = 5$$

$$r_2=\sum_{k=1}^{3} f_2(X_k)Y_k = f_2(X_1)Y_1+f_2(X_2)Y_2+f_2(X_3)Y_3 = 1x1+2.6x2+2.8x2 = 11.8$$

Hence, the system of two linear algebraic equations for finding a_1 and a_2 for the straight-line equation is:

$$\begin{bmatrix} 3 & 6.4 \\ 6.4 & 15.6 \end{bmatrix}\begin{bmatrix} a_1 \\ a_2 \end{bmatrix}=\begin{bmatrix} 5 \\ 11.8 \end{bmatrix}$$

The solution can be obtained by application of Cramer's Rule to be $a_1 = 0.42466$ and $a_2 = 0.58219$. More examples will be given after we discuss how computer programs can be written to compute [C] and {R} and then solve for {A}.

Program **LeastSqG** provided in both **FORTRAN** and **QuickBASIC** versions is developed using the above two equations. It can be readily applied for calculating the coefficients $c_{1\ to\ M}$. Both QucikBASIC and **FORTRAN** versions are listed and sample applications are presented below.

QuickBASIC Version

```
' * Program LeastSqG - General Least-Squares Curve Fitting *
            DECLARE SUB FS (I, X, F)
            CLEAR : CLS : KEY OFF: SCREEN 2: W = 8: h = 8
PRINT "Program LeastSqG - Least-Squares curve-fit of N given points
by a linear "
PRINT "                       combination of M chosen functions"
   INPUT "Enter total number of chosen functions, M (<10)"; M
   INPUT "Enter the number of (X,Y) points, N : ", N
25 LOCATE 23,5: PRINT "Have you edited the subroutine FS for the
selected function? "
   LOCATE 24,5
   INPUT ; "Enter Y if already done; otherwise BREAK the program and
edit! ", A$
   IF A$ = "Y" THEN 35 ELSE LINE (0,22*h)-(639,24*h-1),0,BF: GOTO 25
35 DIM X(N),Y(N)
   LOCATE 5,1: PRINT "Input the X values:  Press <Enter> key after
each number is entered."
   FOR T = 1 TO N: INPUT ; X(T): NEXT T
PRINT "Input the Y values: Press <Enter> key after each number is
entered."
FOR T = 1 TO N: INPUT ; Y(T): NEXT T
                  DIM A(M, M), C(M), FV(M, N)
Generate matrix [A]
FOR I = 1 TO M: C(I) = 0
    FOR K=1 TO N: CALL FS(I,X(K),FT): C(I)=C(I)+FT*Y(K): FV(I,K)=FT:
NEXT K: NEXT I
FOR I = 1 TO M
    FOR J = 1 TO M  : A(I,J)=0
        FOR K=1 TO N: A(I,J)=A(I,J)+FV(I,K)*FV(J,K): NEXT K
        NEXT J: NEXT I: GOSUB 360
PRINT: PRINT: PRINT "The coefficients, c(i), in the ";
             PRINT "equation Y=c(1)F(1)+c(2)F(2)+...+c(M)F(M) are
: "
    FOR T=1 TO M: PRINT USING "  c(#)=##.####^^^^  ";T,C(T): NEXT T:
PRINT
    END
360 ' Solves AX=C by Gauss-Jordan Elimination X and C share same
memory space.
    FOR I=1 TO M: IF I=M AND A(I,I)=0 THEN 435
        IF I < M AND A(I, I) = 0 THEN 395
        IF I < M THEN 390 ELSE C(M)=C(M)/A(M,M): A(M,M)=1: GOTO 410
390 FOR J=I+1 TO M: A(I,J)=A(I,J)/A(I,I): NEXT J: C(I)=C(I)/A(I,I):
GOTO 410
395 FOR J = I + 1 TO M: IF A(J, I) = 0 THEN 405
        FOR K=I TO M: T=A(I,K): A(I,K)=A(J,K): A(J,K)=T: NEXT K
        T=C(I): C(I) = C(J): C(J) = T: GOTO 390
```

```
405      NEXT J
410      FOR K = 1 TO M: IF K = I THEN 425
             IF A(K, I) = 0 THEN 425
             C(K)=C(K)-A(K,I)*C(I)
             FOR J=I+1 TO M: A(K,J)=A(K,J)-A(K,I)*A(I,J): NEXT J
425          NEXT K
430 NEXT I: RETURN
435 PRINT : PRINT "Matrix A is singular!"
    END
    SUB FS (I, X, FVALUE)
    ON I GOTO 100,200,300,400
100 FVALUE = X      : GOTO 999
200 FVALUE = X ^ 2 : GOTO 999
300 FAVLUE = SIN(X): GOTO 999
400 FVALUE = COS(X): GOTO 999
999 END SUB
```

Sample Applications

When four functions are selected as those listed in SUB FS, an interactive application of the program **LeastSqG QuickBASIC** version using the input data entered through keyboard has resulted in a screen display of:

```
Program LeastSqG - Least-Squares curve-fit of N given points by a linear
                   combination of M chosen functions
Enter total number of chosen functions, M (<10) ? 4
Enter the number of (X,Y) points, N : 6
Input the X values:  Press <Enter> key after each number is entered.
? 1? 2? 3? 4? 5? 6
Input the Y values:  Press <Enter> key after each number is entered.
? 2? 4? 7? 11? 23? 45

The coefficients, c(i), in the equation Y=c(1)F(1)+c(2)F(2)+...+c(M)F(M) are
  c(1)=-2.5435E+01
  c(2)= 4.9436E+00
  c(3)= 8.0342E-01
  c(4)=-9.9120E+00
```

If three sinusoidal functions, $\sin(\pi x/20)$, $\sin(3\pi x/20)$, and $\sin(5\pi x/20)$ were selected and replacing those listed in SUB FS, another interactive application of the program LeastSqG QuickBASIC version is shown below.

```
     Have you edited the subroutine FS for the selected functions?
     Enter Y if already done; otherwise BREAK the program and edit! Y

Press any key to continue

Program LeastSqG - Least-Squares curve-fit of N given points by a linear
                   combination of M chosen functions
Enter total number of chosen functions, M (<10) ? 3
Enter the number of (X,Y) points, N : 5
Input the X values:  Press <Enter> key after each number is entered.
? 1.4? 3.2? 4.8? 8? 10
Input the Y values:  Press <Enter> key after each number is entered.
? 2.25? 15? 26.25? 33? 35

The coefficients, c(i), in the equation Y=c(1)F(1)+c(2)F(2)+...+c(M)F(M) are :
  c(1)= 4.8732E+01
  c(2)= 3.3299E+00
  c(3)= 1.0738E+01
```

FORTRAN VERSION

```
C * Program LeastSqG - General Least-Squares Curve Fitting *
C
      DIMENSION A(10,10),C(10),FV(10,10),R(10),X(100),Y(100)
      WRITE (*,2)
    2 FORMAT(' Program LeastSqG - Least-Squares curve-fit of N ',
     *       'given points by a linear'/
     *       ' combination of M chosen functions')
      WRITE (*,4)
    4 FORMAT(' Enter total number of chosen functions, M (<10)')
      READ (*,*) M
      WRITE (*,6)
    6 FORMAT(' Enter the number of (X,Y) points, N : ')
      READ (*,*) N
      WRITE (*,8)
    8 FORMAT(' Have you edited the subroutine FS for the selected',
     *       ' function?'/
     *       ' Press <Enter> if already done; otherwise BREAK the ',
     *       'program and edit!')
      WRITE (*,10)
   10 FORMAT(' Input the X values: ')
      READ (*,*) (X(I),I=1,N)
      WRITE (*,12)
   12 FORMAT(' Input the Y values: ')
      READ (*,*) (Y(I),I=1,N)
C
C     Generate matrix [A] and {R}
C
      DO 25 I=1,M
         R(I) = 0
         DO 25 K=1,N
            CALL FS(I,X(K),FT)
            R(I)=R(I)+FT*Y(K)
   25 FV(I,K)=FT
      DO 30 I=1,M
         DO 30 J=1,M
            A(I, J) = 0
            DO 30 K=1,N
   30 A(I, J) = A(I, J) + FV(I, K) * FV(J, K)
      CALL GAUSS(A,M,10,R)
      WRITE (*,35)
   35 FORMAT(' The coefficients, c(i), in the equation',
     *       ' Y=c(1)F(1)+c(2)F(2)+...+c(M)F(M) are : ')
      WRITE (*,40) (I,R(I),I=1,M)
   40 FORMAT('  c(',I2,') = ',E12.5)
      END

      SUBROUTINE FS(I,X,FVALUE)
      GO TO (100, 200, 300, 400),I
  100 FVALUE = X
      RETURN
  200 FVALUE = X**2
      RETURN
  300 FAVLUE = SIN(X)
      RETURN
  400 FVALUE = COS(X)
      RETURN
      END
```

Sample Application

By selecting the four functions listed in Subroutine FS, an interactive application of the program **LeastSqG** using the input data given below has resulted in a screen display of:

```
  Program LeastSqG - Least-Squares curve-fit of N given points by a linear
                     combination of M chosen functions
Enter total number of chosen functions, M (<10)
4
Enter the number of (X,Y) points, N :
6
Have you edited the subroutine FS for the selected functions?
Pree <Enter> if already done; otherwise BREAK the program and edit!
Input the X values:
1,2,3,4,5,6
Input the Y values:
2,4,7,11,23,45
The coefficients, c(i), in the equation Y=c(1)F(1)+c(2)F(2)+...+c(M)F(M) are :
 c( 1) =  -.25435E+02
 c( 2) =   .49436E+01
 c( 3) =   .80341E+00
 c( 4) =  -.99118E+01
```

MATLAB Application

A **LeastSqG.m** file can be created and added to **MATLAB** m files which will take N sets of X and Y points and fitted by a linear combination of M selected functions in the least-squares sense. The selected functions can be specified in another m file called **FS.m** (using the same name as in the **FORTRAN** and **Quick-BASIC** versions). First, let us look at a version of **LeastSqG.m**:

```
function {C]=LeastSqG(funfcn,X,Y,M,N)
           A=zeros(M,M); R=zeros(M,1);
           for i=1:M
                for j=1:M
                     for k=1:N
                A(i,j)=A(i,j)+feval(funfcn,i,X(k)).*feval(funfcn,j,X(k));
                          end
                     end
                end
           for i=1:M
                for k=1:N
                     R(i)=R(i)+Y(k).*feval(funfcn,i,X(k));
                     end
                end
           A,R,C=A\R;
```

Notice that the coefficients C's is obtained by solving [A]{C} = {R} as in the text. For **MATLAB**, a simple matrix multiplication of the inverse of [A] to and on the left of the vector {R}. Complete execution of **LeastSqG.m** will be indicated by a display of the matrix [A], vector {R}, and the solution vector {C}. The expression feval(funfcn,i,X(k)) in the above program is to evaluate the ith function at X(k) defined in a function file to be specified when **LeastSqG.m** is applied which is to be illustrated next.

Consider the case of given 5 (X,Y) points (N = 5) which are (1.4,2.25), (3.2,15), (4.8,26.25), (8,33), and (10,35). And, the selected functions are sin(πX/20), sin(3πX/20), and sin(5πX/20). That is, M = 3. The supporting function FS.m is simply:

```
function F=FS(i,xv)
         F=sin((2.*i-1).*xv.(pi./20);
```

Having prepared this file **FS.m** on a disk which is in drive A, we can now apply **LeastSqG.m** interactively as follows:

>> X = [1.4,3.2,4.8,8,10]; Y = [2.25,15,26.25,33,35]; C = LeastSqG('A:FS',X,Y,3,5)

The results shown on screen are:

```
A =

     2.6528   -0.4171    1.0752
    -0.4171    3.3109   -0.3202
     1.0752   -0.3202    2.4849
R =
    92.0714
   -72.8215
    30.3922
C =
    35.9251
    -1.1926
    -3.4670
```

If four functions X, X², sin(X), and cos(X) are selected, we may change the **FS.m** file to:

```
function F=FS(i,xv)
         if         i==1, F=xv;
            elseif  i==2, F=xv.^2
            elseif  i==3, F=sin(xv)
            elseif  i==4, F=cos(xv)
            end
```

Same as for the **FORTRAN** and **QuickBASIC** versions, if we are given six (X,Y) points, (1,2), (2,4), (3,7), (4,11), (5,23), and (6,45), **MATLAB** application of **LeastSqG.m** will be:

```
>>X=[1,2,3,4,5,6];Y=[2,4,7,11,23,45];C-LeastSqG('A:FS',X,Y,4,6)
```

The results are:

```
A =                                                R =              C =
   1.0e+003 *                                         1.0e+003 *      -4.7576
                                                                       2.1116
     0.0910    0.4410   -0.0064    0.0013              0.4600          5.7657
     0.4410    2.2750   -0.0404    0.0212              2.4520         -0.9869
    -0.0064   -0.0404    0.0212   -0.0001             -0.0366
     0.0013    0.0212   -0.0001    0.0029              0.0350
```

Notice that the values in [A] and {R} are to be multiplied by the factor 1.0e + 003 as indicated. For saving space, [A], {R}, and {C} are listed side-by-side when actually they are displayed from top-to-bottom on screen. To further utilize the capability of **MATLAB**, the obtained {C} values are checked to plot the fitted curve against the provided six (X,Y) points. The following additional statements are entered in order to have a screen display as illustrated in Figure 3:

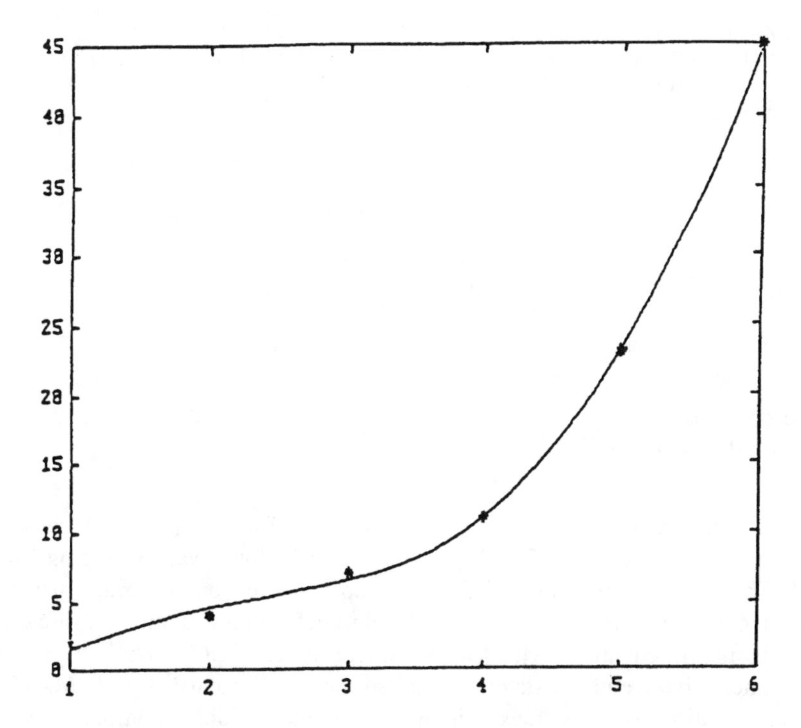

FIGURE 3. The '*' argument in the second plot statement requests that the character * is to be used for plotting the given points,

```
>> XC=[1:0.2:6]; YC=zeros(1,26);
>> for i=1:26
      for k=1:4
          YC(i)=YC(i)+feval('A:FS',k,XC(i)).*C(k);
          end
      end
>> hold; plot(XC,YC); plot(X,Y,'*')
```

The statement XC = [1:0.2:6] generates a vector of XC containing value from 1 to 6 with an increment of 0.2. The command "hold" enables the first plot of XC vs. YC which is the resulting least-squares fitted curve, to be held on screen until the given points (X,Y) are superimposed. The '*' argument in the second plot statement requests that the character * is to be used for plotting the given points, as illustrated in Figure 3.

Next, another example is given to show how **MATLAB** statements can be applied directly with defining a m function, such as **FS** which describes the selected set of functions for least-squares curve fit. Consider the problem of least-squares fit of the points (1,2), (3,5), and (4,13) by the linear combination $Y = C_1f_1(X) + C_2f_2(X)$ where $f_1(X) = X{-}1$ and $f_2(X) = X^2$. An interactive application of **MATLAB** may go as follows:

```
>> X=[1,3,4]; Y=[2,5,13]; T1=X-ones(1,3); T2=X.^2;
>> A(1,1)=T1*T1'; A(1,2)=T1*T2'; A(2,1)=A(1,2); A(2,2)=T2*T2';
>> R(1)=Y*T1';    R(2)=Y*T2';     C=A\R';
>> A,R',C

                    A =
                           13     66
                           66    338
                    R' =
                           49
                          255
                    C =
                        -7.0526
                         2.1316

>>XC=1:0.1:4;YC=C(1)*(XC-ones(1,31))+C(2)*XC.^2;
>>plot(XC,YC,X,Y,'*')
```

The resulting curve is plotted in Figure 4 using 31 points, (XC,YC), calculated based on the equation $Y = -7.0526(X{-}1) + 2.1316X^2$ for X values ranging from 1 to 4. The three given points, (X,Y), are superimposed on the graph using '*' character. Notice from the above statements, the coefficients C(1) and C(2) are solved from the matrix equation [A]{C} = {R} where the elements in [A] are generated using interactively entered statement and so are the elements of {R}. **MATLAB** matrix operations such as transposition, subtraction, multiplication, and inversion are all involved. Also, notice that here no use of **MATLAB** 'hold' as for generating Figure 1, is necessary if X, Y, XC, and YC are all used as arguments in calling the

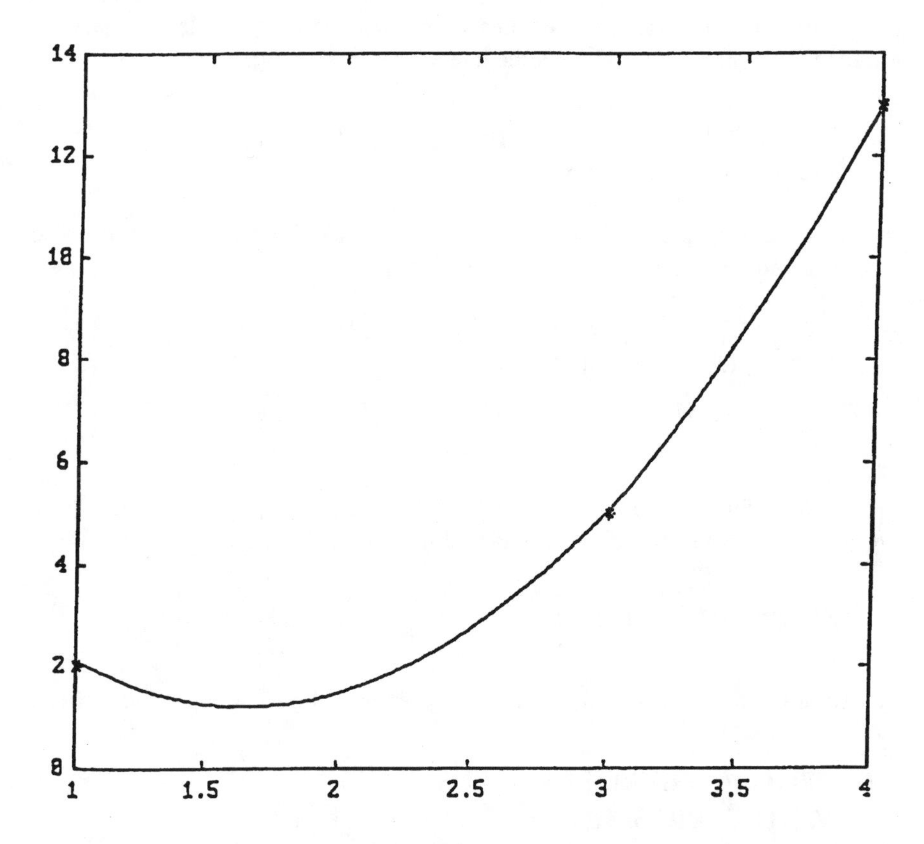

FIGURE 4. The curve is plotted using 31 points, (XC,YC), calculated based on the equation $Y = -7.0526(X-1) + 2.1316X^2$ for X values ranging from 1 to 4.

plot function. The statement XC = 1:0.1:4 generates a XC row matrix having elements starting from a value equal to 1, equally incremented by 0.1 until a value of 4 is reached.

Mathematica Applications

Mathematica has a function called **Fit** which least-squares fits a given set of (x,y) points by a linear combination of a number of selected functions. As the first example of interactive application, let us find the best straight line which gives the least squared errors for fitting a set of five points. This is the case of two selected functions $f_1(x) = 1$ and $f_2(x) = x$. the interactive application goes as follows:

In[1]: = Fit[{{1,2},{2,5},{3,8},{5,11},{8,24}}, {1, x}, x]

Out[1]: = –1.47403 + 3.01948 x

Notice that Fit has three arguments: first argument specifies the data set, second argument lists the selected function, and the third argument is the variable for the derived equation. In case that three points are given and the two selected functions are $f_1(x) = x-1$ and $f_2(x) = x^2$, then the interactive operation goes as follows:

In[2]: = Fit[{{1,2},{3,5},{4,13}}, {x–1, x^2}, x]

Out[2]: = –7.05263 (–1 + x) + 2.13158 x

Two other examples previously presented in the **MATLAB** applications can also be considered and the results are:

In[3]: = (Fit[{{1,2}, {2,4}, {3,7}, {4,11}, {5,23}, {6,45}},
{x, x^2, Sin[x], Cos[x]}, x])

Out[3]: = –4.75756 x + 2.11159 x^2 – 0.986915 Cos[x] + 5.76573 Sin[x]

and

In[4]: = (Fit[{{1.4, 2.25}, {3.2, 15}, {4.8, 26.25}, {8, 33}, {10, 35}},
{Sin[Pi*x/20], Sin[3*Pi*x/20], Sin[5*Pi*x/20]}, x])

$$\textit{Out[4]:}= 35.9251\,\mathrm{Sin}\left[\frac{\mathrm{Pi}\;x}{20}\right] - 1.19261\,\mathrm{Sin}\left[\frac{3\;\mathrm{Pi}\;x}{20}\right] - 3.46705\,\mathrm{Sin}\left[\frac{5\;\mathrm{Pi}\;x}{20}\right]$$

All of the results obtained here are in agreement with those presented earlier.

2.5 PROGRAM CUBESPLN — CURVE FITTING WITH CUBIC SPLINE

If a set of N given points (X_i, Y_i) for i = 1, 2,…,N is to be fitted with a polynomial of N–1 degree passing these points exactly, the polynomial curve will have many fluctuations between data points as the number of points, N, increases. A popular method for avoiding such over-fluctuation is to fit every two adjacent points with a cubic equation and to impose continuity conditions at the point shared by two neighboring cubic equations. This is like using a draftsman's spline attempting to pass through all of the N given points. It is known as cubic-spline curve fit. For a typical interval of X, say X_i to X_{i+1}, the cubic equation can be written as:

$$Y = a_i X^3 + b_i X^2 + c_i X + d_i \tag{1}$$

where a_i, b_i, c_i, and d_i are unknown coefficients. If there are N–1 intervals and each interval is fitted by a cubic equation, the total number of unknown coefficients is equal to 4(N–1). They are to be determined by the following conditions:

1. Each curve passes two specified points. The first and last specified points are used once by the first and N-1st curves, respectively, whereas all interior points are used twice by the two cubic curves meeting there. This gives 2 + 2(N–2), or, 2N–2 conditions.

2. Every two adjacent cubic curves should have equal slope (first derivative with respect to X) at the interior point which they share. This gives N–2 conditions.

(2)

3. For further smoothness of the curve fit, every two adjacent cubic curves should have equal curvature (second derivative with respect to X) at the interior point which they share. This gives N–2 conditions.
4. At the end points, X_1 and X_N, the nature spline requires that the curvatures be equal to zero. This gives 2 conditions.

Instead of solving the coefficients in Equation 1, the usual approach is to apply the above-listed conditions for finding the curvatures, Y″ at the interior points, that is for i = 2,3,…,N–1 since $(Y'')_1 = (Y'')_N = 0$. To do so, we notice that if Y is a cubic polynomial of X, Y″ is then linear in X and can be expressed as:

$$Y'' = AX + B \tag{3}$$

If this is used to fit the ith interval, for which the increment in X is here denoted as $H_i = X_{i+1} - X_i$, we may replace the unknown coefficients A and B with the curvatures at X_i and X_{i+1}, $(Y'')_i$ and $(Y'')_{i+1}$ by solving the two equations:

$$Y_i'' = AX_i + B \quad \text{and} \quad Y_{i+1}'' = AX_{i+1} + B \tag{4,5}$$

By using the Cramer's rule, it is easy to obtain:

$$A = \frac{Y_{i+1}' - Y_i''}{X_{i+1} - X_i} \quad \text{and} \quad B = Y'' - \frac{X_i\left(Y_{i+1}'' - Y_i''\right)}{X_{i+1} - X_i} \tag{6,7}$$

Consequently, Equation 3 can be written as:

$$Y'' = \frac{Y_i''}{H_i}\left(X_{i+1} - X\right) + \frac{Y_{i+1}''}{H_i}\left(X - X_i\right) \tag{8}$$

Equation 8 can be integrated successively to obtain the expressions for Y′ and Y as:

$$Y' + \frac{-Y_i''}{2H_i}\left(X_{i+1} - X\right)^2 + \frac{Y_{i+1}''}{2H_i}\left(X - X_i\right)^2 + C_1 \tag{9}$$

and

$$Y = \frac{Y_i''}{6H_i}\left(X_{i+1} - X\right)^3 + \frac{Y_{i+1}''}{6H_i}\left(X - X_i\right)^3 + C_1X + C_2 \tag{10}$$

The integration constants C_1 and C_2 can be determined by the conditions that at X_i, $Y = Y_i$ and at X_{i+1}, $Y = Y_{i+1}$. Based on Equation 10, the two conditions lead to:

$$Y_i = \frac{Y_i''}{6} H_i^2 + C_1 X_i + C_2 \quad \text{and} \quad Y_{i+1} = \frac{Y_{i+1}''}{6} H_i^2 + C_1 X_{i+1} + C_2 \tag{11,12}$$

Again, Cramer's rule can be applied to obtain:

$$C_1 = \frac{(Y_i'' - Y_{i+1}'')H_i}{6} - \frac{Y_i - Y_{i+1}}{H_i} \tag{13}$$

and

$$C_2 = -\frac{(X_{i+1}Y_i'' - X_i Y_{i+1}'')H_i}{6} - \frac{X_i Y_{i+1}}{H_i} - \frac{X_{i+1}Y_i}{H_i} \tag{14}$$

Substituting C_1 and C_2 into Equations 11 and 12 and rearranging into terms involving the unknown curvatures, the expressions for the cubic curve are:

$$Y = \frac{Y_i''}{6}\left[\frac{(X_{i+1} - X)^3}{H_i} - H_i(X_{i+1} - X)\right] + \frac{Y_{i+1}''}{6}\left[\frac{(X - X_i)^3}{H_i} - H_i(X - X_i)\right]$$

$$+ Y_i\left(\frac{X_{i+1} - X}{H_i}\right) + Y_{i+1}\left(\frac{X - X_i}{H_i}\right) \tag{15}$$

and

$$Y' = Y_i''\left[\frac{H_i}{6} - \frac{(X_{i+1} - X)^2}{2H_i}\right] + Y_{i+1}''\left[\frac{(X - X_i)^2}{2H_i} - \frac{H_i}{6}\right] + \frac{Y_{i+1} - Y_i}{H_i} \tag{16}$$

Equation 15 indicates that the cubic curve for the ith interval is completely defined if in addition to the specified values of Y_i and Y_{i+1}, the curvatures at X_i and X_{i+1}, $(Y'')_i$ and $(Y'')_{i+1}$ respectively, also can be found. To find all of the curvatures at the interior points X_2 through X_{N-1}, we apply the remaining unused conditions mentioned in (2). That is, matching the slopes of two adjacent cubic curves at these interior points. To match the slope at X_i, first we need to have the slope equation for the preceding interval, that is from X_{i-1} to X_i, for which the increment is denoted as H_{i-1}. From Equation 16, we can easily write out that slope equation as:

$$Y' = Y''_{i-1}\left[\frac{H_{i-1}}{6} - \frac{(X_i - X)^2}{2H_{i-1}}\right] + Y''_i\left[\frac{(X - X_{i-1})^2}{2H_{i-1}} - \frac{H_i}{6}\right] + \frac{Y_i - Y_{i-1}}{H_{i-1}} \quad (17)$$

Using Equations 16 and 17 and matching the slopes at the interior point X_i and after collecting terms, we obtain:

$$H_{i-1}Y''_{i-1} + 2(H_{i-1} + H_i)Y''_i + H_iY''_{i+1} = 6\left(\frac{Y_{i-1} - Y_i}{H_{i-1}} - \frac{Y_i - Y_{i-1}}{H_i}\right) \quad (18)$$

This equation can be applied for all interior points, that is, at $X = X_i$ for $i = 2,3,\ldots,N-1$. Hence, we have N–2 equations for solving the N–2 unknown curvatures, $(Y'')_i$ for $i = 2,3,\ldots,N-1$ when the X and Y coordinates of N + 1 points are specified for a cubic-spline curve fit. Upon substituting the calculated curvatures into Equation 15, we obtain the desired cubic polynomial for each interval of X.

If the N given points, (X_i,Y_i) for $i = 1,2,\ldots,N$, has a periodic pattern for every increment of X_N-X_1, we can change the above formation for the open case to suit this particular need by requiring that the points be specified with $Y_N = Y_1$ and that curvatures also should be continuous at the first and last points. That is to remove the 4th rule, and also one condition each for the 2nd and 3rd rules described in (2). Equation 18 is to be used for $i = 1,2,\ldots,N$ to obtain N equations for solving the N curvatures. For obtaining the first and last equations, we utilize the fact that since Y and its derivatives are periodic, in addition to $Y_N = Y_1$, $(Y'')_N = (Y'')_1$, $H_N = H_1$, we also have $Y_{N+1} = Y_2$, $Y_0 = Y_{N-1}$, $(Y'')_{N+1} = (Y'')_2$, $(Y'')_0 = (Y'')_{N-1}$, and $H_0 = H_{N-1}$.

A program called **CubeSpln** has been prepared to handle both the nonperiodic and periodic cases. It formulates the matrix equation $[A]\{Y''\} = \{C\}$ for solving the curvatures at X_i for $i = 1,2,\ldots,N$ based on Equation 18. A Gaussian elimination scheme is needed by this subroutine for obtaining the solutions of Y″. Program **CubeSpln** also has a block of statements for plotting of the spline curves. This subroutine is listed below.

QuickBASIC Version

```
'    Program CubeSpln - Cubic-Spline Curve Fit of N equally spaced points,
'                       (X(I),Y(I)) for I=1 through N.
'    N, (X(I),Y(I)) for I=1 through N, and KK are inputs.
'    KK=1- Open ends, Y(N).NE.Y(1); KK=2- Periodic, Y(N)=Y(1).
'
CLS : KEY OFF: SCREEN 2: H = 8: W = 8: ASPR = 1!: N = 5: KK = 1
      DIM A(N, N), C(N), X(N), Y(N)
      FOR I = 1 TO N: READ X(I): NEXT I
      FOR I = 1 TO N: READ Y(I): NEXT I
                       DATA 1.,2.,3.,4.,5., 2.,4.,7.,8.,11.
```

```
'
' First, solve for the curvatures.
'
    FOR J = 1 TO N: A(1, J) = 0: A(N, J) = 0: NEXT J
    IF KK = 1 THEN  A(1, 1) = 1: A(N, N) = 1: C(1) = 0: C(N) = 0: GOTO 260
    IF KK = 2 THEN 230 ELSE 250
230    A(1,      1) = 2: A(1, 2) = 1
       A(1, N - 1) = 1: A(1, N) = 2
       A(N,      1) = 1
       A(N,      N) =-1
       C(1) = 6 * (Y(2) - Y(1) + Y(N - 1) - Y(N))
       C(N) = 0: GOTO 260
250 PRINT "Error in specifying KK.": GOTO 610
260 FOR I = 2 TO N - 1
        FOR J = 1 TO N
            IF J = I - 1 THEN A(I, J) =  X(I  ) - X(I - 1)      : GOTO 320
            IF J = I     THEN A(I, J) = (X(I+1) - X(I - 1)) * 2: GOTO 320
            IF J = I + 1 THEN A(I, J) =  X(I+1) - X(I)          : GOTO 320
            A(I, J) = 0
320         NEXT J
        C(I) = 6 * ((Y(I+1)-Y(I))/(X(I+1)-X(I)) + (Y(I-1)-Y(I))/(X(I)-X(I-1)))
        NEXT I
    GOSUB 620                        'Solve for (Y")'s using GauJor
365 '
    FOR I = 1 TO N: PRINT C(I); : NEXT I: PRINT
'
'   Next, plot the spline curve from X(1) to X(N).
'
400 W1 = 0        : W2 = 6          : W3 = 0       : W4 = 12
    V1 = 10.5 * W: V2 = V1 + 60 * W: V4 = 20.5 * H: V3 = V4 - 18 * H
    DEF FNXF (X) = V1 + (X - W1) * (V2 - V1) / (W2 - W1)
    DEF FNYF (Y) = V4 + (Y - W3) * (V3 - V4) / (W4 - W3)
    CLS : LINE  (FNXF(6), FNYF(0 ))-(FNXF(0), FNYF(0))
          LINE -(FNXF(0), FNYF(12))
    FOR I = 1 TO N
        LINE (FNXF(X(I)     ), FNYF(Y(I)+.2))-(FNXF(X(I)     ), FNYF(Y(I)-.2))
        LINE (FNXF(X(I)-.05), FNYF(Y(I)    ))-(FNXF(X(I)+.05), FNYF(Y(I)    ))
        NEXT I
480 FOR I = 1 TO 6
        LINE (FNXF(I        ),             V4)-(FNXF(I        ),         V4-3)
        LINE (            V1, FNYF(2*I     ))-(           V1+W, FNYF(2*I    ))
        NEXT I
512 FOR I = 1 TO 7
        LOCATE 22, V1 / W + (I - 1) * 10 - 1   : PRINT I - 1
        LOCATE V4 / H - 3 * I + 3.1, V1 / W - 3: PRINT USING "##"; 2 * (I - 1)
        NEXT I
519 LOCATE 23, 41: PRINT "X": LOCATE 12, 5: PRINT "Y"
    PSET (FNXF(X(1)), FNYF(Y(1)))
    NP = 21: XT = X(1)
    FOR I = 2 TO N
        DD = X(I) - X(I - 1): DX = DD / (NP - 1)
        H1 = C(I - 1) / 6 / DD: H2 = C(I) / (6 * DD)
        H3 = Y(I - 1) / DD - C(I - 1) * DD / 6
        H4 = Y(I) / DD - C(I) * DD / 6
        FOR  J = 2 TO NP
            XT = XT + DX
             F = H1*((X(I)-XT)^3)+H2*((XT-X(I-1))^3)+H3*(X(I)-XT)+H4*(XT-X(I-1))
            LINE -(FNXF(XT), FNYF(F))
            NEXT J
            NEXT I
610 END

620 ' GauJor.Sub - solves a matrix equation A(N,N)X(N)=C(N)
    '              by Gauss-Jordan elimination method.
    ' X and C share same storage space.
      FOR I = 1 TO N
          IF A(I, I) = 0 THEN 660                       'Normalization
650       FOR J = I + 1 TO N
              A(I, J) = A(I, J) / A(I, I)
              NEXT J
          C(I) = C(I) / A(I, I): GOTO 690
```

```
660         FOR J = I + 1 TO N
                IF A(J, I) = 0 THEN 680                               'Pivoting
                FOR K = I TO N
                    T       = A(I, K)
                    A(I, K) = A(J, K)
                    A(J, K) = T
                    NEXT K
                T    = C(I)
                C(I) = C(J)
                C(J) = T
                GOTO 650
680             NEXT J
690         FOR K = 1 TO N
                IF K = I THEN 720                                     'Elimination
                IF A(K, I) = 0 THEN 720
                C(K) = C(K) - A(K, I) * C(I)
                FOR J = I + 1 TO N
                    A(K, J) = A(K, J) - A(K, I) * A(I, J)
                    NEXT J
720             NEXT K
            NEXT I
      RETURN
```

Sample Application

When program **CubeSpln** is run, it gives a plot of the cubic spline curves as shown in Figure 5. The given points are marked with + symbols. Between every two adjacent points, a third-order polynomial is derived. There are two different third-order polynomials for the left and right sides of every in-between points, at which the slopes and curvatures determined by the two third-order polynomials are both continuous.

MATLAB Application

MATLAB has a function file called spline.m which can be applied to perform the curve fit of cubic spline. The function has three arguments: the first and second arguments are for the coordinates of the data points for which a cubic spline curve fit is to be obtained, and the third argument should be an array containing a more finely spaced abscissa at which the curve ordinates should be calculated for plotting the spline curve. Let us redo the problem for which Figure 5 has been obtained. The **MATLAB** application and the resulting display are as follows:

```
>> X=[1,2,3,4,5]; Y=[2,4,7,8,11]; XC=[1:0.2:5]; YC=spline(X,Y,XC)

   YC =

   Columns 1 through 7
   2.0000     2.0920     2.3760     2.8140     3.3680     4.0000     4.6720

   Columns 8 through 14
   5.3460     5.8940     6.5480     7.0000     7.3160     7.5280     7.6820

   Columns 15 through 21
   7.8240     8.0000     8.2560     8.6380     9.1920     9.9640    11.0000

>> plot (X,Y,'*',XC,YC)
```

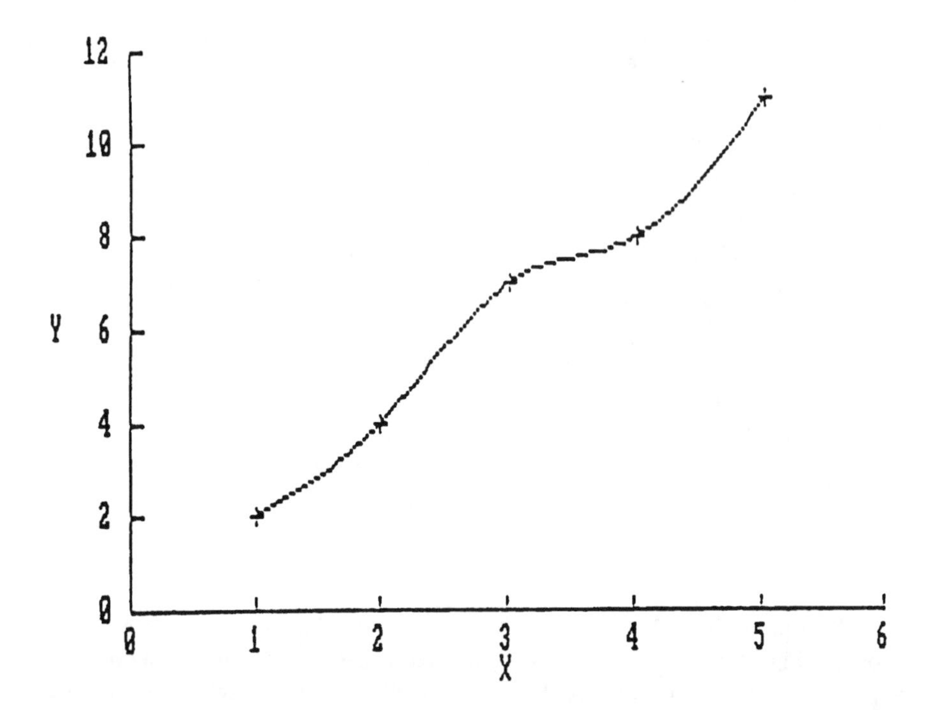

FIGURE 5. When program CubeSpln is run, it gives a plot of the cubic spline curves.

The plot of the spline curve using XC and YC data superimposed with the given 5 (X,Y) points marked by the * character is shown in Figure 6 which is identical to Figure 5 except different in the ranges of the axes.

For dealing with data sets, more features of plot.m of **MATLAB** can be utilized. Figure 7 shows how different data sets can be marked with different characters, axes can be labeled, and various text can be added. The interactive **MATLAB** commands entered are:

```
>> X =[1,2,3,4]; Y =[2,4,7,13]; X1=[0.5 1.2 2.5 3.7]; Y1=[3 6 5 11];
>> X2=[3.0 3.6 4.2 5.1]; Y2=[3 6 8 11];
>> plot(X,Y,'*',X1,Y1,'+',X2,Y2,'.')
>> xlabel('X'), ylabel('Y')
>> text(0.7,12,'X,Y - *')
>> text(0.7,11,'X1,Y1 - +')
>> text(0.7,10,'X2,Y2 - .')
```

Notice the commands xlabel, ylabel, and text are adding labels for the x and y axes, and text, respectively. The specific content string of label or text is to be spelled out inside the two single quotation signs. The first two arguments for text are the coordinates, at which the left lower corner of the first character of that string.

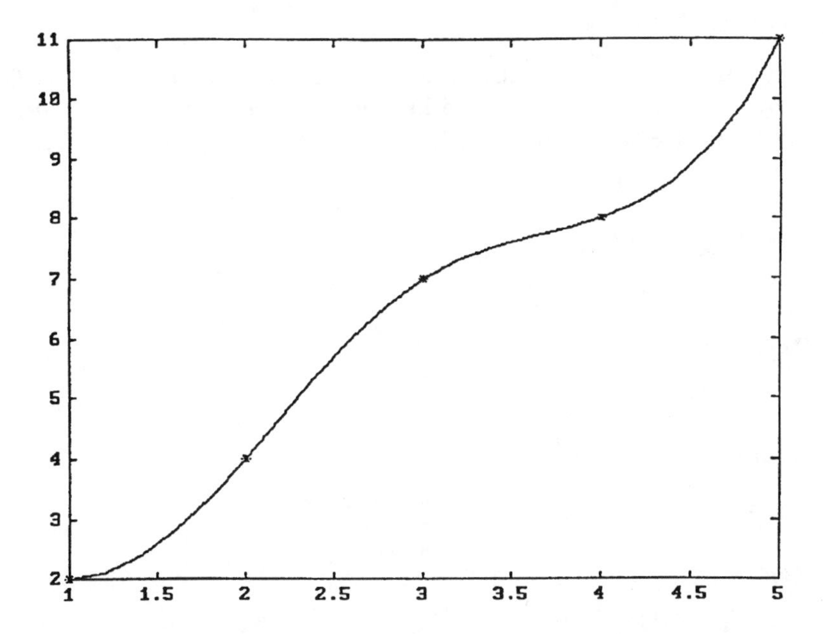

FIGURE 6. The plot of the spline curve using XC and YC data superimposed with the given 5 (X,Y) points marked by the * character is identical to Figure 5 except different in the ranges of the axes.

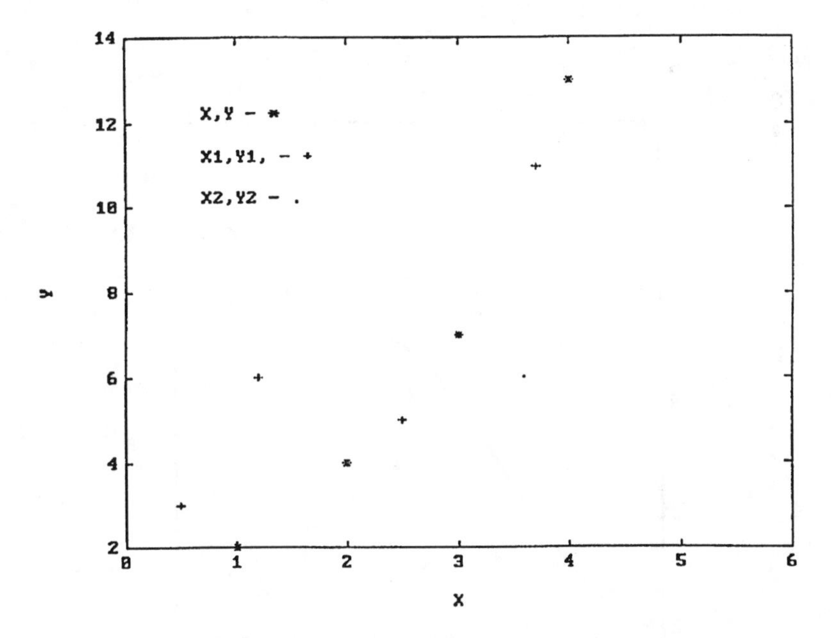

FIGURE 7. How different data sets can be marked with different characters, axes can be labeled, and various text can be added.

MATHEMATICA APPLICATIONS

Mathematica has an interpolation function which fits a collection of specified points by a polynomial. The command **InterpolationOrder** specifies the order of polynomial and the default order is 3. Here are some examples of applications and plots.

Input[1]: =

x = {1,2,3,4,5}

Output[1] =

{1, 2, 3, 4, 5}

Input[2]: =

y = {2,4,7,8,11}

Output[2] =

{2, 4, 7, 8, 11}

Input[3]: =

Plot[Evaluate[Interpolation[y]][x], {x,1,5},
Frame->True, AspectRatio->1]

Output[3] =

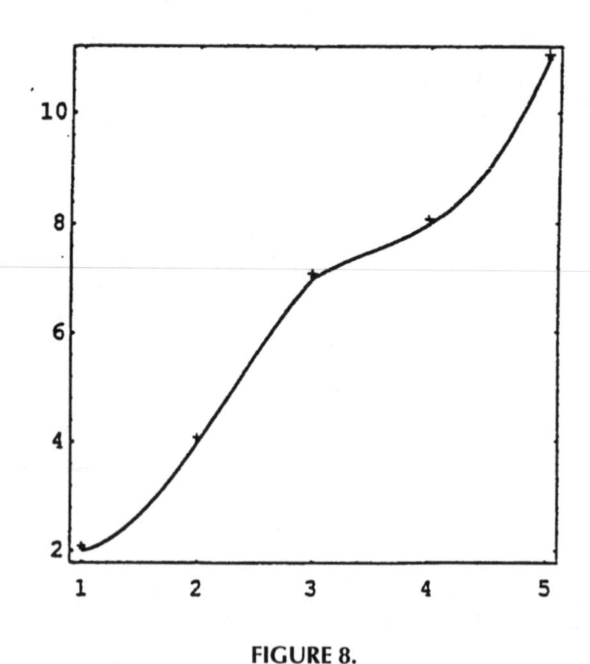

FIGURE 8.

Evaluate calculates the values of the fitted polynomial within the specified interval. In this case, the y values are interpolated using the cubic polynomial.

To add more features to a plot, **Mathematica** has commands Text allowing a string of characters to be placed at a desired location, and FrameLabel allowing labels to be added to the axes. In fact, Figure 8 is not the result of *Input[3]* but the addition of the five + markers with the statement

Input[4]: =

```
Show[%,Graphics[Table[Text["+",{x[[i]],y[[i]]}],{i,1,5}]]]
```

Output[4] =

—Graphics—

% refers to the previous output and **%%** refers to the next-to-the-last output, and so on. Show and Graphics enable additional graphics to be shown. Table lists a series of entries specified by the running index variable which in this case is I having a starting value of 1 and incremented by 1 (omitted), and ending at a value equal to 5. Notice that the four * markers are not exactly located at the coordinates specified, for the reason that the character itself is not centered but offsets like a superscript.

As another example, consider the plot shown in Figure 9 which is resulted from of the following statements:

Input[1]: =

X = {1,2,3,4}

Output[1] =

{1, 2, 3, 4}

Input[2]: =

Y = {2, 4, 7, 13}

Output[2] =

{2, 4, 7, 13}

Input[3]: =

X1 = {0.5, 1.2, 2.5, 3.7}

Output[3] =

{0.5, 1.2, 2.5, 3.7}

Input[4]: =

```
Y1 = {3, 6, 5, 11}
```

Output[4] =

```
{3, 6, 5, 11}
```

Input[5]: =

```
X2 = {3.0, 3.6, 4.2, 5.1}
```

Output[5] =

```
{3.0, 3.6, 4.2, 5.1}
```

Input[6]: =

```
Y2 = {3, 6, 8, 11}
```

Output[6] =

```
{3, 6, 8, 11}
```

Input[7]: =

```
g1 = Show[Graphics[Table[Text["*",{X[[i]],Y[[i]] }],{i,1,5}],
Table[Text[" + ",{X1[[i]],Y1[[i]]}],{i,1,5}],
Table[Text[".",{X2[[i]],Y2[[i]]}],{i,1,5}]]]
```

Output[7] =

—Graphics—

Input[8]: =

```
g2 = Show[g1, Frame->True, AspectRatio->1,
FrameLabel->{"X","Y"}]
```

Output[8] =

—Graphics—

Input[9]: =

```
Show[g2,Graphics[Text["X–Y — *",{0.7,12},{-1,0}],
Text["X1–Y1 — + ",{0.7,11},{-1,0}],
Text["X2–Y2 — .",{0.7,10}],{-1,0}]]]
```

Output[9] =

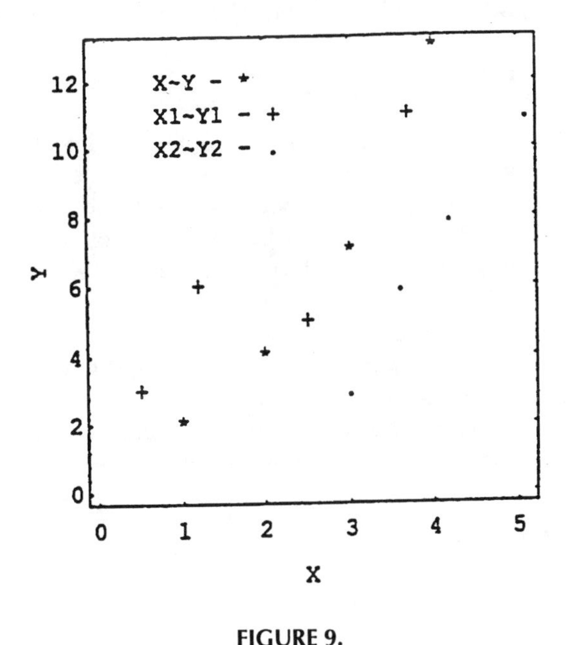

FIGURE 9.

The two intermediate plots designated as g1 and g2 are actually displayed on screen but not presented here. Only the final plot showing all of the ingredients is presented in Figure 9. Giving plot names facilitates the later referral; a better arrangement than using the % option. In Figure 9, it is also illustrated that the label for the vertical axis is rotated by 90 degrees.

Mathematica has a package called SplineFit can also be called into service for the need of spline curve-fit. For creating Figure 9, we may enter the request as follows:

Input[1]: = <<NumericalMath\`SplineFit\`

Input[2]: = XYS = {{1,2},{2,4},{3,7},{4,8},{5,11}};

Input[3]: = Spline = SplineFit[XYS, Cubic]

Output[3] = SplineFunction[Cubic, {0.,4.}, <>]

Input[4]: = ParametricPlot[Spline[x], {x,0,4}, Frame->True, AspectRatio->1]

Output[4] = —Graphics—

Input[5]: = Show[%,Graphics[Table[Text["X",{XYS[[i]]}],{i,1,5}]]]

Output[5] = —Graphics—

In *Input[1]*, << loads the specified package `SplineFit`. *Input[2]* illustrates how the coordinates of points can be entered as pairs and ; can be used to suppress the display of the output. "Cubic" spline fit is requested in *Input[3]* and in *Input[5]* the ParametricPlot command is applied so that the coordinates of the points on the cubic spline curves are generated using a third parameter. *Input[5]* also demonstrates how a matrix should be called in a looping situation.

Notice that two plots for *Output[4]* and *Output[5]* are not presented here, for the reason that Figure 9 already shows all of the needed details. However, it should be mentioned that use of "X" as the character in Text for marking the five XYS points in *Input[5]* will enable it to be centered at the coordinates specified instead of being upwardly offset as is the case of using the character "*" in Figure 9.

2.6 PROBLEMS

EXACT CURVE-FIT

1. Modify the program ExactFit so that the given points (1,3), (3,8), and (4,23) can be exact-fitted by the equation $y = c_1x + c_2x^2 + c_3x^4$.
2. Modify the program ExactFit so that the given points (0,0.2), (2,0.5), (5,–0.4), and (7,–0.2) can be exact-fitted by the equation $y = c_1\sin x + c_2\sin 2x + c_3\sin 3x + c_4\sin 4x$.

LEASTSQ1

1. Given five points (1,1), (2,3), (3,2), (4,5), and (5,4), calculate the coefficients c_1 and c_2 in the linear equation $y = c_1 + c_2x$ which fits the five points by the least-squares method.
2. For a given set of data (1,–2), (2,0), (3,1), and (4,3), two equations have been suggested to fit these points. They are $Y = X-2$ and $Y = (-X^2 + 7X-10)/2$. Based on the least-squares criterion, which equation should be chosen to provide a better fit? Explain why?
3. During a tensile-strength test of a metallic material the following data (X_i,Y_i) for $i = 1,2,\ldots,7$ where X and Y represent strain (extension per unit length) and stress (force per unit area), respectively, have been collected:

X	.265	.395	.695	.955	1.35	2.05	2.45	($\times10^{-3}$)
Y	1.03	1.41	1.71	2.09	2.42	2.76	3.01	($\times10^{3}$)

Fit these data by the least-squares method with an equation $Y = b_1 X Exp(b_2 X)$ and calculate the values of the coefficients b_1 and b_2. (Note: $\ell n(Y/X) = \ell n(b_1) + b_2 X$)

4. Apply polyfit of **MATLAB** to the data given in Problem 1 and fit them linearly. Compare the answer with that of Problem 1.
5. Same as Problem 4 except the data are to be fitted by a quadratic equation.
6. Same as Problem 4 except the data are to be fitted by a cubic equation.
7. Same as Problem 4 except the data are to be fitted by a quartic equation.
8. Use the results of Problems 4 through 7 and enter **MATLAB** statements to compare their errors.
9. Apply polyfit of **MATLAB** to the data given in Problem 3 and fit them linearly. Compare the answer with that of Problem 3.
10. Same as Problem 9 except the data are to be fitted by a quadratic equation.
11. Same as Problem 9 except the data are to be fitted by a cubic equation.
12. Same as Problem 9 except the data are to be fitted by a quartic equation.
13. Use the results of Problems 9 through 12 and enter **MATLAB** statements to compare their errors.
14. Save the results of Problems 4 through 7 and generating the four curves determined by the least-squares method. Obtain a composite graph of these four curves superimposed with the given points marked using the character * by application of plot.
15. Same as Problem 14 but for the curves determined by Problems 9 through 12.
16. Try **Mathematica** to obtain and compare results for the above problems.

LeastSqG

1. Given four points (1,0.5), (2.5,0.88), (3.2,1.35), and (4.5,2.76), they are to be least-squares fitted by a linear combination of two selected functions, $f_1(x) = e^{-x}$ and $f_2(x) = e^{-2x}$, in the form of $f(x) = c_1 f_1(x) + c_2 f_2(x)$. Find c_1 and c_2.
2. For a given set of 10 points, (x_i, y_i) for i = 1,2,...,10, the least-squares method has been applied to fit these data by two students. Student A selects 3 functions $f_1(x) = x$, $f_2(x) = x^3$, and $f_3(x) = x^5$ to obtain the coefficients a_1, a_2, and a_3 for the expression$y = a_1 x + a_2 x^3 + a_3 x^5$. Student B selects 3 other functions $f_1(x) = \sin x$, $f_2(x) = \sin 3x$, and $f_3(x) = \sin 5x$ to obtain the coefficients b_1, b_2, and b_3 for the expression $y = b_1 \sin x + b_2 \sin 3x + b_3 \sin 5x$. Write a program to calculate the least-squares errors E_A and E_B for the curve-fit approaches taken by the students A and B, respectively.
3. Apply the least-squares method to curve-fit the following three given points by a linear combination of two selected functions $f_1(x) = x$ and $f_2(x) = x^3$, namely, $y = c_1 x + c_2 x^3$:

x	0	1	2
y	-1	0	2

Find c_1 and c_2 by use of Cramer's Rule.

4. Given eight data points (X_i,Y_i) for i = 1,2,…,8 as listed, fit them by the least-squares method with the equation $Y = a_1 + a_2X + a_3X^2$. Find a_{1-3} by applying the computer program LeastSqG.

X	1	2	3	4	5	6	7	8
Y	1.13	1.45	1.76	2.19	2.43	2.79	3.51	4.88

5. A set of three points are provided as listed: (1,0.2), (2,0.5), and (3,0.6). These points are to be fitted by application of the least-squares method using a *linear* combination of (a) two functions x and x^2, or (b) two other functions x^0 and x^3. Which fit will be better, a or b? Back up your answer with detailed calculations.
6. Given three points (1,2), (3,5), and (4,13), two selected functions $f_1(x) = x-1$ and $f_2(x) = x^3$ are to be linearly combined to fit these points by the equation $y = a_1f_1(x) + a_2f_2(x) = a_1(x-1) + a_2x^3$. Derive two equations needed for solving a_1 and a_2 by use of the Least-Squares method.
7. Given 7 points of which the coordinates are X(i) and Y(i) for i = 1 to 7, a least-squares fit of these points with a linear combination of 3 selected functions $f_1(X) = X$, $f_2(X) = \sin 2X$, and $f_3(X) = e^{-X}$ in the form of $Y(X) = C(1)f_1(X) + C(2)f_2(X) + C(3)f_3(X)$ has been conducted and the coefficients C(1) to C(3) have been found. Complete the following segment of **FORTRAN** program to calculate the total error E which is the sum of the squares of the differences between Y(X(i)) and Y(i) for i = 1 to 7.

```
DIMENSION X(7),Y(7),C(3)
DATA X,Y,C/(17 real numbers separated by,)/
          insert statements for
       <--- calculation of E involving
          C, X, Y, and f1, f2, and f3.
WRITE (*,*) E
STOP
END
```

8. For a given set of data (1,–2), (2,0), (3,1), and (4,3), two equations have been suggested to fit these points. They are $Y = X-2$ and $Y = (-X^2 + 7X-10)/2$. Based on the least-squares criterion, which equation should be chosen to provide a better fit? Explain why.
9. Given 12 points of which the coordinates are X(i) and Y(i) for i = 1 to 7, a least-squares fit of these points with a linear combination of 4 selected functions $f_1(X) = X$, $f_2(X) = \sin 2X$, $f_3(X) = \cos 3X$, and $f_4(X) = e^{-X}$ in the form of $Y(X) = C(1)f_1(X) + C(2)f_2(X) + C(3)f_3(X) + C(4)f_4(X)$ has been conducted and the coefficients C(1) to C(4) have been found. Complete the following segment of **FORTRAN** program to calculate the total error E which is the sum of the squares of the differences between Y(X(i)) and Y(i) for i = 1 to 12, using a FUNCTION subprogram F(I,X) which evaluate the Ith selected function at a specified X value for i = 1 to 4.

```
DIMENSION X(12),Y(12),C(4)
DATA X,Y,C/(28 real numbers separated by,)/
         insert statements for
      <--- calculation of E involving
         C, X, Y, and f1, f2, and f3.
WRITE (*,*) E
STOP
END
```

10. Any way one can solve the above-listed problem by application of **MATLAB**? Compare the computed results obtained by **QuickBASIC**, **FORTRAN**, and **MATLAB** approaches.
11. Try **Mathematica** and compare results for the above problems.

Cubic Spline

1. Presently, program CubeSpln is not interactive. Expand its capability to allow interactive input of the number of points, N, and coordinates (X_i,Y_i) for i = 1 to N. Also, user should be able to specify the KK value so that both periodic or nonperiodic data points can be fitted. Call this program CubeSpln.X and rerun the case used in Sample Application.
2. Change the program CubeSpln slightly to allow a sixth point to be considered. Add a sixth point whose Y value is equal to that of the first point then run it as a periodic case by changing KK equal to 2. The resulting plot for X(6) = 5.5 should be as shown below.

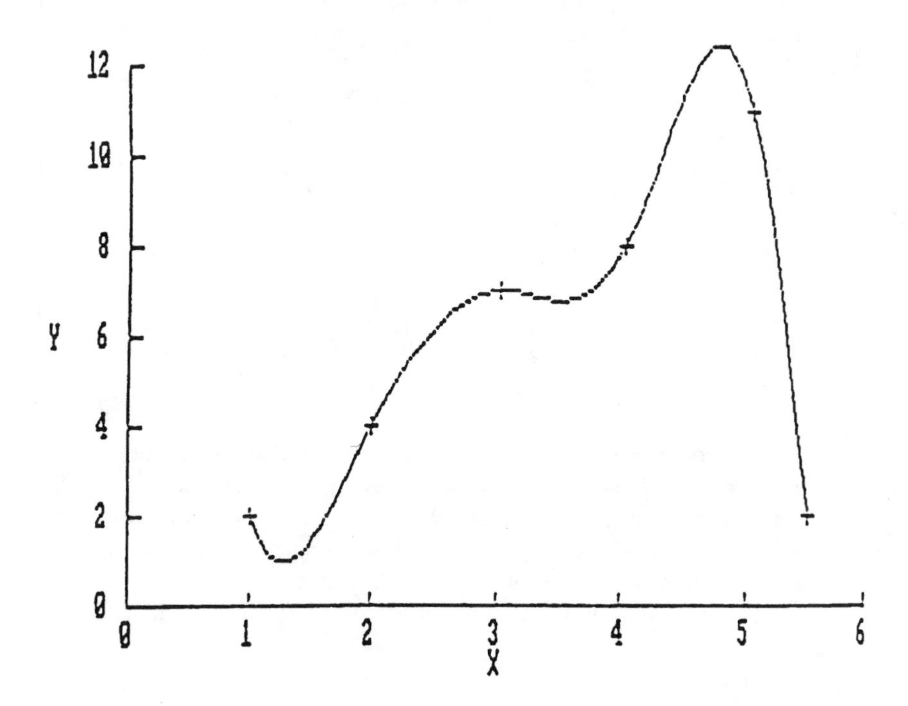

3. Use the program **CubeSpln.X** to run Problem 2.
4. Apply spline.m of **MATLAB** to fit the points (1,2), (2,4), (3,7), and (4,13) and then plot the curve by using plot.m. Mark the points by the character *.
5. Apply spline.m of **MATLAB** to fit the points (0.5,3), (1.2,6), (2.5,5), and (3.7,11) and then plot the curve by using plot.m. Mark the points by the character + .
6. Apply spline.m of **MATLAB** to fit the points (3,3), (3.6,6), (4.2,8), and (5.1,11) and then plot the curve by using plot.m.
7. Combine the curves obtained in Problems 4 to 6 into a composite graph by using solid, broken, and center lines which in use of plot.m require to specify with '-', '- - -', and '-.', respectively. The resulting composite graph should look like the figure below.

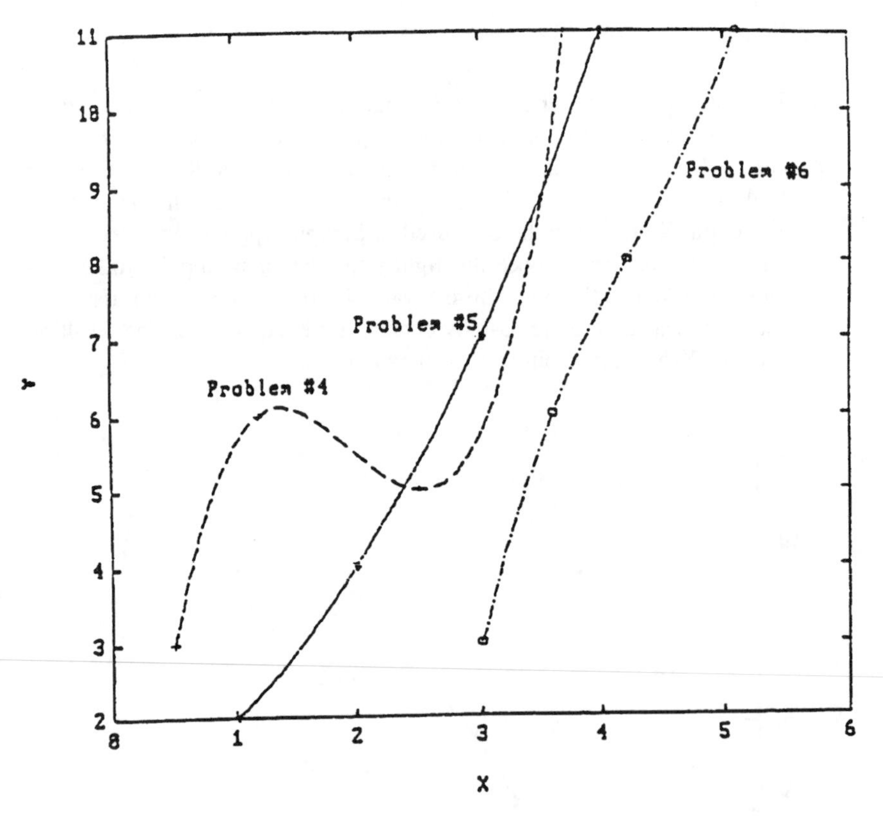

8. Use text command of **MATLAB** to add texts 'Problem 4', 'Problem 5', and 'Problem 6' near the respective curves already drawn in Problem 7.
9. The temperature data in °F, collected during a period of seven days are (2,75), (3,80), (4,86), (5,92), (6,81), and (7,90). Cubic-spline fit these data, plot the curve, and label the horizontal axis with 'Days' and vertical axis with 'Temperature, in Fahrenheit' by use of xlabel and ylabel commands of **MATLAB**.

10. Add to the graph obtained for Problem 9 by marking the data points with the character * and also a text 'Cubic Spline of Temperature Data' at an appropriate location not touching the spline curve.

2.7 REFERENCES

1. Y. C. Pao, "On Development of Engineering Animation Software," in *Computers in Engineering*, K. Ishii, editor, ASME Publications, New York, 1994, pp. 851–855.
2. A. Ralston, *A First Course in Numerical Analysis,* McGraw-Hill, New York, 1965.
3. H. Flanders, R. R. Korfhage, and J. J. Price, *A First Course in Calculus with Analytic Geometry,* Academic Press, New York, 1973.

3 Roots of Polynomials and Transcendental Equations

3.1 INTRODUCTION

In the preceding chapter, we derive equations which fit a given of data either exactly, or, by using a criterion such as the least-squares method. Once such equations have been obtained in the form of $y = C(x)$ when the data are two-dimensional, or, $z = S(x,y)$ when the data are three-dimensional. It is next of common interest to find where the curve $C(x)$ intercepts the x-axis, or, where the surface $S(x,y)$ intercepts with the x-y plane. Mathematically, these are the problems of finding the *roots* of the equations $C(x) = 0$ and $S(x,y) = 0$, respectively. The equation to be solved could be a *polynomial* of the form $P(x) = a_1 + a_2x + \dots + a_ix^{i-1} + \dots + a_{N+1}x^N$ which is of Nth order, or, a *transcendental equation* such as $C(x) = a_1\sin x + a_2\sin 2x + a_3\sin 3x$.

As it is well known, a polynomial of Nth order should have N roots which could be real, or, complex conjugate pair if the coefficients of the polynomial are all real. Geometrically speaking, only those real roots really pass the x-axis. For a transcendental equation, there may be infinite many roots. In this chapter, we shall introduce computational methods for finding the roots of polynomials and transcendental equations. Beginning with the very primitive approach of incremental and half-interval searches, the approximate location of a particular root is to be located. More refined, systematic methods such as the linear interpolation and Newton-Raphson methods are then followed to determine the more precise location of the root. A program called **FindRoot** incorporating the four methods is to be presented for interactive solution of a particular root of a given polynomial or transcendental equation when the upper and lower bounds of the root are provided.

Also discussed is a method called *Successive Substitution.* A transcendental equation derived from analysis of a four-bar linkage problem is used to demonstrate how roots are to be found by application of this method. Another transcendental equation has been derived for the unit-step response analysis of a mechanical vibration system and its roots solved by application of the Newton-Raphson method to illustrate how the design specifications are checked in the time domain.

Since the Newton-Raphson method for solving $F(x) = 0$ which can be a polynomial, or, transcendental equation of one variable is based on the Taylor's series involving the derivatives of $F(x)$, it can be extended to the solution of two-equations $F_1(x,y) = 0$ and $F_2(x,y) = 0$ by application of Taylor's series involving partial derivatives of both F_1 and F_2 with respect to x and y. A program called **NewRaphG** has been developed for this purpose. Also, this generalized Newton-Raphson method allows the quadratic factors of a higher order polynomial to be iteratively and continuously extracted and their quadratic roots solved by the so-called **Bairstow** method. For that, a program called Bairstow is made available for interactive application.

Both **QuickBASIC** and **FORTRAN** versions for the above-mentioned programs are presented. Both the application of the **roots** m-file of **MATLAB** in place of the program Bairstow and direct conversion of the program **FindRoot** into **MATLAB** version are also presented. The **Mathematica**'s function **NSolve** is introduced in place of the program **Bairstow** if the user prefers. Also the linear interpolation method used in the program **FindRoot** has been translated into **Mathematica** version. In fact. **Mathematica** has its own **FindRoot** based on the Newton-Raphson method.

3.2 ITERATIVE METHODS AND PROGRAM FINDROOT

Program FindRoot is developed for interactive selection of an iterative method among the four made available: (1) Incremental Search, (2) Bisection Search, (3) Linear Interpolation, and (4) Newton-Raphson Iteration. Polynomials are often encountered in engineering analyses such as the characteristic equations in vibrational and buckling problems. The roots of a polynomial are related to some important physical properties of the systems being analyzed, such as the frequencies of vibration or buckling loads. A nth degree polynomial can be expressed as:

$$P(x) = a_1 + a_2x + a_3x^2 + \ldots + a_nx^{n-1} + a_{n+1}x^n$$
$$= \sum_{k=1}^{n+1} a_k x^{k-1} = 0 \qquad (1)$$

For n = 1,2,3, there are formulas readily available in standard mathematical handbooks[1] for finding the roots. But for large n values, computer methods are then necessary to help find the roots of a given polynomial. The methods to be discussed here are simple and direct and are applicable to not only polynomials but also transcendental equations such as $5 + 7\cos x - \cos 60° - \cos(60° - x) = 0$ related to a linkage design problem[2] or $x = 40000/\{1–0.35\sec[40(x/10^7)^{0.5}]\}$ arisen from buckling study of slender rods.[3]

INCREMENTAL SEARCH

For convenience of discussion, let us consider a cubic equation:

$$P(x) = 1 + 2x + 3x^2 + 4x^3 = 0 \qquad (2)$$

To find a root of P(x), we first observe that $P(x = -\infty)<0$, $P(x = 0) = 1$, and $P(x = -\infty)>0$. This indicates that the P(x) curve must cross the x axis, possibly once or an odd number of times. Also, the curve may remain above the x-axis or cross it an even number of times. To further narrowing down the range on the x-axis, in which the root is located, we can begin to check the sign of P(x) at x = –10 and search toward the origin using an increment of x equal to 2. That is, we may construct a list such as:

X	-10	-8	-6	-4	-2	0
P(x)	–	–	–	–	–	+

Since P(x) changes sign from x = –2 to x = 0, this incremental search can be continued using an increment of x equal to 0.2 and the left bound x = –10 by replaced by x = –2 to obtain:

X	-1.8	-1.6	• • •	-0.8	-0.6
P(x)	–	–	–	–	+

The search continues as follows:

x	-0.78	-0.76	• • •	-0.62	-0.60
P(x)	–	–	–	–	+

x	-0.618	-0.616	• • •	-0.606	-0.604
P(x)	–	–	–	–	+

x	-0.6058
P(x)	+

If only three significant figures accuracy is required, then x = 0.606 is the root and it has taken 23 incremental search steps to arrive at this answer. If better accuracy is required, the root should then be sought between x = –0.6060 and x = –0.6058.

Program **FindRoot** prepared both is **QuickBASIC** and **FORTRAN** has one of the options using the above-explained incremental search method, it also has other methods of finding the roots of polynomials and transcendental equations to be introduced next.

Bisection Search

The above example of incremental search shows that if we search from left to right of the x-axis for the root of $4x^3 + 3x^2 + 2x + 1 = 0$ between x = –2 and x = 0, it would be longer than if we search from right to left because the root is near x = 0. Rather than using a fixed incremental in the incremental search method, the bisectional method uses the mid-point of the two bounds of x in search of the root. It involves the testing of the signs of the polynomial at the bounds of the root and replacing the bounds. The two search methods follow the same procedure. So, the bisection method would go as follows:

x	-10	0	-5	-2.5	-1.25	-0.625	-0.3125
P(x)	–	+	–	–	–	–	+

x	-0.46875	-0.546875	-0.5859375	-0.6054688
P(x)	+	+	+	+

x	-0.6152344	-0.6103516	-0.6079102	-0.6066895
P(x)	–	–	–	–

x	-0.6060792	-0.6057741	-0.6059266
P(x)	–	+	-2.68817E-04

If we require only three significant figures accuracy, then –0.606 can be considered as the root after having taken 18 bisection search steps.

Linear Interpolation

Notice that both the incremental and bisection search methods make no use of the values of the polynomials at the guessed x values. For example, at x = –10 and x = 1, the polynomial P(x) has values equal to –3719 and 1, respectively. Since P(x = 1) has a smaller value than P(x = –10), we would certainly expect the root to be closer to x = 1 than to x = –10. The linear interpolation makes use of the values of P(x) at the bounds and calculates a new guessing value of the root using the following formulas derived from the relationship between two similar triangles:

$$(x - x_L)/[-P(x_L)] = (x_R - x)/P(x_R) \tag{3}$$

where x_L and x_R are the left and right bounds of the root, which in this case are equal to –10 and 1, respectively. Based on Equation 3 and $P(x_L) = -3719$ and $P(x_R) = 1$, we can have x = –0.002688 and P(x) = 0.9946. Since P(x)>0, we can therefore replace $x_R = 1$ with $x_R = 0.002688$. Linear interpolation involves the continuous use of Equation 3 and updating of the bounds.

Newton-Raphson Iterative Method

Linear interpolation method uses the value of the function, for which the root is being sought; Newton-Raphson method goes one step farther by involving with the derivative of the function as well. For example, the polynomial $P(x) = 4x^3 + 3x^2 + 2x + 1 = 0$ has its first-derivative expression $P'(x) = 12x^2 + 6x + 2$. If we guess the root of P(x) to be $x = x_g$ and $P(x_g)$ is not equal to zero, the adjustment of x_g, calling Δx, can be obtained by application of the Taylor's series:

$$P(x_g + \Delta x) = P(x_g) + [P'(x_g)/1!]\Delta x + [P''(x_g)/2!](\Delta x)^2 \ldots$$

Since the intention is to find an adjustment Δx which should make $P(x_g + \Delta x)$ equal to zero and Δx itself should be small enough to allow higher order of Δx to be dropped from the above expression. As a consequence, we can have $0 = P(x_g) + P'(x_g)\Delta x$, or

$$\Delta x = -P(x_g)/P'(x_g) \tag{4}$$

Equation 4 is to be continuously used to make new guess, $(x_g)_{new} = x_g + \Delta x$, of the root, until $P(x = (x_g)_{new})$ is negligibly small.

The major shortcoming of this method is that during the iteration, if the slope at the guessing point becomes too small, Equation 4 may lead to a very large Dx so that the x_g may fall outside the known bounds of the root. However, this method has the advantage of extending the iterative procedure to solving multiple equations of multiple variables (see program **NewRaphG**).

An interactive program called **FindRoot** has been developed in both **QuickBASIC** and **FORTRAN** languages with all four methods discussed above. User can select any one of theses methods, edits the equation to be solved, specifies the bounds of the root, and gives the accuracy tolerance for termination of the root finding. The programs are listed below along with sample applications.

QUICKBASIC VERSION

```
'  Program FindRoot - Finds roots of polynomial and transcendental equations
'                     using incremental and bisection search, linear
'                     interpolation, or Newton-Raphson iteration.
CLS : PRINT "  Program FindRoot - Finds roots of polynomial or transcendental equations."
      PRINT : PRINT "Four options available : "
              PRINT "  (1) Incremental Search, (2) Bisection Search,"
              PRINT "  (3) Linear Interpolation, and"
              PRINT "  (4) Newton-Raphson Iteration."
              DEF fnx (x) = 4 * x ^ 3 + 3 * x ^ 2 + 2 * x + 1
              DEF fnd (x) = 12 * x ^ 2 + 6 * x + 2
     INPUT "Select a method, ½/3/4 : ", ms
     INPUT "Have you defined F(x) by editing DEF fnx(x)?  Enter Y/N : ", A$
           IF A$ <> "Y" THEN END
     IF ms <> 4 THEN 5
     INPUT "Have you defined F'(x) by editing DEF fnd(x)?  Enter Y/N : ", A$
     IF A$ <> "Y" THEN END
5    INPUT "Enter the lower and upper bounds of the root : ", xl, xr
     INPUT "How small should F(X) be for termination : ", epsilon
     PRINT SPC(3); "Trial #"; SPC(14); "X"; SPC(11); "F(X)"
     nt = 1 : ON ms GOTO 10, 20, 30, 40
10   fl = fnx(xl)
15   dx = (xr-xl)/10: x =xl+dx : GOTO 50
20   fr = fnx(xr)
25    x = .5 * (xl + xr)        : GOTO 50
30   fl = fnx(xl)               : fr = fnx(xr)
35    x = (xr*fl-xl*fr)/(fl-fr): GOTO 50
40    x = xl: fv = fnx(x)       : fpv = fnd(x)
45    x = x - fv / fpv          : GOTO 50
50   fv = fnx(x)
     PRINT USING "     #####    ##.####^^^^    ##.####^^^^"; nt, x, fv
     IF (ABS(fv) < epsilon) THEN 900
     nt = nt + 1: ON ms GOTO 110, 120, 130, 140
110  IF (fl * fv <  0!) THEN xr = x: xl = x - dx: GOTO 15 ELSE  x=x+dx: GOTO 50
120  IF (fr * fv >= 0!) THEN xr = x: fr = fv    : GOTO 25 ELSE xl=x   : GOTO 25
130  IF (fr * fv >= 0!) THEN xr = x: fr = fv    : GOTO 35
                             xl = x: fl = fv    : GOTO 35
140  fpv = fnd(x): GOTO 45
900  END
```

Sample Application

All four methods have been applied for searching the roots of the equation $x^2 - \sin(x) - 1 = 0$ in the intervals (–1,–0.5) and (1,1.5). The negative root equal to –0.63673 was found after 27, 15, 5, and 3 iterations and the positive root equal to 1.4096 after 29, 15, 4, and 4 iterations by the incremental search, bisection search, linear interpolation, and Newton-Raphson methods, respectively. An accuracy tolerance of

1.E–5 was used for all cases. For solving this transcendental equation, Newton-Raphson therefore is the best method.

FORTRAN Version

```
C  Program FindRoot - Finds roots of polynomial and transcendental
C                     equations using incremental and bisection search,
C                     linear interpolation, or Newton-Raphson iteration.
      CHARACTER*1 AS
      F (x) = 4 * x ** 3 + 3 * x ** 2 + 2 * x + 1
      FP(x) = 12 * x ** 2 + 6 * x + 2
      WRITE (*,1)
    1 FORMAT(' Program FindRoot - Finds roots of polynomial or'
     *       ' transcendental equations.')
      WRITE (*,*) ' Four options available : '
      WRITE (*,*) '    (1) Incremental Search,'
      WRITE (*,*) '    (2) Bisection Search,'
      WRITE (*,*) '    (3) Linear Interpolation, and'
      WRITE (*,*) '    (4) Newton-Raphson Iteration.'
      WRITE (*,*) ' Select a method, ½/3/4 : '
      READ  (*,*) ms
      WRITE (*,3)
    3 FORMAT(' Have you defined F(x) by editing?  Enter Y/N :')
      READ  (*,4) AS
    4 FORMAT(A1)
      IF (AS.NE.'Y') GOTO 900
      IF (ms.NE.4) GOTO 5
      WRITE (*,*) ' Have you defined d[F(x)]/dx by editing?  Enter Y/N :'
      READ  (*,4) AS
      IF (AS.NE.'Y') GOTO 900
    5 WRITE (*,*) ' Enter the lower and upper bounds of the root : '
      READ  (*,*) xl, xr
      WRITE (*,*) ' How small should F(X) be for termination :'
      READ  (*,*) epsilon
      WRITE (*,*) '    Trial #              X             F(X)'
      nt = 1
      GOTO (10, 20, 30, 40),ms
   10 fl = f(xl)
   15 dx = (xr - xl) / 10
       x = xl + dx
      GOTO 50
   20 fr = f(xr)
   25  x = .5 * (xl + xr)
      GOTO 50
   30 fl = f(xl)
      fr = f(xr)
   35  x = (xr * fl - xl * fr) / (fl - fr)
      GOTO 50
   40  x = xl
      fv = f(x)
      fpv= fp(x)
   45  x = x - fv / fpv
      GOTO 50
   50 fv = f(x)
      WRITE (*,52) nt, x, fv
   52 FORMAT(I10,2E15.4)
      IF (ABS(fv).LT.epsilon) GOTO 900
      nt = nt + 1
      GOTO (110, 120, 130, 140),ms
  110 IF ((fl * fv) .LT. 0) GOTO 115
      x = x + dx
      GOTO 50
```

```
115 xr = x
    xl = x - dx
    GOTO 15
120 IF ((fr * fv).GE.0.) GOTO 125
    xl = x
    GOTO 25
125 xr = x
    fr = fv
    GOTO 25
130 IF ((fr * fv).GE.0.) GOTO 135
    xl = x
    fl = fv
    GOTO 35
135 xr = x
    fr = fv
    GOTO 35
140 fpv = fp(x)
    GOTO 45
900 END
```

Sample Application

The interactive question-and-answer process in solving the polynomial $4x^3 + 3x^2 + 2x + 1 = 0$ using the Newton-Raphson method and the subsequent display on screen of the iteration goes as follows:

```
Program Findroot - Finds roots of polynomial or transcendental equations.
Four options available :
   (1) Incremental Search,
   (2) Bisection Search,
   (3) Linear Interpolation, and
   (4) Newton-Raphson Iteration.
Select a method ½/3/4 "
4
Have you defined F(x) by editing?  Enter Y/N :
Y
Have you defined d[F(x)]/dx by editing?  Enter Y/N :
Y
Enter the lower and upper bounds of the roots :
-2,0
How small should F(x) be for termination?
1.e-5
  Trial #           x              F(x)
        1     -.1395E+01      -.6806E+01
        2     -.9938E+00      -.1951E+01
        3     -.7465E+00      -.4852E+00
        4     -.6312E+00      -.7308E-01
        5     -.6068E+00      -.2668E-02
        6     -.6058E+00      -.3815E-05
```

Notice that in the **FORTRAN** program **FindRoot**, the *statement functions* F(X) and FP(X) are defined for calculating the values of the given function and its derivative at a specified X value. Also, a *character variable* AS is declared through a CHARACTER*N with N being equal to 1 in this case when AS can have only one character as opposed to the general case of having N characters.

MATLAB APPLICATION

A **FindRoot.m** file can be created and added to **MATLAB** m files for the purpose of finding a root of a polynomial or transcendental equation. In this file, the four methods discussed in the **FORTRAN** or **QuickBASIC** versions can all be incorporated. Since some methods require that the left and right bounds, x_l and x_r, be provided, the m file listed below includes as arguments these bounds along with the tolerance and the limited number of iterations:

```
function FindRoot(FofX,DFDX,xl,xr,Tol,NTlimit)
 % Finds a root of polynomial or transcendental equation, F(X).
 % Four options available : ms=1, Incremental Search,
 %                           ms=2, Bisection Search,
 %                           ms=3, Linear Interpolation, and
 %                           ms=4, Newton-Raphson Iteration.
 % Root is bounded by xl and xr.
 % Limits to NTlimit times or function value < Tol.
   ms=input('Choose a method by entering ½/3/4 : ')
  nt = 1; ExitFlag=0; Newxlxr=1; fl=feval(FofX,xl); fr=feval(FofX,xr); x=xl;
   while nt<NTlimit
     if         ms==1, if Newxlxr==1, dx=(xr-xl)/10; x=xl+dx; Newxlxr=0;
                       end
        elseif ms==2, x=.5*(xl+xr);
        elseif ms==3, x = (xr * fl - xl * fr) / (fl - fr);
        elseif ms==4, x = x - feval(FofX,x)/ feval(DFDX,x);
     end
     fv = feval(FofX,x); ntxandfx=[nt,x,fv]
     if abs(fv) < Tol, ExitFlag=1; break
              elseif ms==1, if fl*fv<0, xr=x; xl=x-dx; fl=feval(FofX,xl);
   Newxlxr=1;
                            else x = x + dx;
                       end
        elseif ms==2, if fr*fv>=0, xr=x; fr=fv;
                            else      xl=x;
                       end
        elseif ms==3, if fr*fv>=0, xr=x; fr = fv;
                            else xl = x; fl = fv;
                       end
        elseif ms==4, fpv = feval(DFDX,x);
     end
     if ExitFlag==1, break
        else nt=nt+1;
     end
   end
   if ExitFlag==0, error('The selected method is not converging!')
   end
```

Notice that the equation for which a root is to be found should be defined in a mile file called **FofX.m**, and that if the Newton-Raphson method, i.e., option 4, is to be used, then the first derivative of this equation should also be defined in a m file called **DFDX.m**. We next present four examples demonstrating when all four methods are employed for solving a root of the polynomial $F(x) = 4x^3 + 3x^2 + 2x + 1 = 0$ between the bounds $x = -1$ and $x = 0$ using a tolerance of 10^{-5}. In addition to **FindRoot.m** file, two supporting m files for this case are:

```
function FP=DFDX(X)
         FP=12.*X.^2+6.*X+2;
function F=FofX(X)
         F=4.*X.^3+3.*X.^2+2.*X+1;
```

The four sample solutions are (some printout have been shortened for saving spaces:

```
>>FindRoot('A:FofX','A:DFDX',-1,0,1.e-5,50)

  Choose a method by entering ½/3/4 : 1

 ms =

      1

 ntxandfx =

    1.0000   -0.9000   -1.2860
    2.0000   -0.8000   -0.7280
    3.0000   -0.7000   -0.3020
    4.0000   -0.6000    0.0160
    5.0000   -0.6900   -0.2657
    6.0000   -0.6800   -0.2305
    7.0000   -0.6700   -0.1964
    8.0000   -0.6600   -0.1632
    9.0000   -0.6500   -0.1310
   10.0000   -0.6400   -0.0998
   11.0000   -0.6300   -0.0695
   12.0000   -0.6200   -0.0401
   13.0000   -0.6100   -0.0116
   14.0000   -0.6000    0.0160
   15.0000   -0.6090   -0.0088
   16.0000   -0.6080   -0.0060
   17.0000   -0.6070   -0.0032
   18.0000   -0.6060   -0.0005
   19.0000   -0.6050    0.0023
   20.0000   -0.6059   -0.0002
   21.0000   -0.6058    0.0001
   22.0000   -0.6059   -0.0002
   23.0000   -0.6059   -0.0001
   24.0000   -0.6059   -0.0001
   25.0000   -0.6059   -0.0001
   26.0000   -0.6059   -0.0001
   27.0000   -0.6058   -0.0000
   28.0000   -0.6058   -0.0000

>>FindRoot('A:FofX','A:DFDX',-1,0,1.e-5,50)

  Choose a method by entering ½/3/4 : 2
 ms =

      2

 ntxandfx =

    1.0000   -0.5000    0.2500
    2.0000   -0.7500   -0.5000
    3.0000   -0.6250   -0.0547
    4.0000   -0.5625    0.1123
    5.0000   -0.5938    0.0328
    6.0000   -0.6094   -0.0099
    7.0000   -0.6016    0.0117
    8.0000   -0.6055    0.0010
    9.0000   -0.6074   -0.0044
   10.0000   -0.6064   -0.0017
   11.0000   -0.6060   -0.0004
   12.0000   -0.6057    0.0003
   13.0000   -0.6058   -0.0000
   14.0000   -0.6058    0.0002
   15.0000   -0.6058    0.0001
   16.0000   -0.6058    0.0000
   17.0000   -0.6058    0.0000
```

```
>>FindRoot('A:FofX','A:DFDX',-1,0,1.e-5,50)

  Choose a method by entering ½/3/4 : 3

ms =

     3

ntxandfx =

    1.0000   -0.3333    0.5185
    2.0000   -0.4706    0.3063
    3.0000   -0.5409    0.1629
    4.0000   -0.5755    0.0802
    5.0000   -0.5919    0.0379
    6.0000   -0.5994    0.0175
    7.0000   -0.6029    0.0080
    8.0000   -0.6045    0.0037
    9.0000   -0.6052    0.0017
   10.0000   -0.6056    0.0008
   11.0000   -0.6057    0.0003
   12.0000   -0.6058    0.0002
   13.0000   -0.6058    0.0001
   14.0000   -0.6058    0.0000
   15.0000   -0.6058    0.0000
   16.0000   -0.6058    0.0000

>>FindRoot('A:FofX','A:DFDX',-1,0,1.e-5,50)

  Choose a method by entering ½/3/4 : 4
ms =

     4

ntxandfx =

    1.0000   -0.7500   -0.5000
    2.0000   -0.6324   -0.0765
    3.0000   -0.6069   -0.0029
    4.0000   -0.6058   -0.0000
```

Notice that incremental search, half-interval search, interpolation, and Newton-Raphson methods take 28, 17, 16, and 4 iterations to arrive at the root $x = -0.6058$, respectively. The last method therefore is the best, but is only for this polynomial and not necessary so for a general case.

Method of Successive Substitution

As a closing remark, another method called successive substitution is sometimes a simple way of finding a root of a transcendental equation, such as for solving the angle in a four-bar linkage problem shown in Figure 1. Knowing the lengths L_{AB}, L_{BC} amd L_{CD}, and the angle of the driving link AB, the angle of the driven link CD, can be found by guessing an initial value of $\gamma^{(0)}$ and then continuously upgraded using the equation:

$$\gamma^{(k+1)} = \cos^{-1}\left\{\frac{1}{L_{BC}}\left[L_{AB}\cos\alpha - L_{CD} + \cos\left(\alpha - \gamma^{(k)}\right)\right]\right\} \tag{5}$$

where the superscript k serves as an iteration counter set equal to zero initially. For α changing from 0 to 360°, it is often required in study of such mechanism to find the change in γ. This is left as a homework for the reader to exercise.

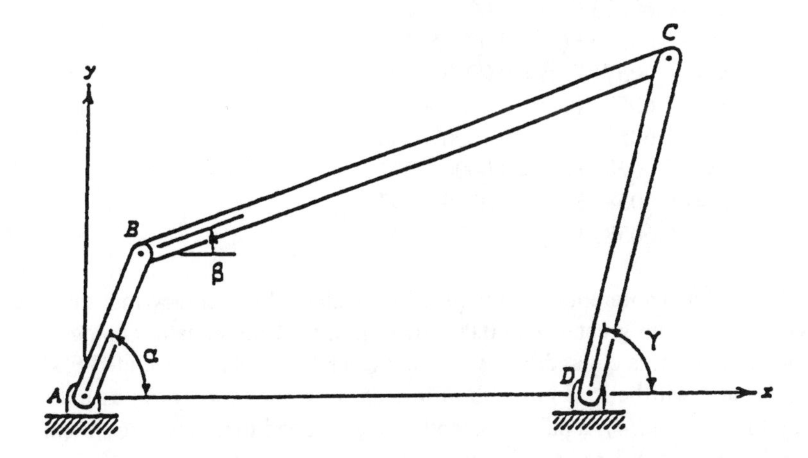

FIGURE 1. Successive substitution sometimes is a simple way of finding a root of a transcendental equation, such as for solving the angle γ in a four-bar linkage problem.

Mathematica Applications

To illustrate how **Mathematica** can be applied to find a root of $F(x) = 1 + 2x + 3x^2 + 4x^3 = 0$ in the interval x = [xl,xr] = [–1,0], the linear interpolation is used below but similar arrangements could be made when the incremental, or, bisection search, or, Newton-Raphson method is selected instead.

Input[1]: = F[x_]: = 1. + 2*x + 3*x^2 + 4*x^3

Input[2]: = xl = –1; xr = 0; fl = F[xl]; fr = F[xr]; fx = fl;

Input[3]: = Print["xl = ",xl," xr = ",xr," F(xl) = ",fl," F(xr) = ",fr]

Output[3]: = xl = –1 xr = 0 F(xl) = –2. F(xr) = 1.

Input[4]: = (While[Abs[fx]>0.00001, x = (xr*fl-xl*fr)/(fl-fr);fx = F[x];

Print["x = ",N[x,5]," F(x) = ",N[fx,5]];
If[fx*fl<0, xr = x;fr = fx;, xl = x;fl = fx;]])

Output[4]: = x = –0.33333 F(x) = 0.51852

x = –0.47059 F(x) = 0.30633
x = –0.54091 F(x) = 0.1629
x = –0.57548 F(x) = 0.080224
x = –0.59185 F(x) = 0.037883
x = –0.59944 F(x) = 0.017521
x = –0.60292 F(x) = 0.0080245
x = –0.60451 F(x) = 0.0036586

x = –0.60523 F(x) = 0.0016646
x = –0.60556 F(x) = 0.00075666
x = –0.60571 F(x) = 0.00034379
x = –0.60577 F(x) = 0.00015618
x = –0.60580 F(x) = 0.000070940
x = –0.60582 F(x) = 0.000032222
x = –0.60582 F(x) = 0.000014635
x = –0.60583 F(x) = $6.6473\text{x}10^{-6}$

Notice that 16 iterations are required to achieve the accuracy that the value of |F(x)| should be no greater than 0.00001. In *Input[1]*, the equation being solved is defined in F[x]. 1. is entered instead of an integer 1 so that all computed F(x) values when printed will be in *decimal form* instead of in *fractional form* as indicated in Output[3]. In *Input[4]*, a pair of parentheses are added to allow *long statements* be entered using many lines and broken and listed with better clarity. Also, N[*exp*,n] is applied to request that the value of expression, *exp*, be handled with n significant figures. The command If is also employed in *Input[4]*. It should be used in the form of If[*condition*, GS1, GS2], which implements the statements in the group GS1 or in the group GS2 when the condition is true or false, respectively. Abs computes the absolute value of an expression specified inside the pair of brackets.

3.3 PROGRAM NEWRAPHG — GENERALIZED NEWTON-RAPHSON ITERATIVE METHOD

Newton-Raphson method[4] has been discussed in the program **FindRoot** in iterative solution of polynomials and transcendental equation. Here, for an extended discussion of this method for solving a set of specified equation, we reintroduce this method in greater detail. This method is based on Taylor's series.[5] Let us start again with the case of one equation of one variable. Let F(X) = 0 be the equation for which a root X_r is to be found. If this root is known to be in the neighborhood of X_g, then based on Taylor's series expansion we may write:

$$F(X_r) = F(X_g) + F'(X_g)\Delta X + F''(X_g)(\Delta X)^2/2! + \ldots \qquad (1)$$

where:

$$\Delta X = X_r - X_g \qquad (2)$$

and the prime in Equation 1 represents differentiation with respect to X. Since X_r is a root of F(X) = 0, therefore $F(X_r) = 0$. And if X_g is sufficiently close to X_r, ΔX is small and the terms involving $(\Delta X)^2$ and higher powers of ΔX in Equation 1 can be neglected. It leads to:

$$X_r = X_g - \left[F(X_g)/F'(X_g)\right] \qquad (3)$$

This result suggests that if we use a projected root value according to Equation 3 as next guess, an iterative process can then be continued until the condition $F(X_g) = 0$ is, if not exactly, *almost* satisfied.

The Newton-Raphson iterative procedure is developed on the above mentioned concept by using the formula:

$$X_g^{(k+1)} = X_g^{(k)} - F\left(X_g^{(k)}\right) / F'\left(X_g^{(k)}\right) \tag{4}$$

where k is an iteration counter. By providing an initial guess, $X_g^{(0)}$, Equation 4 is to be repeatedly applied until $F(X_g^{(k)})$ is almost equal to zero which by using a tolearance can be tested with the condition:

$$\left|F\left(X_g^{(k)}\right)\right| < \varepsilon \tag{5}$$

As an example, consider the case of:

$$F(X) = X^3 - 6X^2 + 11X - 6 = 0 \tag{6}$$

for which

$$F'(X) = 3X^2 - 12X + 11 \tag{7}$$

If we make an initial guess of $X_g^{(0))} = 1.75$ and set a tolerance of = 0.00001, the Newton-Raphson iteration will proceed as follows:

Trial No.	X	F(X)
0	1.7500	0.23438
1	2.0572	-0.05701
2	1.9825	0.01753
3	2.0008	-0.00085
4	2.0001	-0.00011
5	2.0000	0.00001

Program **FindRoot** has a fourth option for Newton-Raphson iteration of a root for a specified equation of *one* variable. The results tabulated above are obtained by the program **FindRoot**.

Transcendental Equations

Not only for polynomials, Newton-Raphson iterative method can also be applied for finding roots of transcendental equations. To introduce a transcendental equation, let us consider the problem of a moving vehicle which is schematically represented by a mass m in Figure 2. The leaf-spring and shock absorber are modelled by k and c, respectively.

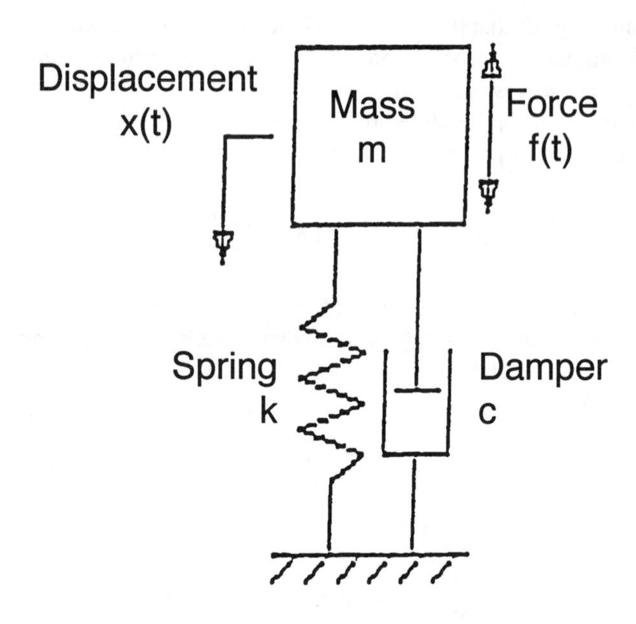

FIGURE 2. Mechanical vibration system with one degree-of-freedom.

If the vehicle is suddenly disturbed by a lift or drop of one of its supporting boundaries by one unit (mathematically, that is a unit-step disturbance), it can be shown[2] that the elevation change in time of the mass, here designated as X(t), is described by the equation:

$$X(t) = 1 - a_1 \text{Exp}(-a_2 t)\sin(a_3 t + a_4) \tag{8}$$

where:

$$a_1 = (k/m)^{0.5}/a_3, \qquad a_2 = c/2m, \tag{9,10}$$

$$a_3 = (4km - c_2)^{0.5}/2m, \quad \text{and} \quad a_4 = \tan^{-1}(a_3/a_2) \tag{11,12}$$

Equation 8 is a transcendental equation.

In actual design of a vehicle, it is necessary to know the lengths of time that are required for the vehicle to respond to the unit-step disturbance and reaching to the amounts equal to 10, 50, and 90 percent of the disturbance. Such calculations are needed to ascertain the delay time, rise time, and other items among the design specifications shown in Figure 3. If one wants to know when the vehicle will rise up to 50 percent of a unit-step disturbance, then it is a problem of finding a root, $t = t_r$, which satisfies the equation:

$$X(t_r) = 1 - a_1 \text{Exp}(-a_2 t_r)\sin(a_3 t_r + a_4) = 0.5$$

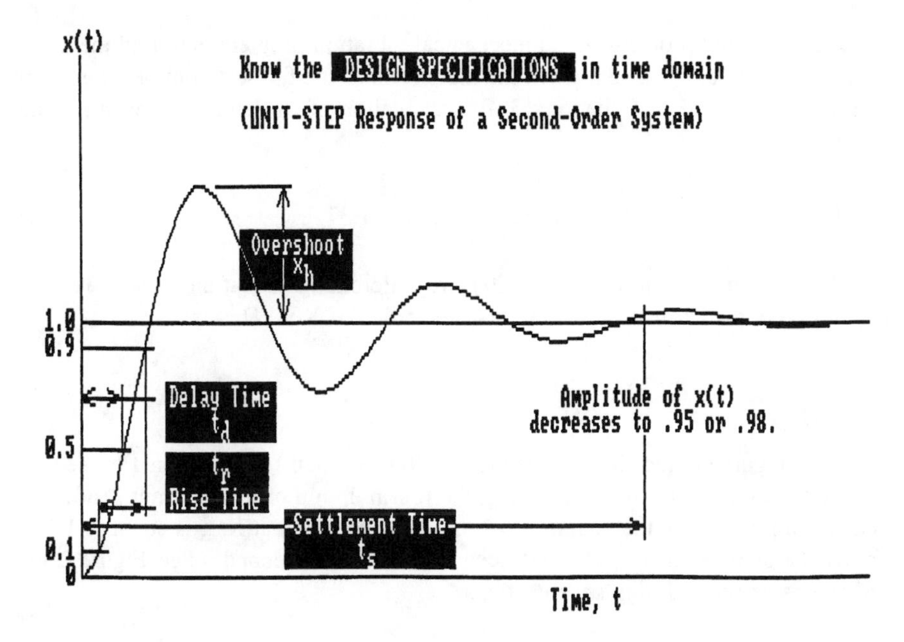

FIGURE 3. Design specifications in time domain: overshoot x_h, delay time t_d, rise time t_r, and settlement time t_s.

Or, the problem can be mathematically stated as solving for t_r from the following transcendental equation by knowing the constants a_{1-4}:

$$a_1 \mathrm{Exp}(-a_2 t)\sin(a_3 t + a_4) - 0.5 = 0$$

As an example, let $a_1 = 1$, $a_2 = 0.2\ \mathrm{sec}^{-1}$, $a_3 = 1\ \mathrm{sec}^{-1}$, and $a_4 = 1.37$ radian then the transcendental equation is:

$$e^{-.2t}\sin(t + 1.37) - 0.5 = 0 \tag{13}$$

To find a root t_r for Equation 3, we select an initial guess $t_r^{(0)} = 0.5$ and apply the fourth option of the program **FindRoot**. The results are listed below. It indicates that the mass reaches 50% of the unit-step disturbance in approximately 1.1 seconds.

```
   Program FindRoot - Finds roots of polynomial or transcendental equations.
Four options available :
  (1) Incremental Search, (2) Bisection Search,
  (3) Linear Interpolation, and
  (4) Newton-Raphson Iteration.
Select a methods, 1/2/3/4 : 4
Have you defined F(x) by editing DEF fnx(x)?  Enter Y/N : Y
Have you defined F'(x) by editing DEF fnd(x)?  Enter Y/N : Y
Enter the lower and upper bounds of the root : 1,2
How small should F(x) be for termination : 1.e-7
   Trial #              X              F(X)
        1       1.1011E+00     -1.4545E-03
        2       1.0991E+00     -3.8743E-07
        3       1.0991E+00     -5.9605E-03
```

An associated problem of the mechanical vibration problem is to find the magnitude and time of overshoot when the mass reaches the farthest point as illustrated in Figure 1. Instead of Equation 13, for calculation of overshoot we examine the equation:

$$X(t) = 1 - e^{-.2t}\sin(t + 1.37) \tag{14}$$

To determine the maximum of X(t), we differentiate Equation 14 with respect to t to derive the expression for the first derivative of X(t). That is:

$$dX(t)/dt = e^{-.2t}\left[.2\sin(t + 1.37) - \cos(t + 1.37)\right] \tag{15}$$

The magnitude and time of maximum X(t) can then be determined by setting Equation 15 equal to zero. In so doing, the fourth option of the program **FindRoot** is again applied using the bounds $t_r = 1$ and $t_r = 2$ to find that X_{max} is equal to 1.523 or overshoot is equal to 53% and occurs at t = 3.145 seconds. See Figure 2 for definitions of these design specifications.

EXTENDED NEWTON-RAPHSON METHOD

The iterative method of Newton-Raphson for solving a either polynomial or transcendental equation of one variable can be extended into solution of multiple equations of multiple variables. Consider the case of two equations of two variables, u(x,y) = 0 and v(x,y) = 0. Let (x_g,y_g) be a guessing solution of these two equations. In that neighborhood, the Taylor's series for f(x,y) and g(x,y) are:

$$u(x_r, y_r) = u(x_g, y_g) + u_{,x}(x_g, y_g)\Delta x + u_{,y}(x_g, y_g)\Delta y + \ldots \tag{16}$$

and

$$v(x_r, y_r) = v(x_g, y_g) + v_{,x}(x_g, y_g)\Delta x + v_{,y}(x_g, y_g)\Delta y + \ldots \tag{17}$$

where $u_{,x} \equiv \partial u/\partial x$ and $v_{,y} \equiv \partial v/\partial y$, and the root location (x_r,y_r) is predicted using the adjustments Δx and Δy. That is,

$$x_r = x_g + \Delta x \quad \text{and} \quad y_r = y_g + \Delta y \tag{18,19}$$

Since it is hoped that $u(x_r,y_r)$ and $v(x_r,y_r)$ would both be equal to zero, Equations 16 and 17 therefore can be expressed, after dropping the higher order terms of Δx and Δy, in the forms of:

$$u_{,x}(x_g, y_g)\Delta x + u_{,y}(x_g, y_g)\Delta y = -u(x_g, y_g) \tag{20}$$

and

$$v_{,x}(x_g, y_g)\Delta x + v_{,y}(x_g, y_g)\Delta y = -v(x_g, y_g) \tag{21}$$

Cramer's rule can then be applied to obtain Δx and Δy as

$$\Delta x = \left(-uv_{,y} + vu_{,y}\right) / \left(u_{,x}v_{,y} - u_{,y}v_{,x}\right) \tag{22}$$

and

$$\Delta y = \left(+uv_{,x} - vu_{,x}\right) / \left(u_{,x}v_{,y} - u_{,y}v_{,x}\right) \tag{23}$$

where u, v, and their derivatives are to be evaluated at (x_g,y_g). Equations 22 and 23 are to be continuously applied to adjust the guessing values of (x_r,y_r) until both $u(x_r,y_r)$ and $v(x_r,y_r)$ are negligibly small.

Program **NewRaphG** has been developed by use of the iterative equations 22 and 23. Both **QuickBASIC** and **FORTRAN** versions of this program are listed below along with a sample application of solving the intercepts of two ellipses, $f(x,y) = (x/3)^2 + (y/4)^2 - 1$ and $g(x,y) = (x/4)^2 + (y/3)^2 - 1$.

QuickBASIC Version

```
' * PROGRAM NewRaphG - Newton-Raphson's method applied for two-equations, two-unknowns case.
    DECLARE SUB FandDFs (X, Y, F, G, FX, FY, GX, GY)
    CLEAR : CLS : E = .00001: N = 100
    PRINT "Program NewRaphG - Solves two-equations, two-variables problems"
    PRINT "                   by Newton-Raphson iteration."
    INPUT "Have you defined the equations by editing the subroutine FandDfs?  Enter Y/N: ",
A$
    IF A$ = "Y" THEN 100 ELSE END
100 INPUT "Enter initial guess, (XO,YO) :", XO, YO
    INPUT "Enter the accuracy tolerance : ", E
    INPUT "Enter the number of iterations allowed : ", N
    PRINT SPC(9); "Iteration #        X              Y              F(X)          F(Y)"
    X = XO: Y = YO
    FOR T = 1 TO N
    PRINT SPC(19); : PRINT USING "##  ##.####^^^^"; T, X;
    PRINT USING "  ##.####^^^^  ##.####^^^^  ##.####^^^^"; Y, F, G
    CALL FandDFs(X, Y, F, G, FX, FY, GX, GY)
    IF ABS(F) + ABS(G) < E THEN END
    DM = FX * GY - FY * GX: X = X + (G * FY - F * GY) / DM: Y = Y + (F * GX - G * FX) / DM
    NEXT T
370 PRINT : PRINT "Iteration has failed after N trials !": END

    SUB FandDFs (X, Y, F, G, FX, FY, GX, GY)
    F = (X / 3) ^ 2 + (Y / 4) ^ 2 - 1!
    G = (X / 4) ^ 2 + (Y / 3) ^ 2 - 1
    FX = 2 * (X / 3) / 3
    FY = 2 * (Y / 4) / 4
    GX = 2 * (X / 4) / 4
    GY = 2 * (Y / 3) / 3
    END SUB
```

FORTRAN Version

```
C     PROGRAM NewRaphG - Newton-Raphson's method applied for solving
C                        two-equations, two-unknowns case.
      Character*1 AS
      WRITE (*,5)
    5 FORMAT(' Program NewRaphG - Solves two-equations, two-variables',
     *       ' problems'/20X,'by Newton-Raphson iteration.')
      WRITE (*,10)
   10 FORMAT(' Have you defined the equations by editing the',
     *       ' subroutine FandDfs?  Enter Y/N: ')
      READ  (*,15) AS
```

```
   15 FORMAT(A1)
      IF (AS.NE.'Y') GOTO 900
      WRITE (*,*) 'Enter initial guess, (X0,Y0) :'
      READ  (*,*) X0, Y0
      WRITE (*,*) 'Enter the accuracy tolerance :'
      READ  (*,*) E
      WRITE (*,*) 'Enter the number of iterations allowed :
      READ  (*,*) N
      WRITE (*,20)
   20 FORMAT(1X,'  Iteration #          X              Y
     *         'F(X,Y)       G(X,Y)')
      X = X0
      Y = Y0
      DO 50 IT = 1, N
      CALL FandDFs(X,Y,F,G,FX,FY,GX,GY)
      WRITE (*,25) IT,X,Y,F,G
   25 FORMAT(8X,I2,4E15.4)
      IF ((ABS(F)+ABS(G)).LT.E) GOTO 900
      DM = FX * GY - FY * GX
      X = X + (G * FY - F * GY) / DM
   50 Y = Y + (F * GX - G * FX) / DM
   55 WRITE (*,*) 'Iteration has failed after N trials !'
  900 END

      SUBROUTINE FandDFs (X, Y, F, G, FX, FY, GX, GY)
      F = (X / 3) ** 2 + (Y / 4) ** 2 - 1
      G = (X / 4) ** 2 + (Y / 3) ** 2 - 1
      FX = 2 * (X / 3) / 3
      FY = 2 * (Y / 4) / 4
      GX = 2 * (X / 4) / 4
      GY = 2 * (Y / 3) / 3
      RETURN
      END
```

Sample Application

```
Program NewRaphG - Solves two-equations, two-variables problems.
                   by Newton-Raphson iteration.
Have you defined the equations by editing the subroutine FandDfs?  Enter Y/N:
Y
Enter initial guesses, (X0,Y0) :
-2,2
Enter the accuracy tolerance :
1.e-5
Enter the number of iterations allowed :
100
  Iteration #          X              Y          F(X,Y)          G(X,Y)
        1       -.2000E+01      .2000E+01     -.3056E+00      -.3056E+00
        2       -.2440E+01      .2440E+01      .3361E-01       .3361E-01
        3       -.2400E+01      .2400E+01      .2733E-03       .2733E-03
        4       -.2400E+01      .2400E+01      .5960E-07       .5960E-07
```

MATLAB Applications

Here, we provide a m file called NewRaphG.m as a companion of the **FORTRAN** and **QuickBASIC** versions:

```
   function X=NewRaphG(Functns,Derivatf,N,X0,TOL,NTlimit)
% Solves N simultaneous equations defined in Functns and their first-
%  derivative functions in Derivatf near the neighborhood {X0}.
% Generalized Newton-Raphson method is used to iterate NTlimit time
%  or when the sum of the absolute values of the functions < TOL.
 NT=1; FS=0; ExitFlag=0; FXS=feval(Functns,N,X0);
 for i=1:N
     FS=FS+abs(FXS(i));
 end
 NT,Xvalues=[X0]',SumOfFsA
 while FS>TOL
       NT=NT+1; DS=feval(Derivatf,N,X0);
       X=X0-DS\FXS; FXS=feval(Functns,N,X);
       FS=0; for i=1:N
                FS=FS+abs(FXS(i));
             end
       X0=X; NT,Xvalues=[X]',SumOfFsA=FS
       if NT==NTlimit, ExitFlag=1; break
       end
 end
 if ExitFlag==1, error('Iteration fails after NTlimit trials.')
 end
```

For using this function, the problem to be solved needs to be defined by creating two m files, in which the equations involved and the expressions for their first derivatives should be spelled out. In case of solving the sample problem used in **FORTRAN** and **QuickBASIC** versions, first we may define the equation as:

```
function FXS=Functns(N,X)
         for i=1:N
             if i==1, FXS(i,1)=(X(1,1)/3).^2+(X(2,1)/4).^2-1;
               elseif i==2, FXS(i,1)=(X(1,1)/4).^2+(X(2,1)/3).^2-1;
             end
         end
```

Next, the expressions for their first derivatives may then be written as:

```
function DFDXS=derivatf(N,X)
         for i=1:N
             if i==1, for j=1:N
                          if j==1, DFDXS(i,j)=2*X(1,1)/3/3;
                           elseif j==2, DFDXS(i,j)=2*X(2,1)/4/4;
                          end
                      end
               elseif i==2, for j=1:N
                               if j==1, DFDXS(i,j)=2*X(1,1)/4/4;
                            elseif j==2, DFDXS(i,j)=2*X(2,1)/3/3;
                                   end
                            end
             end
         end
```

To solve this problem, the interactive application of **MATLAB** proceeds as follows (some displays have been rearranged for saving spaces):

```
>> X0=[-2,2]'; X=NewRaphG('a:Functns','a:Derivatf',2,X0,1.e-5,20)

        NT =          Xvalues =                  SumOfFsA =

            1          -2          2                0.6111
            2          -2.4400     2.4400           0.0672
            3          -2.4003     2.4003           5.4649e-004
            4          -2.4000     2.4000           3.7361e-008

       X =

          -2.4000
           2.4000
```

Notice that the initial values are taken as X(1) = –2 and X(2) = 2, a tolerance of 10^{-5} and the iteration is limited to 20 trials. The solutions are found after four Newton-Raphson trials when the sum of the absolute values of the two equations is equal to 3.7361×10^{-8}.

Mathematica Applications

Mathematica applies the Newton's method in its function **FindRoot** which can be applied for solving a polynomial, or, transcendental equation, and also for multiple equations. We illustrate its applications by using the examples discussed earlier. First, the root near X = 1.75 for a third-order polynomial is found:

In[1]: = FindRoot[{X^3–6*X^2 + 11*X–6 = = 0}, {X,1.75}]

Out[1] = {X -> 2.}

The solution is X = 2. The second example is for finding a root near T = 0.5 for a transcendental equation described inside the first pair of braces:

In[2]: = FindRoot[{Exp[-.2*T]*Sin[T + 1.37] = = 0.5},{T,0.5}]

Out[2] = {T -> 1.09911}

For solving two simultaneous transcendental equations, two examples are presented below. The first is to find one of the intercepts of two ellipses and the second is to find one of the intercepts of a circle of radius equal to 2 and a sine curve.

In[3]: = (FindRoot[{(X/3)^2 + (Y/4)^2 = = 1,(X/4)^2 + (Y/3)^2 = = 1},
{X,–2}, {Y,2}]

Out[3] = {X -> –2.4, Y -> 2.4}

In[4]: = FindRoot[{x = = Sqrt[4y^2],y = = Sin[2*x]},{x,1.95},{y,–0.6}]

Out[4] = {x -> 1.90272, y -> –0.616155}

3.4 PROGRAM BAIRSTOW — BAIRSTOW'S METHOD FOR FINDING POLYNOMIAL ROOTS

Program **Bairstow** is developed for finding the roots of polynomials based on the Newton-Raphson's iterative method for two variables (see program **NewRaphG**). Let a nth-order polynomial be denoted as:

$$P(x) = x^N + a_1x^{N-1} + a_2x^{N-2} + \ldots + a_{N-2}x^2 + a_{N-1}x + a_N \quad (1)$$

Notice that the highest term x^N has a coefficient equal to 1; otherwise the entire equation must be normalized by dividing by that coefficient. The Bairstow's method consists of first selecting a trial divider $D(x) = x^2 + d_1x + d_2$, and to obtain the quotient $Q(x) = x^{N-2} + q_1x^{N-3} + q_2x^{N-4} + \cdots + q_{N-4}x^2 + q_{N-3}x + q_{N-2}$ and a remainder $R(x) = r_1x + r_2$. The objective is to continuously adjust the values of d_1 and d_2 until both values of r_1 and r_2 are sufficiently small. It is apparent that both r_1 and r_2 are dependent of d_1 and d_2. Taylor's series expansions of r_1 and r_2 can be written as:

$$\begin{aligned} r_1(d_1 + \Delta d_1, d_2 + \Delta d_2) &= r_1(d_1, d_2) + r_{1,d_1}(d_1, d_2)\Delta d_1 \\ &\quad + r_{1,d_2}(d_1, d_2)\Delta d_2 + \ldots \end{aligned} \quad (2)$$

and

$$\begin{aligned} r_2(d_1 + \Delta d_1, d_2 + \Delta d_2) &= r_2(d_1, d_2) + r_{2,d_1}(d_1, d_2)\Delta d_1 \\ &\quad + r_{2,d_2}(d_1, d_2)\Delta d_2 + \ldots \end{aligned} \quad (3)$$

where

$$r_{1,d_1} \equiv \partial r_1/\partial d_1, \quad r_{2,d_2} \equiv \partial r_2/\partial d_2$$

and so on.

The adjustments Δd_1 and Δd_2 are to be calculated so as to make the left-hand side of Equations 2 and 3 both equal to zero and these adjustments are expected to be small enough (if the guessed values of d_1 and d_2 values are sufficiently close to those which make both r_1 and r_2 equal to zero) so that the second and higher derivative terms in Equations 2 and 3 can be dropped. This leads to:

$$r_{1,d_1}(d_1, d_2)\Delta d_1 + r_{1,d_2}(d_1, d_2)\Delta d_2 = -r_1(d_1, d_2) \quad (4)$$

and

$$r_{2,d_1}(d_1, d_2)\Delta d_1 + r_{2,d2}(d_1, d_2)\Delta d_2 = -r_2(d_1, d_2) \quad (5)$$

Cramer's rule can then be applied to obtain Δd_1 and Δd_2 as:

$$\Delta d_1 = \left(-r_1 r_{2,d_2} + r_2 r_{1,d_2}\right) \Big/ \left(r_{1,d_1} r_{2,d_2} - r_{1,d_2} r_{2,d_1}\right) \tag{6}$$

and

$$\Delta d_2 = \left(-r_1 r_{2,d_1} + r_2 r_{1,d_1}\right) \Big/ \left(r_{1,d_1} r_{2,d_2} - r_{1,d_2} r_{2,d_1}\right) \tag{7}$$

where r_1, r_2, and their partial derivatives are to be evaluated at (d_1,d_2). Equations 6 and 7 are to be continuously applied to adjust the guessing values of (d_1,d_2) until both $r_1(d_1,d_2)$ and $r_2(d_1,d_2)$ are negligibly small.

To calculate the adjustments Δd_1 and Δd_2 based on Equations 6 and 7, we need to find the partial derivatives $\partial r_1/\partial d_1$, $\partial r_1/\partial d_2$, $\partial r_2/\partial d_1$, and $\partial r_2/\partial d_2$. These derivatives are, however, depend on the d_1 and d_2, and the coefficients q's in the quotient Q(x). This can be shown by actually carried out the division of P(x) by D(x). The results are:

$$q_1 = a_1 - d_1 \quad \text{and} \quad q_2 = a_2 - q_1 d_1 - d_2 \tag{8,9}$$

and

$$q_k = a_k - q_{k-1} d_1 - q_{k-2} d_2, \quad \text{for } k = 3,4,\ldots,N-2 \tag{10}$$

It can also be shown that the coefficients in the remainder R(x) are:

$$r_1 = a_{N-1} - q_{N-2} d_1 - q_{N-3} d_2 \quad \text{and} \quad r_2 = a_N - q_{N-2} d_2 \tag{11,12}$$

We notice that Equations 11 and 12 can be included in Equation 10 if k is extended to N and if the remainder is redefined as:

$$R(x) = (x + d_1) q_{N-1} + q_N \tag{13}$$

That is, r_1 is renamed as q_{N-1} and r_2 is equal to $d_1 q_{N-1} + q_N$. As a consequence, we need to replace r_1 and r_2 in Equations 6 and 7 by q_{N-1} and q_N. For calculation of the adjustments Δd_1 and Δd_2, Equation 10 should be used for q_{N-1} and q_N and to derive their partial derivatives respect to d_1 and d_2. Since all q's are functions of d_1 and d_2, to derive the partial derivatives of the last two q's we must find the partial derivatives for all q's starting with q_1. From Equations 8 to 10, we can have:

$$\partial q_1/\partial d_1 = -1, \quad \partial q_1/\partial d_2 = 0, \tag{14,15}$$

$$\partial q_2/\partial d_1 = (-\partial q_1/\partial d_1) d_1 - q_1 = d_1 - q_1, \tag{16}$$

$$\partial q_2/\partial d_2 = -(\partial q_1/\partial d_2)q_1 - 1 = - \quad (17)$$

and for k = 3,4,…,N

$$\partial q_k/\partial d_1 = -(\partial q_{k-1}/\partial d_1)d_1 - q_{k-1} - (\partial q_{k-2}/\partial d_1)d_2 \quad (18)$$

$$\partial q_{k+1}/\partial d_2 = -(\partial q_k/\partial d_2)d_1 - q_{k-1} - (\partial q_{k-1}/\partial d_2)d_2 \quad (19)$$

It can be concluded from the above results that:

$$\partial q_{k+1}/\partial d_2 = \partial q_k/\partial d_1 \quad \text{for} \quad k = 1,2,\ldots,N-1 \quad (20)$$

Now, we can summarize the procedure of Bairstow's method for factorizing a quadratic equation from an Nth-order polynomial as follows: (Some changes of variables are made in the computer programs to be presented next, such as q's are changed to b's, d_1 and d_2 are changed to u and v, respectively, and c's are introduced to represent the derivatives of q's.)

(1) Specify the values of N, a_1 through a_N, and a tolerance ϵ.
(2) Assume an initial guessing values for d_1 and d_2 for the divider D(x).
(3) Calculate the coefficients q_1 through q_{N-2} for the quotient Q(x) using Equations 8 to 10.
(4) Also use Equation 10 to calculate the coefficients q_{N-1} and q_N for the remainder R(x).
(5) Test the absolute values of q_{N-1} and q_N. If they are both less than ϵ, two root of P(x) are to be calculated by use of the quadratic formulas. The order of P(x), N, is to be reduced by 2, and q_1 through q_{N-2} are to become a_1 through a_{N-2}, respectively, and return to Step 2. This looping continues until the quotient Q(x) is of order two or one, for which the root(s) easily can be calculated.
(6) If the absolute value of either q_{N-1} or q_N is greater than ε, calculate the partial derivatives of q_k with respect to d_1, c's using Equations 14, 16, and 18 for k = 3,4,…,N. The derivatives of q_k with respect to d_2 are already available due to Equation 20.
(7) Use Equations 6 and 7 to calculate the adjustments Δd_1 and Δd_2, noticing that r_1 and r_2 are to be replaced by q_{N-1} and q_N, respectively. The iteration is resumed by returning to Step 3.

Both **QuickBASIC** and **FORTRAN** versions of the program **Bairstow** coded following the steps described above are to be presented next.

QuickBASIC Version

```
' * Program BAIRSTOW - Applies Bairstow's method for finding polynomial roots
    SCREEN 2: CLS : CLEAR : KEY OFF : PRINT "* Program BAIRSTOW *"  : PRINT
    PRINT "  - Solving polynomial roots by quadratic factorization.": PRINT
    PRINT : INPUT "Order of the polynomial"; N : NP1 = N + 1
    DIM A(N), A1(NP1), C(NP1), D(2), Q(NP1) : PRINT
    PRINT "Enter the coefficients of the polynomial starting from the";
    PRINT " highest order."
    PRINT "  Press <Enter> key after entering a number."
           FOR I = 1 TO NP1: INPUT ; A1(I): NEXT I
           FOR I = 1 TO N: A(I) = A1(I + 1) / A1(1): NEXT I
    PRINT "Enter the initial guesses of d(1) and d(2), and Epsilon, e.g.,";
    INPUT " .005,.005,.00001 : "; D10, D20, Epsilon
    PRINT : PRINT "The Roots are :"
    PRINT : PRINT "          REAL PART    IMAGINARY PART    ITERATIONS"
160                    IF N > 1 THEN 175
                     PReal = -A(1) : PImag = 0: I = 1
    PRINT USING "############.#####  ##########.#####"; PReal, PImag;
    PRINT USING "  ############"; I: GOTO 385
175                     IF N > 2 THEN 240
                    D(1) = A(1): D(2) = A(2): I = 1
190                     PReal = -D(1) / 2: R = D(1) ^ 2 - 4 * D(2)
                       IF R > 0 THEN 215
195                    R = -R: PImag = SQR(R) / 2
    PRINT USING "############.#####  ##########.#####"; PReal, PImag;
    PRINT USING "  ############"; I
                        N = N - 1: PImag = -PImag
    PRINT USING "############.#####  ##########.#####"; PReal, PImag;
    PRINT USING "  ############"; I: GOTO 230
215                       PImag = SQR(R) / 2
    PRINT USING "############.#####  ##########.#####";PReal+PImag,0;
    PRINT USING "  ############"; I: N = N - 1
    PRINT USING "############.#####  ##########.#####";PReal-PImag,0;
    PRINT USING "  ############"; I
230                 N = N - 1: IF N <= 0 THEN 385
        FOR K = 1 TO N: A(K) = Q(K): NEXT K: GOTO 160
240                     D(1) = D10: D(2) = D20
'
'           * Limit iteration to 200 trials *
'
    FOR I = 1 TO 200: Q(1) = A(1) - D(1): Q(2)=A(2)-Q(1)*D(1)-D(2)
      FOR K = 3 TO N: Q(K) = A(K)-Q(K-1)*D(1)-Q(K-2)*D(2): NEXT K
      C(1) = Q(1) - D(1): C(2) = Q(2) - C(1) * D(1) - D(2)
      FOR K=3 TO N- 1: C(K) = Q(K)-C(K-1)*D(1)-C(K-2)*D(2): NEXT K
      IF N > 3 THEN 305
      Determ = C(N - 1) - C(N - 2) ^ 2: IF Determ <> 0 THEN 295
290   PRINT : PRINT "Denominator is zero!": GOTO 385
295   DD1 = (Q(N) - Q(N - 1) * C(N - 2)) / Determ
      DD2 = (C(N-1)*Q(N-1)-C(N-2)*Q(N))/Determ: GOTO 320
305   Determ = C(N-1)*C(N-3)-C(N-2)^2: IF Determ = 0 THEN 290
      DD1 = (Q(N) * C(N - 3) - Q(N - 1) * C(N - 2)) / Determ
      DD2 = (C(N - 1) * Q(N - 1) - C(N - 2) * Q(N)) / Determ
320   D(1) = D(1) + DD1: D(2) = D(2) + DD2: S = ABS(DD1) + ABS(DD2)
      IF S < Epsilon THEN 190
      IF I <= 1 THEN S0 = S: GOTO 370
340   IF I <> 50 OR S < S0 THEN 355
      PRINT : PRINT "The process is diverging!"
      PRINT "d(1)= ";D(1);"d(2)= ";D(2);"Deltad(1) = ";DD1;
      PRINT "Delta d(2) = ";DD2: GOTO 385
355   IF I <> 100 THEN 370
      PRINT : PRINT "The process is slow in converging!"
      PRINT "d(1)= "; D(1); "d(2)= "; D(2); "Delta d(1) = "; DD1;
      PRINT "Delta d(2) = "; DD2
370 NEXT I
    PRINT : PRINT "Iteration is terminated after 200 trials."
    PRINT "d(1)= ";D(1);"d(2)=";D(2);"Delta d(1) =";DD1;
    PRINT "Delta d(2) = ";DD2
385 END
```

Sample Application

As an example, the polynomial $P(x) = x^4 - 5x^3 + 13x^2 - 19x + 10 = 0$ is solved by application of the **QuickBASIC** version of the program **Bairstow**. The response on screen is:

```
Enter the order of the polynomial :
4
Enter the coefficients of the polynomial starting from the highest
      order and press Enter key after entering each number.
1
-5
13
-19
10
The roots are :

        REAL PART     IMAGINARY PART      ITERATIONS
         2.00000           .00000              7
         1.00000           .00000              7
         1.00000          2.00000              1
         1.00000         -2.00000              1
```

The quotient in this case is a quadratic equation:

$$Q(x) = [x-(1+2i)][x-(1-2i)] = (x-1)^2 - (2i)^2 = x^2 - 2x + 5 = 0.$$

FORTRAN Version

```
C   * Program Bairstow - Applies Bairstow's method for finding polynomial roots
      DIMENSION A(100),A1(101),C(101),D(2),Q(98)
      WRITE (*,3)
    3 FORMAT (1X,'Enter the order of the polynomial')
      READ (*,*) N
      NP1=N+1
      WRITE (*,5)
    5 FORMAT(1X,'Enter the coefficients of the polynomial starting',
     *' from the highest order'/5X,' and press Enter key after ',
     *'entering each number.')
      DO 7 I=1,NP1
    7 READ (*,*) A1(I)
      DO 9 I=1,N
    9 A(I)=A1(I+1)/A1(1)
      WRITE (*,10)
   10 FORMAT(1X,'Enter the initial guesses of d(1) and d(2)',
     *          ' and Epsilon, e.g.,.005,.005,.00001:')
      READ (*,*) D10,D20,Epsilon
      WRITE (*,11)
   11 FORMAT(1X,'The Roots are :'//'          REAL PART     IMAGINARY',
     *' PART      ITERATIONS')
  165 IF (N.GT.1) GOTO 180
      PReal=-A(1)
      PImag=0.
      I=1
      WRITE (*,170) PReal,PImag,I
  170 FORMAT(F17.5,F18.5,7X,I7)
      GOTO 390
```

```
  180 IF (N.GT.2) GO TO 245
      D(1)=A(1)
      D(2)=A(2)
      I=1
  190 PReal=-D(1)/2.
      R=D(1)**2-4.*D(2)
      IF (R.GT.0.) GO TO 220
      R=-R
      PImag=SQRT(R)/2.
      WRITE (*,170) PReal,PImag,I
      N=N-1
      PImag=-PImag
      WRITE (*,170) PReal,PImag,I
      GOTO 235
  220 PImag=SQRT(R)/2.
      TR=PReal+PImag
      TI=0.
      WRITE (*,170) TR,TI,I
      N=N-1
      TR=PReal-PImag
      WRITE (*,170) TR,TI,I
  235 N=N-1
      IF ((N.LT.0).OR.(N.EQ.0)) GO TO 390
      DO 242 K=1,N
  242    A(K)=Q(K)
      GOTO 165
  245 D(1)=D10
      D(2)=D20
C             * Limit iteration to 200 trials *
      DO 375 I=1,200
         Q(1)=A(1)-D(1)
         Q(2)=A(2)-Q(1)*D(1)-D(2)
         DO 272 K=3,N
  272       Q(K)=A(K)-Q(K-1)*D(1)-Q(K-2)*D(2)
         C(1)=Q(1)-D(1)
         C(2)=Q(2)-C(1)*D(1)-D(2)
         NM1=N-1
         DO 282 K=3,NM1
  282       C(K)=Q(K)-C(K-1)*D(1)-C(K-2)*D(2)
         IF (N.GT.3) GO TO 310
         Determ=C(N-1)-C(N-2)**2
         IF (Determ.NE.0.) GO TO 300
  295    WRITE (*,298)
  298    FORMAT(/5X,'Denominator is zero.!')
         GOTO 390
  300    DD1=(Q(N)-Q(N-1)*C(N-2))/Determ
         DD2=(C(N-1)*Q(N-1)-C(N-2)*Q(N))/Determ
         GOTO 325
  310    Determ=C(N-1)*C(N-3)-C(N-2)**2
         IF (Determ.EQ.0.) GO TO 295
         DD1=(Q(N)*C(N-3)-Q(N-1)*C(N-2))/Determ
         DD2=(C(N-1)*Q(N-1)-C(N-2)*Q(N))/Determ
  325    D(1)=D(1)+DD1
         D(2)=D(2)+DD2
         S=ABS(DD1)+ABS(DD2)
         IF (S.LT.Epsilon) GO TO 190
         IF (I.GT.1) GO TO 345
         S0=S
         GO TO 375
  345    IF ((I.NE.50).OR.(S.LT.S0)) GO TO 360
         WRITE (*,352)
  352    FORMAT(/5X,'The process is not converging!')
         GO TO 390
  360    IF (I.NE.100) GO TO 375
         WRITE (*,368)
  368    FORMAT(/5X,'The process is slow in converging!')
  375    CONTINUE
      WRITE (*,382)
  382 FORMAT(/5X,'Iteration is terminated after 200 trials.')
  390 END
```

Sample Application

Consider the polynomial $P(x) = x^3 + 2x^2 + 3x + 4 = 0$. When the **FORTRAN** version of the program **Bairstow** is run, the response on screen is:

```
Enter the order of the polynomial :
3
Enter the coefficients of the polynomial starting from the highest
      order and press Enter key after entering each number.
1
2
3
4
The roots are :

        REAL PART     IMAGINARY PART     ITERATIONS
         -.17469          1.54687            31
         -.17469         -1.54687            31
        -1.65063           .00000             1
```

When the ITERATIONS column indicates 1, it signals that the quotient is of order one or two. In this case, the quotient is $Q(x) = x + 1.65063$. In fact, no iteration has been performed for solving Q(x).

MATLAB Application

MATLAB has a file called **roots.m** which can be applied to find the roots of a polynomial $p(x) = 0$. To do so, the coefficients of an nth-order p(x) should be ordered in descending powers of x into a row matrix of order n + 1. For example, to solve $p(x) = x^3 + 2x^2 + 3x + 4 = 0$, we enter:

```
>> p= [1,2,3,4]; x= roots(p)
```

and obtain a screen display

```
x =
      -1.6506
      -0.1747 + 1.5469i
      -0.1747 - 1.5469i
```

As a second example of solving $x^4 – 5x^3 + 13x^2 – 19x + 10 = 0$, **MATLAB** interactive entries indicated by the leading >> signs and the resulting display are:

```
>> p=[1,-5,13,-19,10]

   p =

         1     -5     13    -19     10
```

```
>> x=roots(p)

   x =
             1.0000 + 2.0000i
             1.0000 - 2.0000i
             2.0000
             1.0000
```

Comparing the two examples, we notice that by placing ; after a statement suppresses the display of the computed value(s). The elements of the first p matrix (a single row) is not displayed!

It is of interest to introduce the plot capability of **MATLAB** by use of the results presented above which involve a polynomial P(x) and its roots. From graphical viewpoint, the roots are where the polynomial curve crossing the x-axis. **MATLAB** has a **plot.m** file which can be readily applied here. Let us again consider the polynomial $P(x) = x^4 - 5x^3 + 13x^2 - 19x + 10 = 0$ and plot it for $0 \leq x \leq 3$. For adequate smoothness of the curve, an increment of x equal to 0.1 can be selected for plotting. The interactive **MATLAB** commands entered for obtaining Figure 4 are:

```
>> p=[1,-5,13,-19,10]; x=[0:0.1:3]; y=polyval (p,x);
>> plot(x,y), hold
>> XL=[0 3]; YL=[0 0]; plot(XL,YL)
```

Notice that another m file **polyval** of **MATLAB** has been employed above. The statement y = polyval(p,x) generates a array of y values using the polynomial defined by the coefficient vector p and calculated at the values specified in the x array. The **hold** statement put the current plot "on hold" so that an additional horizontal line connecting the two points defined in the XL and YL arrays can be superimposed. The first plot statement draws the curve and axes and tic marks while the second plot statement draws the horizontal line.

The horizontal line drawn at y = 0 help to show the intercepts of the polynomial curve and the x-axis, by observation near x = 1 and x = 2 which confirm the result found by the **MATLAB** file roots.m.

Mathematica Applications

For finding the polynomial roots, **Mathematica**'s function **NSolve** can be applied readily.

Keyboard input (and then press shift and Enter keys)

NSolve[x^3 + 2x^2 + 3x + 4 = = 0,x]

The **Mathematica** response is:

Input[1]: =

NSolve[x^3 + 2x^2 + 3x + 4 = = 0,x]

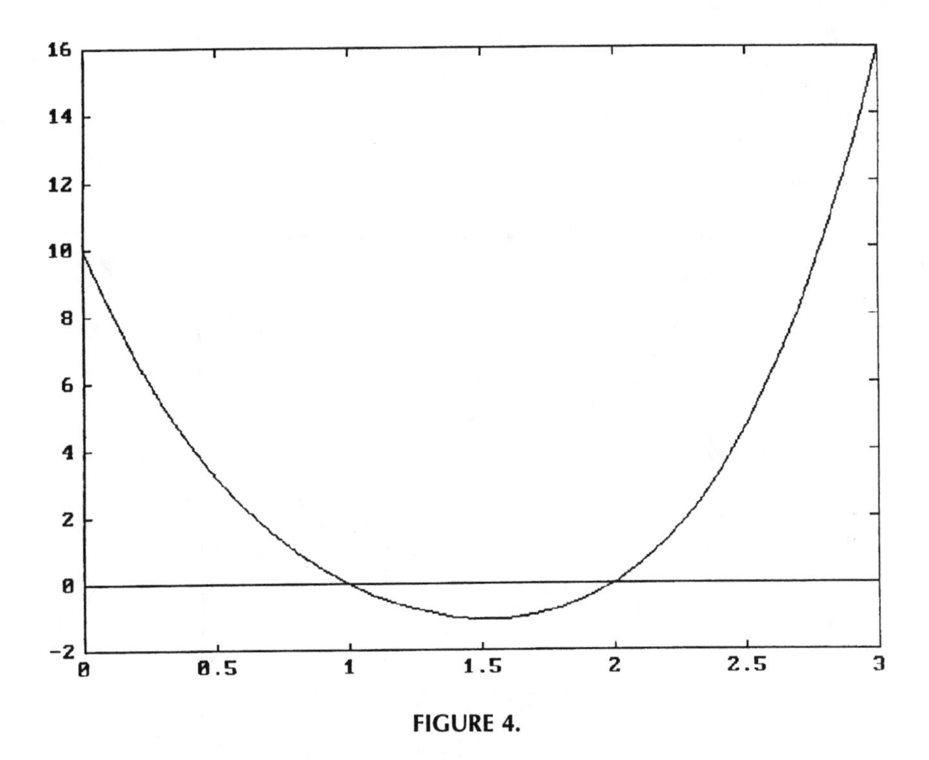

FIGURE 4.

Output[1] =

{{x -> –1.65063}, {x -> –0.174685 — 1.54687 I},
{x -> –0.174685 — 1.54687 I}}

Keyboard input (and then press **Shift** and **Enter** keys)

NSolve[x^4–5x^3 + 13x^2–19x + 10 = = 0,x]

The **Mathematica** response is:

Input[2]: =

NSolve[x^4–5x^3 + 13x^2–19x + 10 = = 0,x]

Output[2] =

{{x -> 1. – 2. I}, {x -> –1 + 2. I}, {x -> 1.}, {x -> 2.}}

To show the locations of the roots of a polynomial, **Mathematica**'s function Plot can be applied to draw the polynomial. The following statements (Keyboard input will hereon be omitted since it is always repeated in the Input response) enable Figure 5 to be generated:

Input[3]: =

```
Plot[x^4–5x^3 + 13x^2–19x + 10,{x,0,3},
  Frame->True, AspectRatio->1]
```

Output[3] =

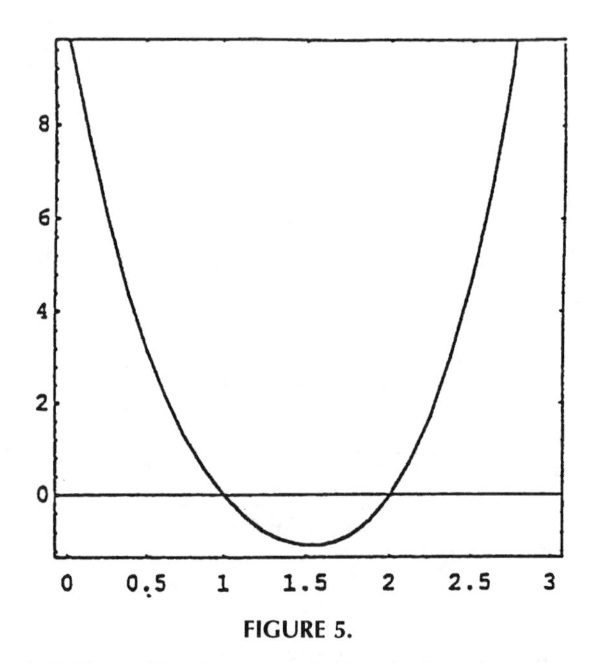

FIGURE 5.

Notice that {x,0,3} specifies the range of x for plotting, Frame->True requests that the plot be framed, and AspectRatio-> requests that the scales in horizontal and vertical directions be equal. The graph clearly shows that there are two roots at x = 1 and x = 2.

3.5 PROBLEMS

FINDROOT

1. A root of $F(x) = 3x–2e^{0.5x} = 0$ is known to exist between x = 1 and x = 2. Calculate the guessed locations of this root *twice* by application of the *linear interpolation* method.
2. A root is known to exist between x = 0 and x = 1 for the polynomial $P(x) = x^3–4.5x^2 + 5.75x–1.875 = 0$ because $P(x = 0) = –1.875$ and $P(x = 1) = 3.75$. What will be the next two guessed root values if *linear interpolation* method is used? Show details of your calculation.
3. A root is known to exist between x = 1 and x = 2 for the polynomial $x^3 + 0.5x^2 + 3x–9 = 0$.
   Based on the *linear interpolation* method, make *two* successive guesses of the location of the root. Show details of the calculations.

4. For finding a root of the polynomial $x^3–8.9x^2–21.94x + 128.576 = 0$ within the bounds $x = 0$ and $x = 4$, the *linear interpolation* method is to be applied. Show only the details involved in computation of *two* successive trial guesses of the root.
5. Use the Newton-Raphson iterative method to find the root of $2X^3–5 = 0$ between $X = 1$ and $X = 2$.
6. Complex roots of a polynomial can be calculated by application of the program **FindRoot** simply by treating the variable X in the polynomial F(X) as a complex variable. Using a complex number which has a real part and an imaginary part as an initial guess for X to evaluate F(X) and its derivatives, both values will also be complex. The Newton-Raphson iterative process is to be continued until both the real and imaginary parts of F(X) are sufficiently small. According to this outline, modify program **FindRoot** to generate a new program **NewRaphC** for determining a complex root for the polynomial $X^4 + 5X^2 + 4 = 0$.
7. In solving eigenvalue problems (see programs **CharacEq** and **Eigen-ODE**), the characteristic equation of an engineering system is in the form of a polynomial. Physically, the roots of this polynomial may have the meaning of frequency, or, buckling load, or others. In the program **Eigen-ODE**, a vibrational problem leads to a characteristic equation of $\lambda^3–50\lambda^2 + 600\lambda – 1000 = 0$. Apply the program FindRoot to find a root between λ equal to 1 and 2 accurate to three significant figures. This root represents the lowest frequency squared.
8. Apply the Newton-Raphson method to find a root of the polynomial $f(x) = 3x^3 + 2x^2–x–30 = 0$ by first guessing it to be equal to 3.0. Carry out *two* iterative steps by hand calculation to obtain the adjustments that need to be made in guessing the value of this root.
9. Apply the program FindRoot to solve Problem 8 given above.
10. Apply the *linear interpolation* method to find a root of the polynomial $f(x) = 3x^3 + 2x^2–x–30 = 0$ between $x = 1$ and $x = 3$. Carry out *two* iterative steps by hand calculation to obtain the new bounds.
12. The well known secant formula for column bucking[3] relating the average unit load P/A to the eccentricity ratio ec/r^2 is:

$$P/A = \sigma_{max}/\{1 + (ec/r^2)\sec[(L/r)(P/EA)^{1/2}/2]\}$$

where σ_{max} is the proportional limit of the column, L/r is the slenderness ratio, and E is Young's modulus of elasticity. Solve the above transcendental equation by using $\sigma_{max} = 620$ MPa and E = 190 GPa to find P/A for $ec/r^2 = 0.1$ and $L/r = 20$.
13. Solve the friction factor f from the Colebrook and White equation[6] for the flow in a pipe $(1/f)^{1/2} = 1.74–0.868\{(2K/D) + [18.7/Re(f)^{1/2}]\}$ where Re is the Reynold's number and K/D is the relative roughness parameter. Plot a curve of f vs. Re, and compare the result with the Moody's diagram.
14. Find the first five positive solution of the equation $XJ_0(X)–2J_1(X) = 0$ where J_0 and J_1 are the Bessel functions of order 0 and 1, respectively.[7]

15. Write a program **SucceSub** for implementing the successive substitution method and apply it to Equation 5 for solving the angle γ for α changing from 0° to 360° in equal increment of 15°.
16. Apply the program **SucceSub** to solve Problem 1.
17. Apply the program **SucceSub** to solve Problem 12.
18. Rise time is defined as the time required for the response X(t) to increase its value from 0.1 to 0.9, referring to Equation 8 and Figure 1. For a second-order system with $a_1 = 1$, $a_2 = 0.2\ sec^{-1}$, $a_3 = 1\ sec^{-1}$, and $a_4 = 1.37$ radians, use Equation 8 to calculate the rise time of the response X(t) by applying the computer program **FindRoot**.
19. Write a **m** file for **MATLAB** and name it **FindRoot.m** and then apply it for solving Problem 12.
20. Apply **FindRoot.m** for solving Problem 14.
21. Apply **FindRoot.m** for solving Problem 18.
22. Apply **Mathematica** to solve Problem 12.
23. Apply **Mathematica** to solve Problem 14.
24. Apply **Mathematica** to solve Problem 18.

NewRaphG

1. Shown below are two ellipses which have been drawn using the equations:

$$\left(\frac{x}{28}\right)^2 + \left(\frac{y}{24}\right)^2 = 1 \quad \text{and} \quad \left(\frac{x'}{9.5}\right)^2 + \left(\frac{y'}{32}\right)^2 = 1$$

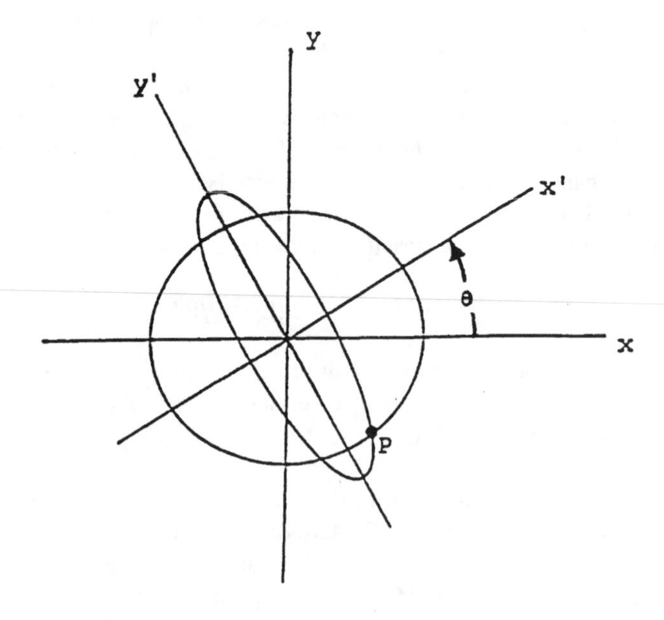

where the coordinate axes x′ and y′ are the result of rotating the x and y axes by a counterclockwise rotation of = 30°. The new coordinates can be expressed in term of x and y coordinates as:

$$x' = x\cos\theta + y\sin\theta \quad \text{and} \quad y' = -x\sin\theta + y\cos\theta$$

Use an appropriate pair of values for x_P and y_P as initial guesses for iterative solution of the location of the point P which the two ellipses intercept in the fourth quadrant by application of the program NewRaphG.

2. The circle described by the equation $x^2 + y^2 = 2^2$ and the sinusoidal curve described by the equation y = sin2x intercepted at two places as shown below. This drawing obtained using the **MATLAB** command axis('square') actually is having a square border when it is shown on screen but distorted when it is printed because the printer has a different aspect ratio. Apply the Newton-Raphson iterative method to find the intercept of these two curves near (x,y) = (2,- 0.5).

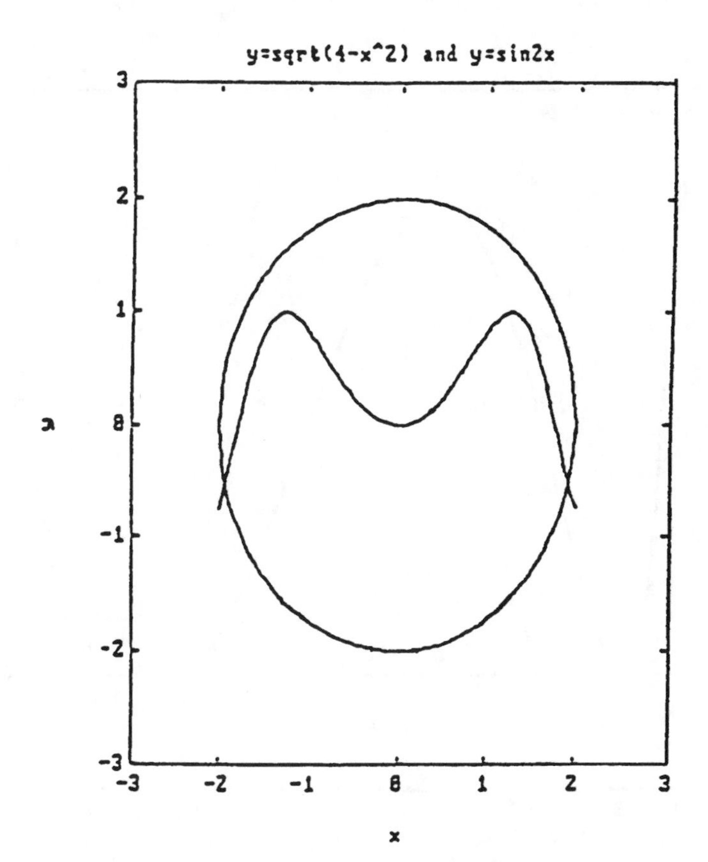

3. Apply program NewRaphG for finding a root near X = 0.4 and Y = 0.6 from the equations SinXSinY + 5X–7Y = –0.77015 and $e^{0.1X}$-X^2Y + 3Y = 2.42627. The solutions should be accurate up to 5 significant digits.
4. If one searches near x = 2 and y = 3 for a root of the equations f(x,y) = $(x-1)y + 2x^2(y-1)^2-35$ and $g(x,y) = x^3-2x^2y + 3xy^2 + y^3-65$ what should be the adjustments of x and y based on the Newton-Raphson method?
5. Write a **MATLAB** m file and call it NewRaph2.m and then apply it to solve Problems 1 to 4.
6. Apply **Mathematica** to solve Problems 1 to 4.

Bairstow

1. Is $x^2-x + 1$ a factor of $4x^4-3x^3 + 2x^2-x + 5$? If not, calculate the adjustments for u and v which are equal to –1 and 1, respectively, based on the Bairstow's method.
2. Apply **plot.m** of **MATLAB** to obtain a plot of $P(x) = x^6 + x^5-8x^4 + 14x^3 + 13x^2-111x + 90 = 0$ vs. x for $-4 \le x \le 3$ as shown below.

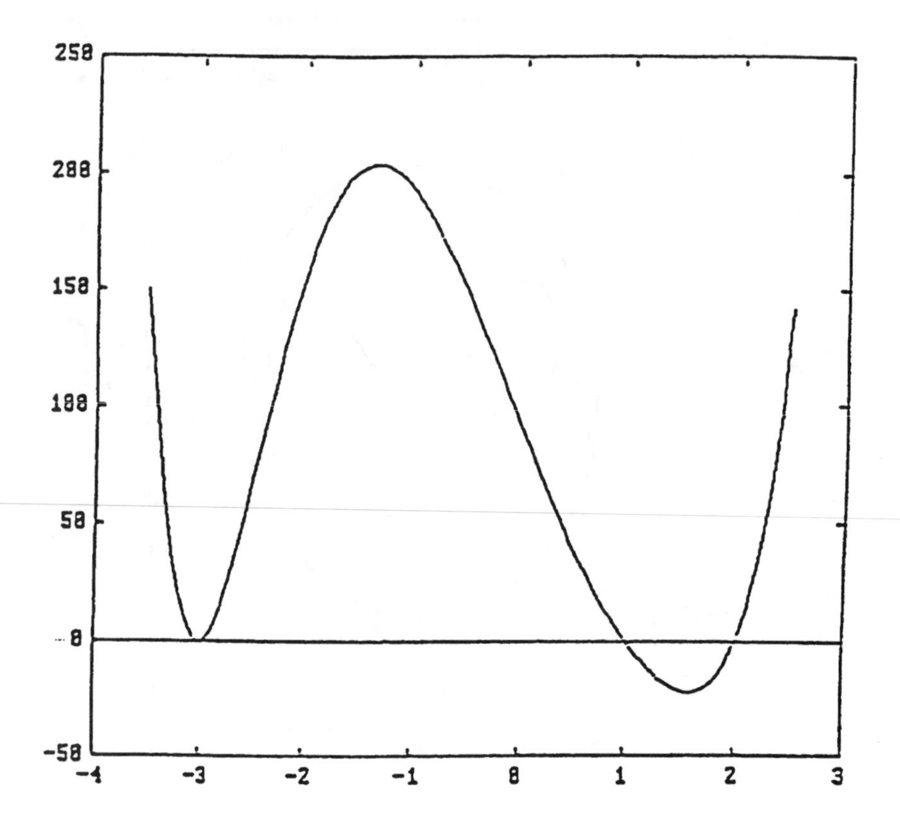

3. Find the roots of $P(x) = x^6 + x^5-8x^4 + 14x^3 + 13x^2-111x + 90 = 0$ by application of the program **Bairstow**. Are the real roots graphically verified in the plot shown in Problem 2?
4. Apply **root.m** of **MATLAB** to find the roots of $P(x) = x^6 + x^5-8x^4 + 14x^3 + 13x^2\ 111x + 90 = 0$.
5. Apply the program **Bairstow** to solve the characteristic equations derived by the program **CharacEq**: (1) $\lambda^3-15\ \lambda^2-18\lambda = 0$, (2) $\lambda^3-18\lambda^2 + 109\lambda -222 = 0$, and (3) $\lambda^3-12\lambda^2 + 47\lambda -60 = 0$.
6. Apply the program **Bairstow** to solve for the characteristic equations derived for the program **EigenODE.Stb**:

$$\lambda^2 - 36\lambda + 243 = 0,$$

$$\lambda^3 - 96\lambda^2 - 2560\lambda - 16384 = 0,$$

$$\lambda^4 - 200\lambda^3 + 13125\lambda^2 - 312500\lambda + 1953125 = 0,$$

and

$$\lambda^5 - 360\lambda^4 + 46656^3 - 2612736\lambda^2 + 5.878656x10^7\lambda - 3.62791x10^8 = 0$$

7. Apply **roots.m** of **MATLAB** to find the roots of the polynomials given in Problem 5 and then graphically verify the locations of these roots by plotting the polynomial curves using **plot.m**.
8. Repeat Problem 7 except for those polynomials given in Problem 6.
9. Apply NSolve of **Mathematica** to solve Problems 2 to 6.

3.6 REFERENCES

1. M. Abramowitz and I. A. Stegun, editors, "Handbook of Mathematical Functions with Formulas, Graphs, and Mathematical Tables," *Applied Mathematics Series 55*, National Bureau of Standards, Washington, DC, 1964.
2. Y. C. Pao, *Elements of Computer-Aided Design and Manufacturing, CAD/CAM*, John Wiley & Sons, New York, 1984.
3. A. Higdon, E. H. Ohlsen, W. B. Stiles, J. A. Weese, and W. F. Riley, *Mechanics of Materials*, John Wiley & Sons, New York, Fourth Edition, 1985.
4. J. F. Traub, Iterative Methods for Solution of Equation, Prentice-Hall, Englewood Cliffs, NJ, 1964.
5. A. E. Taylor, Advanced Calculus, Ginn & Co., Boston, 1955.
6. W. H. Li, *Fluid Mechanics in Water Resources Engineering*, Allyn & Bacon, Boston, 1983.
7. C. R. Wylie, Jr., *Advanced Engineering Mathematics*, McGraw-Hill, New York, 1960.

4 Finite Differences, Interpolation, and Numerical Differentiation

4.1 INTRODUCTION

Linear interpolation is discussed in the preceding chapter as a method for finding a particular root of a polynomial, or, transcendental equation when the upper and lower bounds of the interval for search are provided. To continue the discussion of the general topic of interpolations which not necessarily linear could be quadratic (parabolic, cubic, quartic, and so on, we in this chapter present methods for this general need of interpolation in engineering analyses by treating not only equations but also a set of N tabulated data, (x_i,y_i) for i = 1–N. *Finite difference table* will be introduced and constructed for the equally-spaced data, that is $x_2-x_1 = x_3-x_2 = \ldots = x_N-x_{N-1}$. This table can be utilized as a *forward-difference*, *backward- difference*, or, *central-difference* table depending on how its is applied for the interpolation use.

Taylor's series and a *shifting operator* are to be used in derivation of the interpolation formulas in terms of the forward-difference, backward-difference, and central-difference operators. A program **DiffTabl** has been developed for printing out a difference table of a set of equally-spaced data.

Differentiation operator will also be introduced for the derivation of the numerical differentiation needs. When a set of equally-spaced data, (x_i,y_i) for i = 1–N, are given, formulas in terms of the forward-difference, backward-difference, and central-difference operators are derived for the need of calculating the value of dy/dx at a listed x value or unlisted. If x is not equal to one of the x_i, interpolation and differentiation have to be done combinedly through a modification of the Taylor's series expansion.

For curve-fit by polynomials and for interpolation, applications of the versatile *Lagrangian interpolation formula* are also discussed. A program called **LagrangI** is made available for this need.

QuickBASIC, **FORTRAN**, and **MATLAB** versions of the above-mentioned programs are to be provided. Application of the **Mathematica**'s function **Interpolating Polynomial** in place of **LagrangI** is demonstrated.

In solution of the problems governed by a system of ordinary differential equations with either some initial and/or boundary conditions specified, the finite differences will be applied. In Chapter 6, such method for finding the approximate answer to the problem is discussed. Accuracy of such approximate solution will depend on the increment of the independent variable, stepsize, adopted and on which approximate method is employed.

Because numerical differentiation is highly inaccurate, whenever possible numerical integration should be preferred over numerical differentiation. In case that one needs to find the velocity of a certain motion study and has the option of collecting the displacement or acceleration data, then the acceleration data should be taken not the displacement data. The reason is that one has the choice of applying numerical differentiation to the displacement data or numerical integration to the acceleration data to obtain the velocity results. The numerical integration which is the topic of Chapter 5 has the smoothing effect and hence is more accurate! Graphically, differentiation is of a *local* evaluation of determining the slope at a selected point on a curve which could be the result of fitting a number of data points discussed in Chapter 3 while integration is of a global evaluation of finding the area under the curve between two specified limits of the independent variable. For a set of three given points fitted linearly by two linear segments and quadratically by a parabola, the slope at the mid-point could have very different slope values while the areas under the linear segments and under the parabola would not differ too significantly. Hence, it is worthy of emphasizing that learning the computational methods is easier when compared to making decision of which method is best to solve the problem at hand.

4.2 PROGRAM DIFFTABL — APPLICATIONS OF FINITE-DIFFERENCE TABLE

Program **DiffTabl** has been developed for the need of constructing a table of finite differences of a given set of N two-dimensional points, (x_i,y_i) for i = 1–N. The x values are assumed to be equally spaced, i.e., , $x_2-x_1 = x_3-x_2 = \cdots = x_N-x_{N-1} = h$, h being called the *increment*, or, *stepsize*. This so-called *difference table* can be applied for interpolation of the y value for a specified, unlisted x value inside the range of $x = x_1$ and $x = x_N$ (extrapolation if outside the range), and differentiation. Table 1 shows a typical difference table.

The symbol Δ used in Table 1 is called Forward Difference Operator. If we refer the numbers listed in the x and y columns as x_1 to x_6 and y_1 to y_6, respectively, the first number listed under Δy, 1.9495, is obtained from the calculation of y_2-y_1 and is identified as Δy_1. The last number listed in the Δy column, 5.3015, is equal to y_6-y_5 and referred to as Δy_5. Or, we may write the general formula as, for i = 1 to 5,

$$\Delta y_i = y_{i+1} - y_i \tag{1}$$

Δy_i is called the first forward difference of y at x_i. The higher order forward differences listed in Table 1 are obtained by extended application of Equation 1. That is,

$$\Delta^2 y_i = \Delta(y_{i+1} - y_i) = \Delta y_{i+1} - \Delta y_i = y_{i+2} - 2y_{i+1} + y_i \tag{2}$$

TABLE 1
Difference Table (y = 1 to 2x + 3x² to 4x³ + 5x⁴).

x	y	Δy	Δ²y	Δ³y	Δ⁴y	Δ⁵y
1.1	4.4265					
		1.9495				
1.2	6.3760		0.637			
		2.5865		0.126		
1.3	8.9625		0.763		0.012	
		3.3495		0.138		0.000
1.4	12.3120		0.901		0.012	
		4.2505		0.150		
1.5	16.5625		1.051			
		5.3015				
1.6	21.8640					

$$\begin{aligned}\Delta^3 y_i &= \Delta^2\left(y_{i+1} - y_i\right) = \Delta^2 y_{i+1} - \Delta^2 y_i \\ &= \left(y_{i+3} - 2y_{i+2} + y_{i+1}\right) - \left(y_{i+2} - 2y_{i+1} + y_i\right) \\ &= y_{i+3} - 3y_{i+2} + 3y_{i+1} - y_i\end{aligned} \tag{3}$$

and so on. We shall show later how the third through seven columns of Table 1 can be interpreted differently when the backward and central difference operators are introduced. First, we will demonstrate how Table 1 can be applied for interpolation of the y value at an unlisted x value, say y(x = 1.24). To do that, the *shifting operator*, E, needs to be introduced. The definition of E is such that:

$$Ey_i = y_{i+1} \tag{4}$$

That is, if E is operating on y_i, the y value is shifted down to the next provided y value. Interpolation is a problem of not shifting a full step but a fractional step. For the need of finding y at x = 1.24, the x value falls between $x_2 = 1.2$ and $x_3 = 1.3$. Since the stepsize, h, is equal to 0.1, a full shift from $y_2 = 6.3760$ would lead to y_3 which is equal to 8.9625. We expect the value of y(x = 1.24) to be between y_2 and y_3. Instead of $E^1 y_2$, the value of $E^{0.4} y_2$ is to be calculated by shifting only 40%.

To find the meaning of $E^{0.24}$, or, more generally E^r for $0<r<1$, we substitute Equation 4 into Equation 1 to obtain:

$$\Delta y_i = y_{i+1} - y_i = Ey_i - y_i = (E-1)y_1$$

$$\Delta = E - 1, \qquad \text{or,} \qquad E = 1 + \Delta \tag{5,6}$$

By application of *binomial expansion*, we can then have:

$$E^r = (1+\Delta)^r = \sum_{k=0}^{\infty} \binom{r}{k} \Delta^k \tag{7}$$

where the binomial coefficients are defined as:

$$\binom{r}{0} = 1 \quad \text{and} \quad \binom{r}{k} = \frac{r(r-1)\ldots[r-(k-1)]}{1 \cdot 2 \cdots k} \tag{8}$$

We can now use Equation 7 to obtain:

$$\begin{aligned} y(x=1.24) &= E^{0.4} y(x=1.2) = E^{0.4} y_2 = (1+\Delta)^{0.4} y_2 \\ &= \left[1 + 0.4\Delta + \frac{0.4(0.4-1)}{1 \cdot 2}\Delta^2 + \ldots\right] y_2 \\ &= \left(1 + 0.4\Delta - 1.12\Delta^2 + 0.064\Delta^3 - 0.0416\Delta^4 + 0.022952\Delta^5\right) y_2 \end{aligned} \tag{9}$$

Equation 9 can be applied for linear interpolation if up to the Δy_2 terms are adopted; for parabolic interpolation if up to the $\Delta^2 y_2$ terms are adopted; and so on. Since Table 1 has up to the fifth order forward differences available but the last column contains a zero value, Equation 9 can therefore be effectively up to the fourth-order forward-difference interpolation. The numerical results of $y(x = 1.24)$ using linear, parabolic, cubic, and fourth-order are 7.4106, 7.3190, 7.3279, and 7.3274, respectively. Since we know $y = 1-2x + 3x^2-4x^3 + 5x^4$, the exact value of $y(x = 1.24)$ is equal to 7.3274.

An explanation for discrepancies in all of these four attempts of interpolations, relative to the exact value, is provided in a homework exercise given in the Problems set.

Backward-Difference Operator

Notice that the first numbers listed in columns three through seven in Table 1 are the five forward differences of y_1, and that only four forward differences (the second numbers in columns three through six) of y_2 are available. Lesser and lesser forward differences are available for later y's until there is only Δy_5 for y_5. That is to say, to interpolate y(x) for an x value between $x_5 = 1.5$ and $x_6 = 1.6$, Equation 9 can only be used up to the Δy_5 term. To remedy this situation and to make most use of the provided set of 6 (x,y) data, it is appropriate at this time to introduce the backward-difference operator, ∇, which is defined as:

$$\nabla y_i = y_i - y_{i-1} \tag{10}$$

By combining Equations 1, 7, and 10, we notice that:

$$\nabla y_{i+1} = y_{i+1} - y_i = \Delta y_i \tag{11}$$

and

$$\nabla y_{i-1} = y_{i+1} - y_i = (1 - E^{-1}) y_{i+1} \tag{12}$$

So,

$$\nabla = 1 - E,^{-1} \quad \text{or,} \quad E^{-1} = 1 - \nabla \tag{13,14}$$

Equation 12 is an important result because it indicates that the last numbers listed in columns three through seven of Table 1 are the first through five backward differences of y_6. If we could derive an interpolation formula in terms of ∇, there are up to fifth-order backward difference of y_6 available. Toward that end, let us consider the need of interpolating the value of $y(x = 1.56)$. This y value can be reached by shifting backward by 0.4 step from $x = 1.6$ since the stepsize for Table 1 is $h = 0.1$. By using Equation 14 and noticing Equations 7 and 8, we can have:

$$y(x = 1.56) = E^{-0.4} y(x = 1.6) = E^{-0.4} y_6 = (1 - \nabla)^{0.4} y_6$$

$$= \left[1 + 0.4(-\nabla) + \frac{0.4(0.4-1)}{1 \cdot 2}(-\nabla)^2 + \ldots \right] y_6$$

$$= \left(1 - 0.4\nabla - 0.12\nabla^2 - 0.064\nabla^3 - 0.0416\nabla^4 - 0.02295\nabla^5\right) y_6 \tag{15}$$

One can then apply Equation 15 to obtain the interpolated $y(x = 1.56)$ values using up to the fifth order backward differences. This is left as a homework exercise given in the Problems set.

Central-Difference Operator

For the interest of completeness and later application in numerical solution of ordinary differential equations, we also introduce the *central difference operator*, δ. When it is operating on y_i, the definition is:

$$\delta y_i = y\left(x_i + \frac{h}{2}\right) - y\left(x_i - \frac{h}{2}\right) \tag{16}$$

The first-order central difference is seldom used and the second-order central difference is frequently applied, which is:

$$\delta^2 y_i = \delta\left[y\left(x_i + \frac{h}{2}\right)\right] - \delta\left[y\left(x_i - \frac{h}{2}\right)\right] = y(x_i + h) - 2y(x_i) + y(x_i - h) \tag{17}$$

$$= y_{i+1} - 2y_i + y_{i-1}$$

Differentiation Operator

Another important operator that needs to be introduced in connection with the application of difference table is the *differentiation operator*, D, which is defined as:

$$Dy(x = x_i) \equiv \left.\frac{dy(x)}{dx}\right|_{x=x_i} \equiv Dy_i \tag{18}$$

As it is our intention to apply an available difference table for numerical differentiation at one of the listed x values, or, at an unlisted x value, by using either the forward or backward differences of y values. For example, we may want to find Dy at x = 1.2, or, at x = 1.24. To derive an expression for D in terms of Δ, we recall the Taylor's series of a function y(x = a + h) near the neighborhood of x = a for a small increment of h:

$$y(a+h) = y(a) + \frac{y'(a)}{1!}h + \frac{y''(a)}{2!}h^2 + \ldots + \frac{y^{(j)}(a)}{j!}h^j + \ldots \tag{19}$$

where $y^{(j)}$ is the jth derivative with respect to x. Using the notation of differentiation operator D and the shifting operator E, the above expression can be written as:

$$\begin{aligned} Ey(a) &= y(a) + \frac{hDy(a)}{1!} + \frac{h^2D^2y(a)}{2!} + \ldots + \frac{h^jD^jy(a)}{j!} + \ldots \\ &= \left[1 + hD + h^2D^2 + \ldots + h^jD^j + \ldots\right]y(a) = e^{hD}y(a) \end{aligned} \tag{20}$$

or,

$$E = e^{hD} \quad \text{and} \quad D = \frac{1}{h}\ell nE \tag{21}$$

In order to use the difference table for numerical differentiation, we substitute Equation 6 into Equation 21 to obtain:

$$D = \frac{1}{h}\ell n(1+\Delta) \tag{22}$$

By substituting the logarithmic function in Equation 21 with an infinite series[1] and applying the D operator for y_i, the result is:

$$Dy_i = \frac{1}{h}\left(\Delta - \frac{1}{2}\Delta^2 + \frac{1}{3}\Delta^3 - \ldots\right)y_i \tag{23}$$

Hence, to find Dy(x = 1.2) by using the finite differences in Table 1, Equation 23 can be applied to obtain:

$$Dy_2 = \frac{1}{0.1}\left(2.5865 - 0.5x0.763 + \frac{1}{3}0.138 - \frac{1}{4}0.012\right) = 22.48$$

Notice that the above result is when up to the fourth-order forward differences of y_2 are all utilized. Linear, parabolic, and cubic numerical differentiations at x = 1.2 could also be calculated by taking only one, two, and three terms inside the parentheses of the above expression. The respective results are 25.865, 22.05, and 22.51. Since $y(x) = 1–2x + 3x^2–4x^3 + 5x^4$ and $y'(x) = –2 + 6x–12x^2 + 20x^3$, the exact value of y(x = 1.2) = –2 + 7.2–17.28 + 34.56 = 22.48 indicates that the fourth-order calculation is the best.

When Dy_i is needed for x_i near the end of x list, it is better to express D in terms of the backward-difference operation ∇, which based on Equations 14 and 21 is:

$$\begin{aligned} Dy(x_i) &= \frac{1}{h}(\ell n\ E)y_i = \frac{1}{h}\left[\ell n(1-\nabla)^{-1}\right]y_i \\ &= \frac{-1}{h}\left(-\nabla - \frac{1}{2}\nabla^2 - \frac{1}{3}\nabla^3 - \ldots\right)y_i = \frac{1}{h}\left(\nabla + \frac{\nabla^2}{2} + \frac{\nabla^3}{3} + \frac{\nabla^4}{4} + \ldots\right)y_i \end{aligned} \tag{24}$$

The shifting operator E and differentiation operator D can be combined to derive formulas for numerical differentiation of y(x) at x values unlisted in the difference table either in terms of forward-difference operator or backward-difference. First, let recall Equations 7 and 23 and apply them to find $y'(x = x_i + rh)$ in terms of the forward-difference operator as follows:

$$\begin{aligned} y'(x_i + rh) &= \frac{1}{h}DE^r y_i = \frac{1}{h}\ell n(1+\Delta)(1+\Delta)^4 y_i \\ &= \frac{1}{h}\left(\Delta - \frac{\Delta^2}{2} + \frac{\Delta^3}{3} - \ldots\right)\left[1 + r\Delta + \frac{r(r-1)}{2}\Delta^2 + \ldots\right]y_i \\ &= \frac{1}{h}\left(\Delta + \frac{2r-1}{2}\Delta^2 + \frac{3r^2-6r+2}{6}\Delta^3 + \frac{2r^3-9r^2+11r-3}{12}\Delta^4 + \ldots\right)y_i \end{aligned} \tag{25}$$

Similarly, $y'(x_i\text{-}rh)$ can be expressed in terms of backward-differential operator, as:

$$\begin{aligned} y'(x_i - rh) &= \frac{1}{h}DE^{-r} y_i = \frac{1}{h}\left[-\ell n(1-\nabla)\right](1-\nabla)^r y_i \\ &= \frac{1}{h}\left(\nabla + \frac{\nabla^2}{2} + \frac{\nabla^3}{3} + \ldots\right)\left[1 - r\nabla + \frac{r(r-1)}{2}\nabla^2 + \ldots\right]y_i \\ &= \frac{1}{h}\left(\nabla - \frac{2r-1}{2}\nabla^2 + \frac{3r^2-6r+2}{6}\nabla^3 - \frac{2r^3-9r^2+11r-3}{12}\nabla^4 + \ldots\right)y_i \end{aligned} \tag{26}$$

It should be particularly pointed out that in using Equation 26 for finding $y'(x)$ where $x_{i-1}<x<x_i$, r is to be calculated as $(x_i-x)/h$ and not as $(x-x_{i-1})/h$. For example, in using Table 1, to calculate $y'(x = 1.56)$ based on Equation 26 r should be equal to $(1.6–1.56)/0.1 = 0.4$ and not equal to $(1.56–1.5)/0.1 = 0.6$, and i equal to 6 not 5 because in Table 1 $x_6 = 1.6$ and $x_5 = 1.5$.

Program **DiffTabl** has been prepared for interactive interpolation and differentiation using a difference table such as Table 1. User can interactively specify the data points and where the interpolation or differentiation is to be calculated and also up to what order of finite differences should the computation be performed. Both **QuickBASIC** and **FORTRAN** versions of the program are made available. Listings are given below along with some sample applications. At present, the highest order of finite difference allowed is the fourth.

QuickBASIC Version

```
' Program DiffTabl - Generates and applies Differences Table for interpolation and
'                    differentiation.
                     CLEAR : CLS : DIM C(4), YandDY(100, 5)
PRINT "Program DiffTabl - Generates differences table up to fourth-order difference"
PRINT "                        based on N pairs of data points, (Xi,Yi)."
PRINT : INPUT "Input number of data points, N (>4) : ", N
PRINT "Input X values and press <Enter> key after entering each number :"
       FOR I = 1 TO n: INPUT ; x(I): NEXT I: PRINT
PRINT "Input Y values and press <Enter> key after entering each number :"
       FOR I = 1 TO n: INPUT ; YandDY(I, 1): NEXT I: PRINT
FOR IC = 2 TO 5                                       'Calculate finite differences
    FOR IR=1 TO N-IC+1: YandDY(IR,IC)=YandDY(IR+1,IC-1)-YandDY(IR,IC-1): NEXT IR
    NEXT IC: DX = X(2) - X(1)
CLS
PRINT : PRINT " I      Xi         Yi        DYi        D2Yi       D3Yi       D4Yi": PRINT
PRINT USING "##"; 1; : PRINT USING "####.#####"; X(1), YandDY(1, 1)
PRINT USING "                      ####.#####" ; YandDY(1, 2)
PRINT USING "##"; 2; : PRINT USING "####.#####"; X(2), YandDY(2, 1);
PRINT USING "##############.#####"             ; YandDY(1, 3)
PRINT USING "                      ####.#####" ; YandDY(2, 2);
PRINT USING "           ####.#####"             ; YandDY(1, 4)
FOR I = 3 TO N - 2
    PRINT USING "##";I;: PRINT USING "####.#####";X(I),YandDY(  I,1);
    PRINT USING "           ####.#####";  YandDY(I-1,3),YandDY(I-2,5)
    PRINT USING "                      ####.#####";    YandDY(  I,2);
    PRINT USING "           ####.#####"          ;    YandDY(I-1,4): NEXT I
PRINT USING "##"; N-1; : PRINT USING "####.#####" ;X(N-1),YandDY(N - 1, 1);
PRINT USING "##############.#####"                ;       YandDY(N - 2, 3)
PRINT USING "                      ####.#####"    ;       YandDY(N - 1, 2)
PRINT USING "##"; N; : PRINT USING "####.#####"   ; X(N), YandDY(N, 1): PRINT
100 PRINT "Want to interpolate or find the derivative of Y(X) for a given X value, "
    INPUT "  or end the program?  Enter I/D/E : ", A$:   IF A$ = "E" THEN END
    INPUT "Input X value where interpolation or derivative is to be calculated : ", XG
    INPUT "Enter F/B for using Forward or Backward differnece : ", M$
    FOR K = 1 TO N: IF (XG > X(K)) THEN 225 ELSE I = K - 1
        IF I > 0 THEN 210 ELSE I = 1: PRINT "** Extrapolation ! **"
210     IF M$ = "F" THEN 220 ELSE R = (X(I+1) - XG) / (X(I+1) - X(I)): GOTO 230
220     R = (XG - X(I)) / (X(I + 1) - X(I)): GOTO 230
225     NEXT K: I = N - 1: PRINT "** Extrapolation ! **": GOTO 210
230 INPUT "Input the highest order of finite differences required : ", IO
    IF A$  = "I" THEN 235
    IF A$ <> "D" THEN END
    IF R = 1  THEN I = I + 1: GOTO 231
    IF R = 0  THEN I = I + 1: GOTO 231
    C(1) = 1: C(2) = R- .5: C(3)=R^2/2-R+1!/3: C(4)=R^3/6-.75*R^2+11*R/12-.25
    IF M$="B" THEN 233
    FX   = 0: FOR K=1 TO IO: FX=FX+C(K)*YandDY(I, K + 1): NEXT K: GOTO 234
231 IF M$="B" THEN 232                                   'Backward Differentiation
    FX   = 0: FOR K=1 TO IO                               'Forward Differentiation
                IF I > N - K THEN 420
                SIGN = (-1) ^ (K + 1): FX = FX + SIGN * YandDY(I, K + 1) / K
                NEXT K: GOTO 234
```

```
232 FX   = 0: FOR K=1 TO IO : IF (I - K) < 1 THEN 420
              FX = FX + YandDY(I - K, K + 1) / K: NEXT K: GOTO 234
233 FX   = 0: FOR K=1 TO IO: IF I + 1 - K < 1 THEN 420
              FX = FX + (-1)^(K+1)*C(K)*YandDY(I + 1 - K, K + 1): NEXT K
234 FX   = FX/DX: PRINT : PRINT "The calculated dY/dX = ", FX: PRINT : GOTO 100
235 IF M$= "B" THEN 335
    FX   = YandDY(I, 1): FOR K = 1 TO IO                 'Forward Interpolate
                          IF I > (N - K) THEN 420
                          GOSUB 405
                          FX = FX + BC * YandDY(I, K + 1)
                          NEXT K: GOTO 400
335 I = I+1: IF I> N  THEN I = N                         'Backward Interpolate
    FX= YandDY(I, 1):  FOR K = 1 TO IO
                        IF (I - K) < 1 THEN 420
                        GOSUB 405
                        FX = FX + (-1)^K*BC * YandDY(I - K, K + 1): NEXT K
400 PRINT : PRINT "Answer is : Y(X) = "; FX: PRINT : GOTO 100
405 '  Sub. Binomial Coefficient of (R)K=R(R-1)...(R-K+1)/K!
410 BC = 1: FOR IT = 1 TO K: BC = BC * (R - IT + 1) / IT: NEXT IT: RETURN
420 PRINT "Improper request (order too high); reenter input data!": END
```

Sample Application

```
Program DiffTabl - Generates differences table up to fourth-order differemce
                   based on N pairs of data points, (Xi,Yi).
Input number of data points, N (>4) : 6
Input X values and press <Enter> key after entering each number :
? 1.1? 1.2? 1.3? 1.4? 1.5? 1.6
Input Y values and press <Enter> key after entering each number :
? 4.4265? 6.3760? 8.9625? 12.312? 16.5625? 21.864

 1   1.10000   4.42650
                         1.94950
 2   1.20000   6.37600             0.63700
                         2.58650             0.12600
 3   1.30000   8.96250             0.76300             0.01200
                         3.34950             0.13800
 4   1.40000  12.31200             0.90100             0.01200
                         4.25050             0.15000
 5   1.50000  16.56250             1.05100
                         5.30150
 6   1.60000  21.86400

Want to interpolate or find the derivative of Y(X) for a given X value,
  or end the program?  Enter I/D/E : D
Input X value where interpolation or derivative is to be calculated : 1.24
Enter F/B for using Forward or Backward difference : F
Input the highest order of finite differences required : 4
The calculated dY/dX =       25.12127

Want to interpolate or find the derivative of Y(X) for a given X value,
  or end the program?  Enter I/D/E : E
```

FORTRAN VERSION

```
C      Program DiffTabl - Generates and applies Differences Table for
C                         interpolation and differentiation.
       DIMENSION C(4),X(100),Y(100),YandDY(100,5)
       Character*1 AS,MS
       WRITE (*,2)
     2 FORMAT(' Program DiffTabl - Generates differences table up to',
      *       ' fourth-order difference'/19X,' based on N pairs of',
      *       ' data points, (Xi,Yi).')
       WRITE (*,*) 'Input number of data points, N (>4) : '
       READ  (*,*) N
       WRITE (*,4)
```

```
    4 FORMAT(' Input X values and press <Enter> key after entering',
     *        ' each number :')
      DO 6 I=1,N
    6 READ  (*,*) X(I)
      WRITE (*,8)
    8 FORMAT(' Input Y values and press <Enter> key after entering',
     *        ' each number :')
      DO 10 I=1,N
   10 READ  (*,*) YandDY(I,1)
C     Calculate finite differences
      DO 20 IC=2,5
      DO 20 IR=1,N-IC+1
   20 YandDY(IR,IC)=YandDY(IR+1,IC-1)-YandDY(IR,IC-1)
      DX=X(2)-X(1)
      WRITE (*,25)
   25 FORMAT(' I       Xi         Yi        DYi        D2Yi       D3Yi',
     *        '       D4Yi')
      WRITE (*,27) 1,X(1),YandDY(1,1)
   27 FORMAT(I2,2F10.5,2F20.5)
      WRITE (*,30) YandDY(1,2)
   30 FORMAT(12X,2F20.5)
      WRITE (*,27) 2,X(2),YandDY(2,1),YandDY(1,3)
      WRITE (*,30) YandDY(2,2),YandDY(1,4)
      DO 35 I=3,N-2
      WRITE (*,27) I,X(I),YandDY(I,1),YandDY(I-1,3),YandDY(I-2,5)
   35 WRITE (*,30) YandDY(I,2),YandDY(I-1,4)
      WRITE (*,27) N-1,X(N-1),YandDY(N-1,1),YandDY(N-2,3)
      WRITE (*,30) YandDY(N-1,2)
      WRITE (*,27) N,X(N),YandDY(N,1)
   37 WRITE (*,40)
   40 FORMAT(/' Want to interpolate or find the derivative of Y(X) for',
     *        ' a given X value,'/'  or end the program?',
     *        '  Enter I/D/E : ')
      READ  (*,45) AS
   45 FORMAT(A1)
      IF (AS.EQ.'E') GOTO 450
      WRITE (*,47)
   47 FORMAT(' Input X value where interpolation or derivative is to',
     *        ' be calculated : ')
      READ  (*,*) XG
      WRITE (*,50)
   50 FORMAT(' Enter F/B for using Forward or Backward differnece :')
      READ (*,45) MS
      DO 125 K=1,N
      IF (XG.GT.X(K)) GOTO 125
      I=K-1
      IF (I.GT.0) GOTO 210
      I=1
      GOTO 150
  125 CONTINUE
      I=N-1
  150 WRITE (*,*) '** Extrapolation ! **'
  210 IF (MS.EQ.'F') GOTO 220
      R=(X(I+1)-XG)/(X(I+1)-X(I))
      GOTO 230
  220 R=(XG-X(I))/(X(I+1)-X(I))
  230 WRITE (*,232)
  232 FORMAT(' Input the highest order of finite differences'
     *        ' required :')
      READ  (*,*) IO
      IF (AS.EQ.'I') GOTO 285
      IF (AS.NE.'D') GOTO 450
      IF ((R.NE.1.).AND.(R.NE.0.)) GOTO 234
      I=I+1
      GOTO 237
```

```
  234 C(1)=1
      C(2)=R-.5
      C(3)=R**2/2-R+1./3
      C(4)=R**3/6-.75*R**2+11*R/12-.25
      IF (MS.EQ.'B') GOTO 255
C     Differentiation using Forward Differences
C       (1) with interpolation
      FX=0
      DO 235 K=1,IO
  235 FX=FX+C(K)*YandDY(I,K+1)
      GOTO 265
  237 IF (MS.EQ.'B') GOTO 245
C       (2) without interpolation
      FX=0
      DO 240 K=1,IO
      IF (I.GT.N-K) GOTO 420
      SIGN=(-1)**(K+1)
  240 FX=FX+SIGN*YandDY(I,K+1)/K
      GOTO 265
C     Differentiation using Backward Differences
C       (1) without interpolation
  245 FX=0
      DO 250 K=1,IO
      IF ((I-K).LT.1) GOTO 420
  250 FX=FX+YandDY(I-K,K+1)/K
      GOTO 265
C       (2) with interpolation
  255 FX=0
      DO 260 K=1,IO
      IF ((I+1-K).LT.1) GOTO 420
  260 FX=FX+(-1)**(K+1)*C(K)*YandDY(I+1-K,K+1)
  265 FX=FX/DX
      WRITE (*,270) FX
      GOTO 37
  270 FORMAT(/' The calculated dY/dX = ',E12.5)
  285 IF (MS.EQ.'B') GOTO 335
C     Interpolation using Forward Differences
      FX=YandDY(I,1)
      DO 290 K=1,IO
      IF (I.GT.(N-K)) GOTO 420
      CALL BinoCoef(R,K,BC)
  290 FX=FX+BC*YandDY(I,K+1)
      GOTO 400
C     Interpolation using Backwrad Differences
  335 I=I+1
      IF (I.GT.N) I=N
      FX=YandDY(I,1)
      DO 340 K=1,IO
      IF ((I-K).LT.1) GOTO 420
      CALL BinoCoef(R,K,BC)
  340 FX=FX+(-1)**K*BC*YandDY(I-K,K+1)
  400 WRITE (*,405) FX
      GOTO 37
  405 FORMAT(/' Answer is : Y(X) = ',E12.5)
  420 WRITE (*,425)
  425 FORMAT(/' Improper request (order too high); reenter input',
     *        ' data!')
  450 END
      SUBROUTINE BinoCoef(R,K,BC)
C     Binomial Coefficient of (R)K=R(R-1)...(R-K+1)/K!
  410 BC=1
      DO 415 IT=1,K
  415 BC=BC*(R-IT+1)/IT
      RETURN
      END
```

Sample Application

Using the input data and difference table as for the **QuickBASIC** version, the interactive application of the **FORTRAN** version gives a sample run as follows:

```
Program DiffTabl - Generates differences table up to fourth-order differemce
                   based on N pairs of data points, (Xi,Yi).

Input number of data points, N (>4) : 6
Input X values and press <Enter> key after entering each number :
? 1.1? 1.2? 1.3? 1.4? 1.5? 1.6
Input Y values and press <Enter> key after entering each number :
? 4.4265? 6.3760? 8.9625? 12.312? 16.5625? 21.864

 1   1.10000   4.42650
                         1.94950
 2   1.20000   6.37600             0.63700
                         2.58650             0.12600
 3   1.30000   8.96250             0.76300             0.01200
                         3.34950             0.13800
 4   1.40000  12.31200             0.90100             0.01200
                         4.25050             0.15000
 5   1.50000  16.56250             1.05100
                         5.30150
 6   1.60000  21.86400

  Want to interpolate or find the derivative of Y(X) for a given X value,
    or end the program?  Enter I/D/E : I
  Input X value where interpolation or derivative is to be calculated : 1.24
  Enter F/B for using Forward or Backward difference : F
  Input the highest order of finite differences required : 4

  Answer is : Y(X) =    .73274E+01

  Want to interpolate or find the derivative of Y(X) for a given X value,
    or end the program?  Enter I/D/E : I
  Input X value where interpolation or derivative is to be calculated : 1.56
  Enter F/B for using Forward or Backward difference : B
  Input the highest order of finite differences required : 4

  Answer is : Y(X) =    .19607E+02

  Want to interpolate or find the derivative of Y(X) for a given X value,
    or end the program?  Enter I/D/E : D
  Input X value where interpolation or derivative is to be calculated : 1.56
  Enter F/B for using Forward or Backward difference : B
  Input the highest order of finite differences required : 4

  The calculated dY/dX =    .54085E+02

  Want to interpolate or find the derivative of Y(X) for a given X value,
    or end the program?  Enter I/D/E : E
```

MATLAB Application

A file **DiffTabl.m can** be created and added to **MATLAB** m files for printing out the difference table. This file may be written as:

```
function DiffTabl(X,Y)
        N=length(X); DY=zeros(N-1,N-1);
        for I=1:N-1
            if I==1, for j=1:N-1
```

```
                              DY(1,j)=Y(j+1)-Y(j);
                              end
                              DY(1,1:N-1)
                   else   for j=1:N-I
                              DY(i,j)=DY(I-1,j+1)-DY(I-1,j);
                              end
                              DY(i,1:N-I)
                   end
                end;
```

This m file then can be applied as illustrated by the following examples:

```
>> X=[1,2,3,4,5]; Y=[2,4,7,11,24]; format compact, DiffTabl(X,Y)
        ans =
             2     3     4    13
        ans =
             1     1     9
        ans =
             0     8
        ans =
             8
>> X=[1,2,3,4,5,6,7]; Y=[2,5,7,10,22,35,48]; DiffTabl(X,Y)
        ans =
             3     2     3    12    13    13
        ans =
            -1     1     9     1     0
        ans =
             2     8    -8    -1
        ans =
             6   -16     7
        ans =
           -22    23
        ans =
            45
```

The statement **format compact** requests the results to be displayed without unnecessary line spaces on screen.

It is appropriate at this time to demonstrate how some graphic capability of **MATLAB** can be effectively utilized here in connection with the difference table. First, the calculation of the first derivatives can be graphically interpreted as the slope of the linear segments connecting the given points as shown in Figure 1 which is obtained with the following interactively entered statements:

```
>> X=[1.1:0.1:1.6]; Y=[4.4265, 6.3760,8.9625,12.312,16.5625,21.864];
>> plot(X,Y,'-',X,Y,'*'), xlabel('X-axis'), ylabel('Y-axis')
>> text(1.2,15,'Linearly Connected Data Points')
```

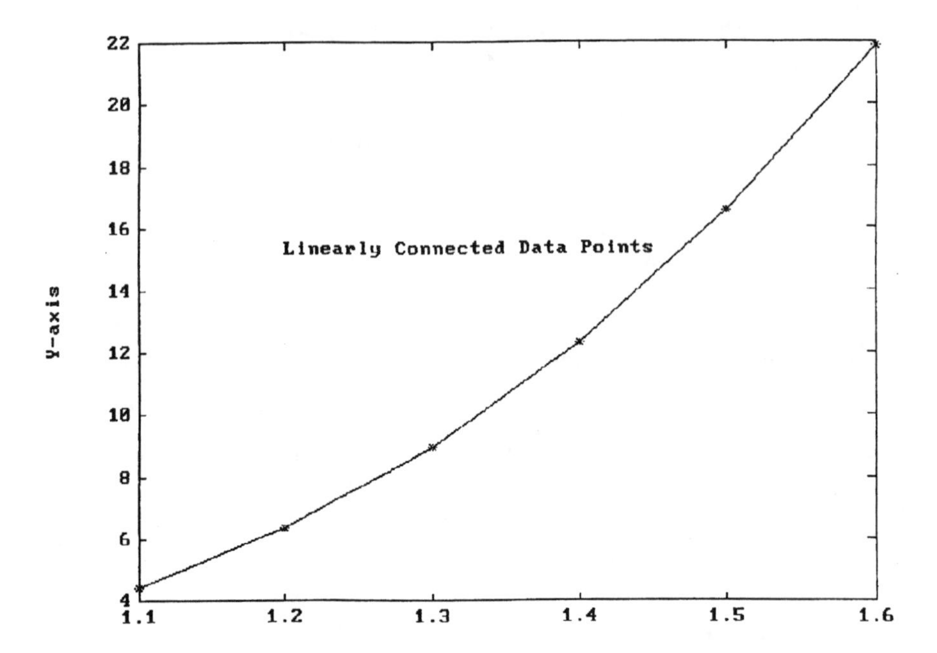

FIGURE 1. The calculation of the first derivatives can be graphically interpreted as the slope of the linear segments connecting the given points.

The first, X = , statement creates an array having 6 elements whose values start at 1.1 and ends at 1.6 and have a uniform increment of 0.1. In the plot statement, the character — inside the first set of single quotation signs requests that the given set of points specified by the coordinates arrays X and Y are to be connected by *solid lines* while the character * inside the second set of single quotation signs is for marking those points.

It also is appropriate at this time to introduce the *bar graph* feature of **MATLAB** when we consider data set and difference table. Figure 2 is presented to show the use of **bar** and **num2str** commands of **MATLAB**. The **bar** command plots a series of vertical bars based on a set of coordinates arrays X and Y where X values must be equally spaced. The **num2str** command converts a numerical value into a string, it often facilitates the display of numerical values in conjunction with the **text** command. The following interactively entered statements have enabled Figure 2 to be displayed:

```
>> X=[1:1:5]; Y=[2,4,7,11,24]; bar(X,Y)
>> for I=1:5
          text(i,Y(I),num2str(Y(I)))
   end
```

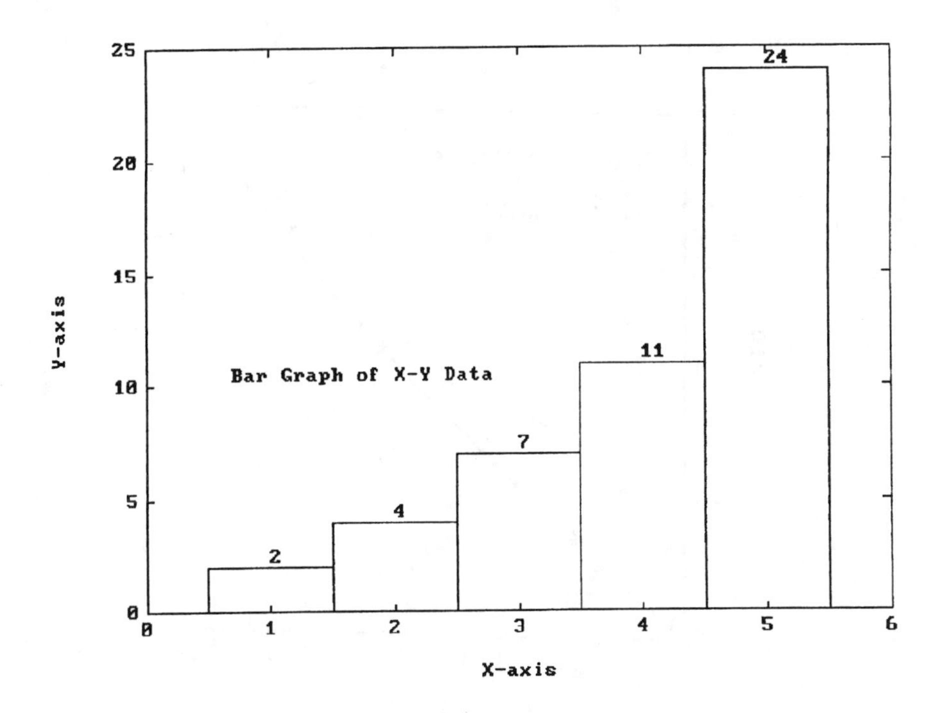

FIGURE 2.

Notice that the first two arguments for text are where the **text** string should be placed whereas the third argument converts the value of Y(I) to be printed as a string. The **for-end** loop allows all Y values to be placed at proper heights.

Mathematica Applications

To produce a plot similar to Figure 3 in the program **DiffTabl** by application of **Mathematica**, we may enter statements and obtain the following:

Input[1]: = X = Table[i,{i,1.1,1.6,0.1}; Y = Exp[X];

Input[2]: = g1 = Show[Graphics[Line[Table[{X[[i]],Y[[i]]},{i,1,6}]]]]

Input[3]: = g2 = Show[g1, Frame->True, AspectRatio->1,
FrameLabel->{"X-axis","Y-axis"}]

Input[4]: = g3 = Show[g2,Graphics[Table[Text["X",{X[[i]],Y[[i]]},
{i,1,16}]]

Input[5]: = Show[g3,Graphics[Text["Linearly Connected",{1.12,4.8},
{–1,0}],Text["Data Points",{1.12,4.6},{–1,0}]]]

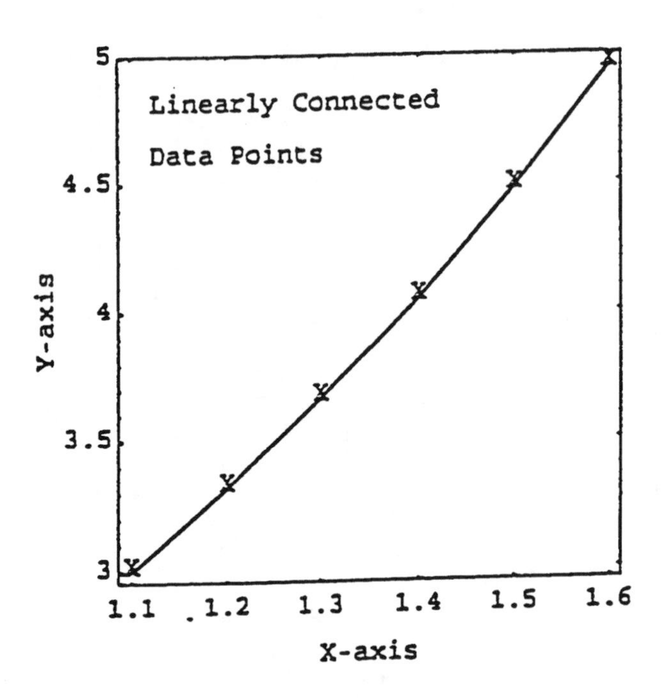

FIGURE 3.

Only the final plot is presented here. The intermediate plots designated as g1, g2, and g3 can be recalled and displayed if necessary. The **Line** command in *Input[3]* directs the specified pairs of coordinates to be linearly connected.

A bar graph can be drawn by application of **Mathematica** command **Rectangle** and their respective values by the command Text. The following statements recreate Figure 4 in the program **DiffTabl.**:

Input[1]: = X = {1,2,3,4,5}; Y = {2,4,7,11,24};

Input[2]: = g1 = Show[Graphics[Table[Rectangle[{X[[i]]–0.4,0},
X[[i]] + 0.4,Y[[i]]}],{i,1,5}]]]

Input[3]: = g2 = Show[g1,Graphics[Table[Text[Y[[i]],
{X[[i]]–0.1,Y[[i]] + 1}],{i,1,5}]]]

Input[4]: = g3 = Show[g2, Frame->True, AspectRatio->1]

Input[5]: = g4 = Show[g3,Graphics[Text["Bar Graph of X–Y Data",
{0.5,18},{–1,0}]]]

Input[6]: = Show[%,FrameLabel->{"X-axis","Y-axis"}]

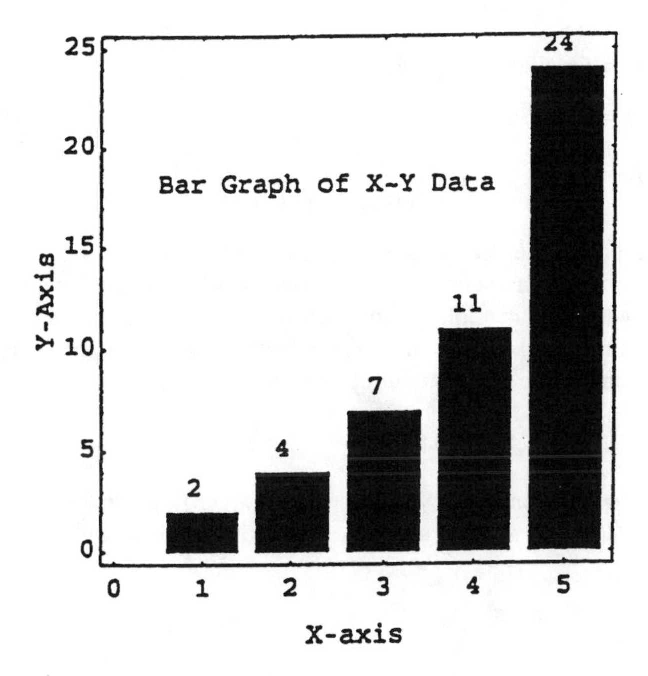

FIGURE 4.

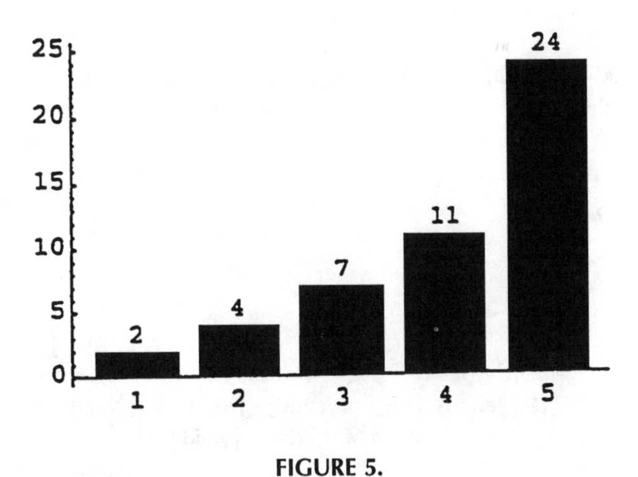

FIGURE 5.

Notice that when no expression inside a pair of doubt quotes is provided for the command **Text**, the value of the specified variable will be printed at the desired location. This is demonstrated in *Input[3]*.

Mathematica also has a function called **BarChart** in its **Graphics** package which can be applied to plot Figure 5 as follows (again, some intermediate Output responses are omitted):

Input[1]: = Y = {2,4,7,11,24};

Input[2]: = <<Graphics`Graphics`

Input[3]: = g1 = BarChart[Y]

Input[4]: = g2 = Show[g1,Graphics[Table[Text[Y[[i]],
{i,Y[[i]] + 1}],{i,1,5}]]

To print out a difference table of a given set of n y values, we can arrange the y values and up to the n-1st order of their differences in a matrix form. The y values are to be listed in the first column and their ith-order diferences are to be listed in the i + 1st column for i = 1,2,...,n–1. The following **Mathematica** input and ouput statements demonstrate the print out of a set of 6 y values:

Input[1]: = y = {1,3,7,12,44,78};

Input[2]: = n = Length[y]; yanddys = Table[x,{i,n},{j,n}];
MatrixForm[yanddys]

Output[2] =

x	x	x	x	x	x
x	x	x	x	x	x
x	x	x	x	x	x
x	x	x	x	x	x
x	x	x	x	x	x
x	x	x	x	x	x

Input[3]: = Do[yanddys[[i,1]] = y[[i]]; MatrixForm[yanddys]

Output[3] =

1	x	x	x	x	x
3	x	x	x	x	x
7	x	x	x	x	x
12	x	x	x	x	x
44	x	x	x	x	x
78	x	x	x	x	x

Input[4]: = Do[Do[yanddys[[i,j]] = yanddys[[i + 1,j–1]]-yanddys[[i,j–1],
{i,n-j + 1}],{j,2,n}]; MatrixForm[yanddys]

Output[4] =

1	2	2	–1	27	–78
3	4	1	26	–51	x
7	5	27	–25	x	x
12	32	2	x	x	x
44	34	x	x	x	x
78	x	x	x	x	x

Notice that in *Input[2]*, the **Mathematica** functions **Length** has been applied to determine the number of components in the array y, **Table** is used to initialize a matrix of n by n with the character x, and **MatrixForm** allows the matrix, yanddys, to be printed in a matrix form. *Input[3]* stores the y array into the first column of the matrix yanddys by application of the **Mathematica** command **Do**. Such looping is extended in *Input[4]* where the higher order differences are generated by using an inner index i and an outer index j. The column number j of the matrix yanddys is increased from 2 to n but the length of each column is continuously decreased to n-j + 1. Such **DoDo** arrangement is made possible by keeping the y values and their differences in a column-by-column form.

4.3 PROGRAM LAGRANGI — APPLICATIONS OF LAGRANGIAN INTERPOLATION FORMULA

Program **LagrangI** is designed to curve-fit a given set of n points, (x_i,f_i) for i = 1, 2,...,n, by a polynomial of n-1st degree based on the *Lagrangian Interpolation Formula:*

$$f(x)=\sum_{i=1}^{n} f_i \left\{ \prod_{\substack{k=1\\k\neq i}}^{n} \left[(x-x_k)/(x_i-x_k)\right] \right\} \tag{1}$$

If only the value of the function f(x) at a specified value of $x = x_s$ is needed, then Equation 1 can be applied to compute

$$f(x_s)=\sum_{i=1}^{n} f_i \left\{ \prod_{\substack{k=1\\k\neq i}}^{n} \left[(x_2-x_k)/(x_i-x_k)\right] \right\} \tag{2}$$

In Equations 1 and 2, the symbol Π is to represent a product of a specified number of factors such as:

$$\prod_{k=1}^{n} F_k = F_1F_2\ldots F_n \tag{3}$$

Equation 1 can be proven if we write the equation which fits the n given points (x_i,f_i) for i = 1 to n by a combination of n functions $L_{1 \text{ to } n}(x)$ as:

$$f(x)=f_1L_1(x)+f_2L_2(x)+\ldots+f_nL_n(x) \tag{4}$$

Notice that the ordinates $f_{1 \text{ to } n}$ are utilized in Equation 4. We expect the functions $L_{1 \text{ to } n}(x)$ to behave in such a way that when $x = x_i$ only the $f_iL_i(x)$ term in Equation 4 will contribute to f(x). That is to say when $x = x_i$, $L_i(x_i)$ should be equal to unity and the other L(x) should be equal to zero. **Mathematica**lly, we write demand that:

$$L_i(x_i)=1 \quad \text{and} \quad L_j(x_i)=0 \quad \text{for} \quad j \neq i \tag{5}$$

The second condition of Equation 5 suggests that $x\text{-}x_k$ are factors of $L_i(x)$ for $k = 1,2,\ldots,n$ but not $x\text{-}x_i$. Therefore, we may write:

$$L_i(x)=c_i(x-x_1)(x-x_2)\ldots(x-x_{i-1})(x-x_{i+1})\ldots(x-x_n) \tag{6}$$

The constant associated with $L_i(x)$, c_i is to be determined by satisfying the first condition of Equation 5. That is:

$$c_i = 1\Bigg/\left(\prod_{\substack{k=1\\k\neq i}}^{n}(x_i-x_k)\right) \tag{7}$$

Consequently, the complete expression for $L_i(x)$ is:

$$L_i(x)=\prod_{\substack{k=1\\k\neq i}}^{n}\left[(x-x_k)/(x_{i-x_k})\right] \tag{8}$$

And, when Equation 8 is substituted into Equation 4, we arrive at Equation 1.

A numerical example will clarify the application of Equation 2. Consider the case of three given points $(x_1,f_1) = (1,2)$, $(x_2,f_2) = (1.5,2.5)$, $(x_3,f_3) = (3,4)$, then $n = 3$. If we need to calculate $f(x = 2)$, Equation 2 can be used to find the equation which passes all three points. That is:

$$f(x)=\frac{(x-x_2)(x-x_3)}{(x_1-x_2)(x_1-x_3)}f_1+\frac{(x-x_1)(x-x_3)}{(x_2-x_1)(x_2-x_3)}f_2$$

$$+\frac{(x-x_1)(x-x_2)}{(x_3-x_1)(x_3-x_2)}f_3$$

$$=\frac{(x-1.5)(x-3)}{(1-1.5)(1-3)}2+\frac{(x-1)(x-3)}{(1.5-1)(1.5-3)}2.5$$

$$+\frac{(x-1)(x-1.5)}{(3-1)(3-1.5)}4$$

$$=2(x-4.5x+4.5)-\frac{10}{3}(x^2-4x+3)+\frac{4}{3}(x^2-2.5x+1.5)$$

$$=x+1$$

When x = 2, f(x = 2) = 3. Actually, the value of f(x = 2) can be specifically calculated as:

$$f(x=2)=\frac{(x-x_2)(x-x_3)}{(x_1-x_2)(x_1-x_3)}f_1+\frac{(x-x_1)(x-x_3)}{(x_2-x_1)(x_2-x_3)}f_2$$

$$+\frac{(x-x_1)(x-x_2)}{(x_3-x_1)(x_3-x_2)}f_3$$

$$=\frac{(2-1.5)(2-3)}{(1-1.5)(1-3)}2+\frac{(2-1)(2-3)}{(1.5-1)(1.5-3)}2.5$$

$$+\frac{(2-1)(2-1.5)}{(3-1)(3-1.5)}4$$

$$=-1+\frac{10}{3}+\frac{2}{3}=3$$

QuickBASIC Version

```
'*** Program LagrangI - Lagrangian interpolation. ***
'
SCREEN 2: CLEAR: CLS: KEY OFF
OPTION BASE 1 : W=8 : H=8
PRINT "Program LagrangI - Interpolates F(X) using N
pairs of data points,"
PRINT "                    (Xi,Fi), based on Lagrangian
interpolation formula."
PRINT
PRINT "Input number of data points, N : ";
INPUT N
DIM F(N),X(N)
PRINT
PRINT "Input X values and press <Enter> key after
entering a number
          FOR I=1 TO N
              INPUT;X(I)
              NEXT I
PRINT    : PRINT
          PRINT "Input F values and press <Enter> key
after entering a number :"
          FOR I=1 TO N
              INPUT;F(I)
              NEXT I
          PRINT
```

```
PRINT    : PRINT "Input X value for which F needs to be
interpolated : " ;
           INPUT XG
           S=0
           FOR I=1 TO N
               P=1
               FOR K=1 TO N
                   IF K=I THEN 190 ELSE
P=P*(XG-X(K))/(X(I)-X(K))
                   NEXT K
               S=S+P*F(I)
               NEXT I
PRINT    : PRINT "Answer is : F(X) = ";S
           PRINT                            : END
```

Sample Application

```
Program LagrangI - Interpolates F(X) using N pairs of data points, (Xi,Fi)
                   based on Lagrangian interpolation formula.
Input number of data points, N :
5
Input X values :
1,2,3,4,5
Input F values :
2,4,6,8,11
Input X value for which F needs to be interpolated :
4.56
Answer is : F(X) =          9.4517350
```

FORTRAN Version

```
C          *** Program LagrangI - Lagrangian interpolation. ***
C
      DIMENSION F(100),X(100)
      WRITE (*,2)
    2 FORMAT(' Program LagrangI - Interpolates F(X) using N pairs of',
     *       ' data points, (Xi,Fi)'/
     *       '              based on Lagrangian interpolation formula.')
      WRITE (*,*) 'Input number of data points, N : '
      READ  (*,*) N
      WRITE (*,4)
    4 FORMAT(' Input X values :')
      READ (*,*) (X(I),I=1,N)
      WRITE (*,6)
    6 FORMAT(' Input F values :')
      READ (*,*) (F(I),I=1,N)
      WRITE (*,8)
    8 FORMAT(' Input X value for which F needs to be interpolated : ')
      READ (*,*) XG
      S=0
      DO 20 I=1,N
      P=1
      DO 15 K=1,N
      IF (K.EQ.I) GO TO 15
      P=P*(XG-X(K))/(X(I)-X(K))
```

```
15 CONTINUE
20 S=S+P*F(I)
   WRITE (*,*) 'Answer is : F(X) = ',S
   END
```

Sample Application

```
Program LagrangI - Interpolates F(X) using N pairs of data points, (Xi,Fi)
                   based on Lagrangian interpolation formula.
Input number of data points, N :
3
Input X values :
1,1.5,3
Input F values :
2,2.5,4
Input X value for which F needs to be interpolated :
2
Answer is : F(X) =          3.0000000
```

MATLAB Application

A **m** file called **LagrangI.m** can be created and added to **MATLAB** m files for interpolating a Y value for a giving X value based on a set of (X,Y) data point using the Lagrangian formula. This file may be written as:

```
function Yvalue=LagrangI(X,Y,XV)
         N=length(X); Yvalue=0;
         for I=1:N
             Term=Y(I); for k=1:N
                          if k~=I, Term=Term.*(XV-X(k))./(X(I)-X(k)); end
                              end
                           Yvalue=Yvalue+Term;
                           end
             end
```

This **m** file can then be applied by specifying the data points, X vs. Y, as illustrated by the following examples:

```
>> X=[1,2,3,4,5]; Y=[2,4,6,8,11]; Yvalue=LagrangI(X,Y,XV)

  Yvalue =

      9.4517
```

The graphic capability of **MATLAB**can also be utilized here to interpret Lagrangian interpolation. In Figure 6, five given points marked with the character * have been exactly fitted with a fourth-order polynomial which is plotted for $1 \le X \le 5$ with a solid line. The interpolation at X = 4.56 using Lagrangian formula is illustrated by the broken line and dotted line. The interactively entered **MATLAB** statements, in addition to those already displayed above, are:

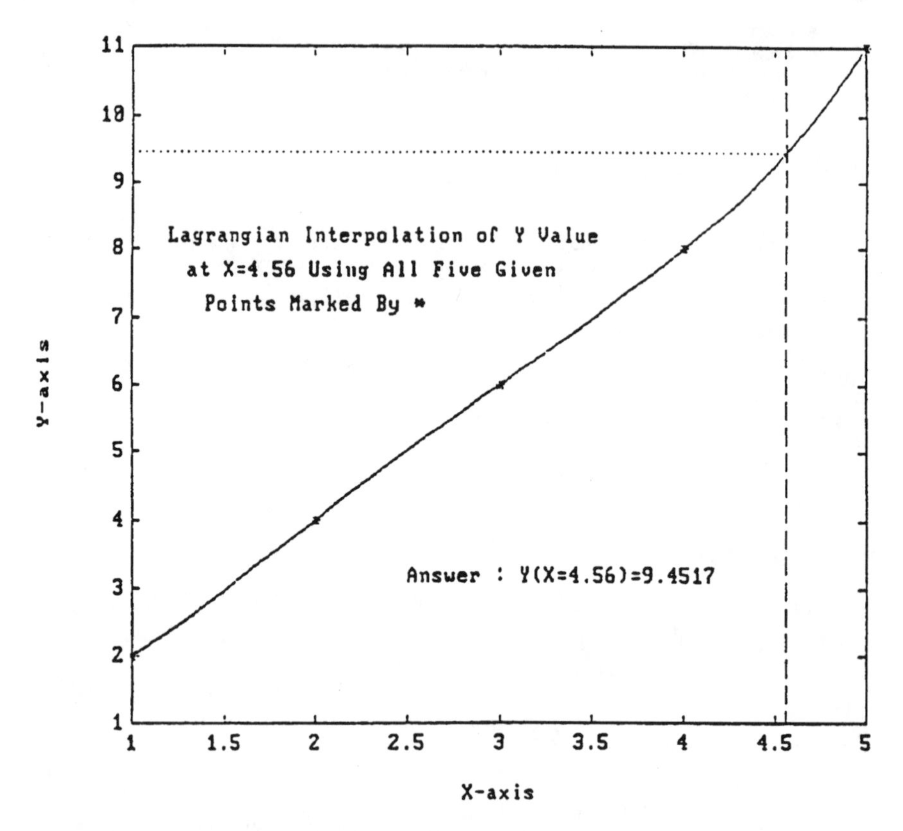

FIGURE 6. Five given points marked with the character * have been exactly fitted with a fourth-order polynomial which is plotted for $1 \leq X \leq 5$ with a solid line. The interpolation at X = 4.56 using Lagrangian formula is illustrated by the broken line and dotted line.

```
>> XV=[4.56 4.56]; YV=[1 11];
>> XH=[1 4.56]; YH=[9.4517 9.4517];
>> C=LeastSqG('A:FSPoly',X,Y,5,5);
>> for I=1:5
        Creverse(I)=C(6-I);
   end
>> XC=[1:0.1:5]; YC=polyval(Creverse,XC);
>> plot(XC,YC,XV,YV,XH,YH,X,Y,'*');
>> xlabel('X-axis'), ylabel('Y-axis')
>> text(1.2,8.0,'Lagrangian Interpolation of Y Value')
>> text(1.3,7.5,'at X=4.56 Using All Five Given')
>> text(1.4,7.0,'Points Marked By *')
>> text(2.5,3.0,'Answer : Y(X=4.56)=9.4517')
```

Notice that **plot.m** automatically uses solid, broken, and dotted lines to plot the four-order polynomial curve based on arrays XC and YC, and the vertical line based on arrays XV, YV, and the horizontal line based on arrays XH, YH, respectively. The details involved in exact curve-fit of the five given point by applying **LeastSqG.m** already has been discussed in the program **Gauss**. The coefficients, {C}, of the fourth-order polynomial determined by **LeastSqG.m** are arranged in descending order. In order to apply **polyval.m** of **MATLAB**, the order of {C} has to be reversed and stored in {Creverse} which is implemented above by the **for** and **end** loop.

MATHEMATICA APPLICATIONS

Derivation of the polynomial which passing through a set of given (x,y) points based on the Lagrangian formula can be achieved by application of the **Interpolating Polynomial** function of **Mathematica**. For example, a fourth-order polynomial can be derived for a given set of 5 (x,y) data points as follows:

In[1]: = pofx = InterpolatingPolynomial [{{1,2},{2,4},{3,6},{4,8},{5,11}},x]

Out[1]: = $2+\left(2+\frac{(-4+x)(-3+x)(-2+x)}{24}\right)(-1+x)$

To interpolate the y value of using the derived polynomial at x equal to 4.56, we replace all x's appearing in the above expression (saved in pofx) with a value of 4.56 by interactively entering

In[2]: = pofx/. x -> 4.56

Out[2]: = 9.45174

Linear and parabolic interpolations can also be implemented by selecting appropriate data points from the given set. For example, to interpolate the y value at x = 1.25 by linear interpolation, we enter:

In[3]: = p1 = InterpolatingPolynomial[{{1,2},{2,4}},x]

Out[3]: = 2 + 2 (–1 + x)

In[4]: = p1/. x -> 1.25

Out[4]: = 2.5

To parabolically interpolate the y value at x = 3.75 using the points (3,6), (4,8), and (5,11), the interactive application of **Mathematica** goes as:

In[5]: = p2 = InterpolatingPolynomial[{{3,6},{4,8},{5,11}},x]

$$Out[5]: = \ 6+\left(2+\frac{-4+x}{2}\right)(-3+x)$$

In[6]: = p2/. x -> 3.75 *Out[6]:* = 7.40625

4.4 PROBLEMS

DiffTabl

1. Construct the difference table based on the following listed data and then find the y value at x = 4.5 by using the *backward-difference* formula up to the *third-order* difference.

x	1	2	3	4	5
y	2	4	7	12	20

2. Explain why interpolations using Equation 9 by the first through fourth orders all fail to match the exact value of y(x = 1.24) = 7.3274 by making 4 plots for x values ranging from 1.2 to 1.3 with an increment of Δx = 0.001. These 4 plots are to be generated with the 4 equations obtained when the first 2, 3, 4, and 5 points are fitted by a first-, second-, third-, and fourth-degree polynomials, respectively. Also, draw a x = 1.24, vertical line crossing all 4 curves.
3. Find the first-, second-, third-, and fourth-order results of y(x = 1.56) by use of Equation 15.
4. Write E^r in terms of binomial coefficient and the backward-difference operator ∇, similar to Equation 7.
5. Find the first-, second-, third-, and fourth-order results of y′(x = 1.24) by use of Equation 24.
6. Find the first-, second-, third-, and fourth-order results of y(x = 1.56) by use of Equation 25.
7. Given 6 (x,y) points (1,0.2), (2,0.4), (3,0.7), (4,1.5), (5,2.9), and (6,4.7), *parabolically* interpolate y(x = 3.4) first by use of forward differences and then by use of backward differences.
8. Modify either the **QuickBASIC** or **FORTRAN** version of the program **DiffTabl** to include the fifth difference for the need of forward or backward interpolation and numerical differentiation.
9. Given 5 (x,y) points (0,0), (1,1), (2,8), (3,27), and (4,64), construct a *complete* difference table based on these data. Compute (1) y value at x = 1.25 using a *forward, parabolic (second-order)* interpolation, (2) y value at x = 3.7 using a *backward, cubic (third-order)* interpolation, and (3) dy/dx value at x = 0 using a *forward, third-order* approximation.
10. Based on Equation 21, derive the forward-difference formulas for D^2y_i and D^3y_i.
11. Use the result of Problem 10 to compute D^2y_2 and D^3y_1 by adopting the forward-difference terms in Table 1 as high as available.

12. Use the data in Table 1 to compute the first derivative of y at $x = 1.155$ by including terms up to the third-order forward difference.
13. Apply **MATLAB** for the points given in Problem 1 to print out the *rows* of x, y, Δy, $\Delta^2 y$, $\Delta^3 y$, and $\Delta^4 y$.
14. Same as Problem 13 but the points in Problem 6.
15. Apply **Mathematica** and DO loops to print out a difference table similar to that shown in **Mathematica** Application of Section 4.2 for the points given in Problem 1.
16. Apply **Mathematica** and **DO** loops to print out a difference table similar to that shown in **Mathematica** Application of Section 4.2 for the points given in Problem 6.
17. Compute the binomial coefficient for $r = 0.4$ and $k = 1,2,3,4,5$ according to Equation (8) in Section 4.2 using **MATLAB**.
18. Rework Problem 17 but using **Mathematica.**

LAGRANGI

1. Given five points (1,1), (2,3), (3,2), (4,5), and (5,4), use the last three points and Lagrangian interpolation formula to compute y value at $x = 6$.
2. A set of 5 (x,y) points is given as (1,2), (2,4), (4,5), (5,2), (6,0), apply the Lagrangian interpolation formulas to find the y for $x = 3$ by parabolic interpolation using the *middle three points*. Check the answer by (a) without fitting the three points by a parabolic equation, and (b) by deriving the parabolic equation and then substituting x equal to 3 to find the y value.
3. Apply the Lagrangian formula to curve-fit the following listed data near $x = 5$ by a cubic equation. Use the derived cubic equation to find the y value at $x = 4.5$.

x	1	2	3	4	5
y	2	4	7	12	20

4. Use the data set given in Problem 3 to exactly curve-fit them by a quartic equation $y(x) = a_1 + a_2x + a_3x^2 + a_4x^3 + a_5x^4$. Do this manually based on the Lagrangian formula.
5. Write a program and call it **ExactFit.Ln5** for computation of the coefficients a_{1-5} in the y(x) expression in Problem 4.
6. Generalize the need in Problem 4 by extending the exact fit of N given (x,y) points by a polynomial $y(x) = a_1 + a_2x + \ldots + a_ix^{i-1} + \ldots + a_Nx^{N-1}$ based on the Lagrangian formula. Call this program **ExactFit.LnN**.
7. Based on the Lagrangian formula, use the first four of the five points given in Problem 1 to interpolate the y value at $x = 2.5$ and then the last four of the five points also at $x = 2.5$.
8. Write a program and call it **Expand.1** which will expand the set of five points given in Problem 2 to a set of 21 points by using an increment of x equal to 0.2 and *linear* interpolation based on the Lagrangian formula. For any x value which is not equal to any of the x values of the five given

points, this x value is to be tested to determine between which two points it is located. These two given points are to be used in the interpolation process by setting N equal to 2 in the program **LagrangI**. This procedure is to be repeated for x values between 1 and 6 in computation of all new y values.

9. As for Problem 8 except *parabolic* interpolation is to be implemented. Call the new program **Expand.2**.
10. Extend the concept discussed in Problems 8 and 9 to develop a general program **Expand.M** for using N given points and Mth-order Lagrangian interpolation to obtain an expanded set.
11. Apply the function **InterpolatingPolynomial** of **Mathematica** to solve Problems 1 and 2.
12. Check the result of Problem 4 by **Mathematica**.
13. Apply LagrangI.m to solve Problem 1 by **MATLAB**.
14. Apply LagrangI.m to solve Problem 2 by **MATLAB**.
15. Apply LagrangI.m to solve Problem 7 by **MATLAB**.

4.5 REFERENCE

1. R. C. Weast, Editor-in-Chief, *CRC Standard Mathematical Tables*, the Chemical Rubber Co. (now CRC Press LLC), Cleveland, OH, 1964, p. 381.

5 Numerical Integration and Program Volume

5.1 INTRODUCTION

Sometimes, one cannot help wonder why π appears so often in a wide range of mathematical problems and why it has a value of approximately equal to 3.1416. One may want to calculate this 16th letter in the Greek alphabet and would like to obtain its value as accurate as 3.14159265358979 achieved by the expert.[1] Geometrically, π represents the ratio of the circumference to the diameter of a circle. It is commonly known that if the radius of a circle is r, the diameter is equal to 2r, the circumference is equal to $2\pi r$, and the area is equal to πr^2. Hence, to calculate the diameter we simply double the radius but to calculate the circumference and the area of a circle is more involved. The transcendental number π is the result of calculating the circumference or area of a circle by numerical integration.

In this chapter, we discuss various methods that can be adopted for the need of numerical integration. Before we elaborate on determining the value of π, let us describe the problem of numerical integration in general.

Consider the common need of finding the area inside a closed contour C_1 such as the one shown in Figure 1A, or the area between the outside contour C_2 and the inside contour C_3 shown in Figure 1B. The latter could be a practical problem of determining the usable land size of a surveyed lot which include a pond. To evaluate the area enclosed by the contour C_1 approximately by application of digital computer, the contour can be treated as two separate curves divided by two points situated at its extreme left and right, denoted as P_L and P_R, respectively. A rectangular coordinate system can be chosen to adequately describe these two points with coordinates (X_L,Y_L) and (X_R,Y_R). Here, for convenience, we shall always place the entire contour C_1 in the first quadrant of the X-Y plane. Such an arrangement makes possible to have the coordinate (X,Y) values any point on C_1 being greater than or equal to zero.

The area enclosed in the contour C_1 can be estimated by subtracting the area A_B between the bottom-branch curve $P_LP_BP_R$ and the X-axis, from the area A_T between the top-branch curve $P_LP_TP_R$ and the X-axis. Approximated evaluation of the areas A_B and A_T by application of digital computer proceeds first with selection of a finite number of points P_i from P_L to P_R. That is, to approximate a curve such as $P_LP_TP_R$ by a series of linear segments. Let N be the number of points selected on the curve $P_LP_TP_R$, then the coordinates of a typical point are (X_i,Y_i) for i ranges from 1 to N and in particular $(X_1,Y_1) = (X_L,Y_L)$ and $(X_N,Y_N) = (X_R,Y_R)$. The area between a typical linear segment P_iP_{i+1} and the X-axis is simply equal to:

$$A_i = (Y_i + Y_{i+1})(X_{i+1} - X_i)/2 \tag{1}$$

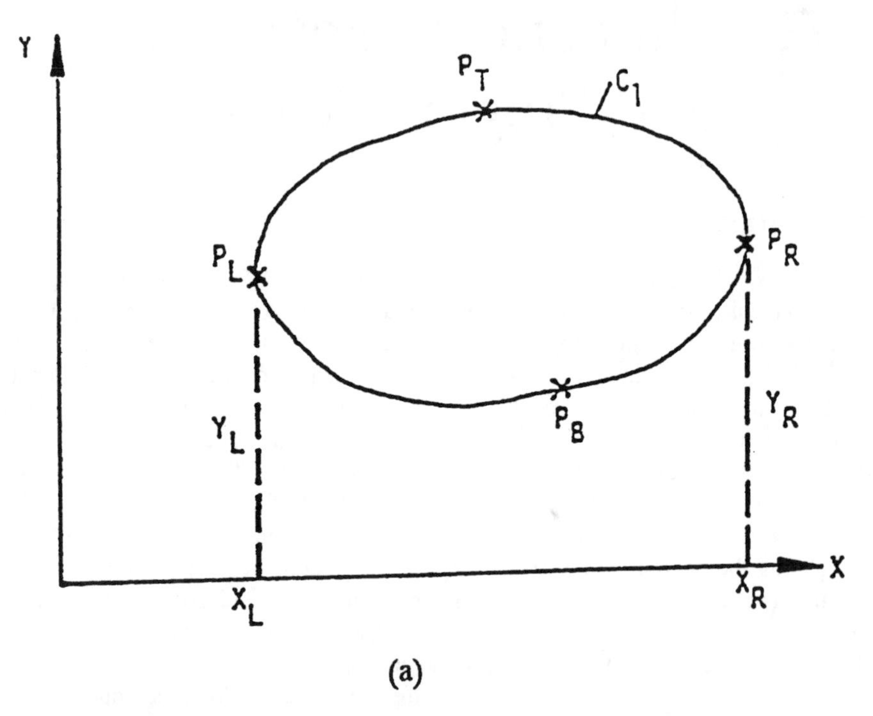

FIGURE 1A. The common need of finding the area inside a closed contour C_1.

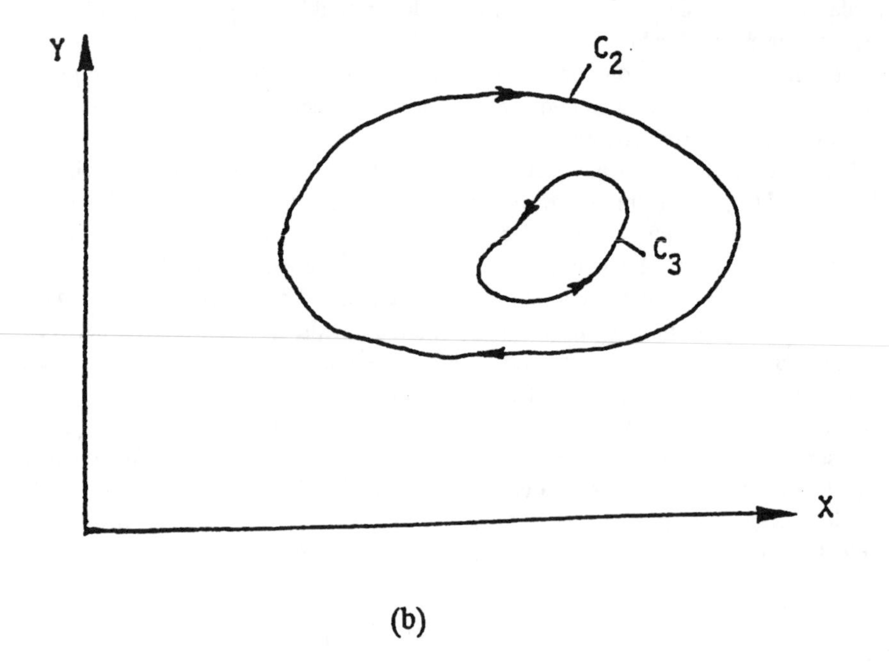

FIGURE 1B. The common need of finding the area between the outside contour C_2 and the inside contour C_3.

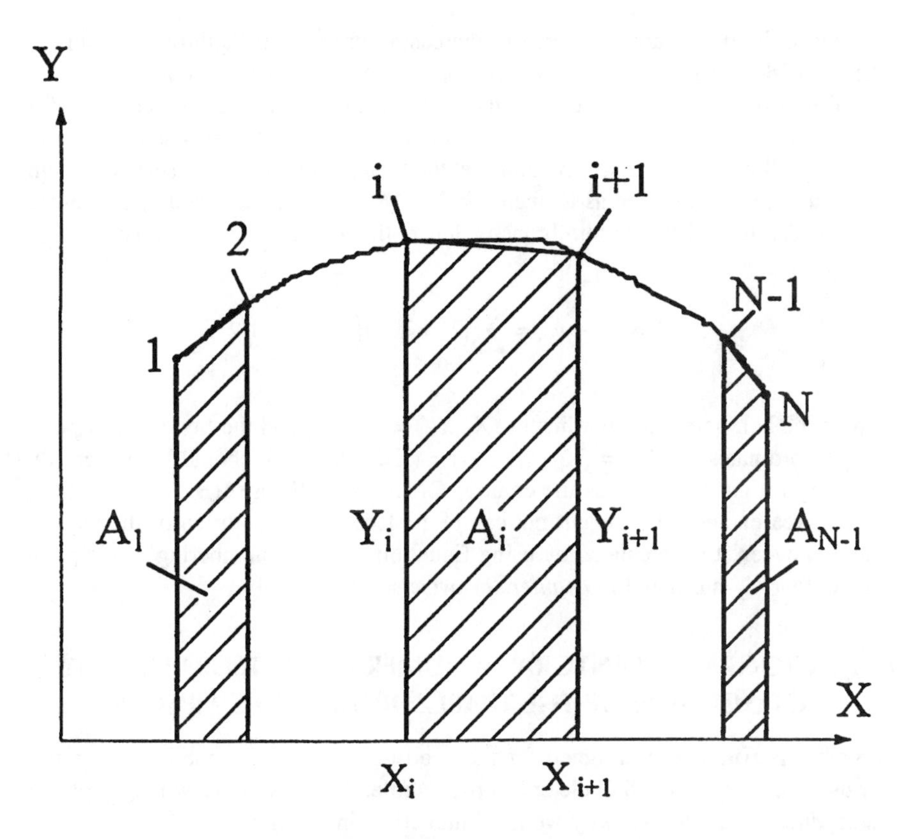

FIGURE 2. $(Y_i + Y_{i+1})/2$ is the average height and $(X_{i+1}-X_i)$ is the width of the shaded strip.

Notice that $(Y_i + Y_{i+1})/2$ is the average height and $(X_{i+1}-X_i)$ is the width of the shaded strip shown in Figure 2. Obviously, the total area A_T between the top branch of contour C_1, $P_LP_TP_R$, and the X-axis is the sum of all strips under the N–1 linear segments P_iP_{i+1} for i = 1,2,…,N. In other words, we may mathematically write:

$$A_T = \sum_{i=1}^{N-1} A_i = \sum_{i=1}^{N-1} (Y_i + Y_{i+1})(X_{i+1} - X_i)/2 \qquad (2)$$

To obtain the area A_B between the bottom branch of contour C_1 and the X-axis, we follow the same procedure as for the area A_T except that the first point is to be assigned to P_R and the last point to P_L. Suppose that there are M points selected along $P_RP_BP_L$, then the coordinates of these points are (X_i,Y_i) for i = 1,2,…,M and in particular $(X_1,Y_1) = (X_R,Y_R)$ and $(X_M,Y_M) = (X_L,Y_L)$. Consequently, the area A_B can be calculated, similar to Equation 2 as:

$$A_B = \sum_{i=1}^{M-1} A_i = \sum_{i=1}^{M-1} (Y_i + Y_{i+1})(X_{i+1} - X_i)/2 \qquad (3)$$

Since the points are numbered in increasing order from P_R through P_B and P_L, it is then clear that X_{i+1} is always less than X_i. A_B thus carries a minus sign.

Based on the above discussion, the area enclosed by contour C_1 can therefore be calculated by *adding* A_T and A_B if the numbering of the points selected on the contour follows a *clockwise* direction. Let the total number of points selected around the contour C_1 be denoted as K, then K = N + (M–2) because P_R and P_L are re-used in consideration of the bottom branch. Hence, the area enclosed in C_1 is:

$$A = A_T + A_B = \sum_{i=1}^{K} \left(Y_i + Y_{i+1}\right)\left(X_{i+1} - X_i\right)/2 \tag{4}$$

where the Nth point has coordinates $(X_N,Y_N) = (X_R,Y_R)$ and the first and last points have coordinates $(X_1,Y_1) = (X_{K+1},Y_{K+1}) = (X_L,Y_L)$. And it should be evident that in case of a cut-out, such as the contour C_2 shown in Figure 1(B), the subtraction of the area enclosed by the cut-out can be replaced by an addition of the value of the area when it is calculated by using Equation 4 but the numbering of the points on contour C_3 is ordered in *counterclockwise* sense.

5.2 PROGRAM NUINTGRA — NUMERICAL INTEGRATION BY APPLICATION OF THE TRAPEZOIDAL AND SIMPSON RULES

Program **NuIntGra** is designed for the need of performing numerical integration by use of either trapezoidal rule or Simpson's rule. These two rules will be explained later. First, let us discuss why we need numerical integration.

Figure 3 shows a number of commonly encountered cross-sectional shapes in engineering and scientific applications. The interactive computer program **NuIntGra** has an option of allowing keyboard input of the coordinates of the vertices of the cross section and then carrying out the area computation of cross-sectional area based on Equation 4.

The following gives some detailed printout of the results for the cross sections shown in Figure 3. It is important to point out that the points on the contours describing the cross-sectional shapes should be numbered as indicated in Figure 3, namely, clockwise around the outer boundary and counterclockwise around the inner boundary.

```
Program NuIntGra - Numerical Integration using Trapezoidal or Simpson Rule *
Enter a description of the problem : Fig. 3©
Are ordinates tabular data or is an integrand function to be defined?
  Enter ½ : 1
This version is limited to using trapezoidal rule only.
How many closed contours are to be considered? 1
Contour No. 1
How many points are involved?  13
Enter X(1) through X(13), press <Enter> after entering each number :
? 0? 0? 1? 1? 0? 0? 3? 3? 2? 2? 3? 3? 0
```

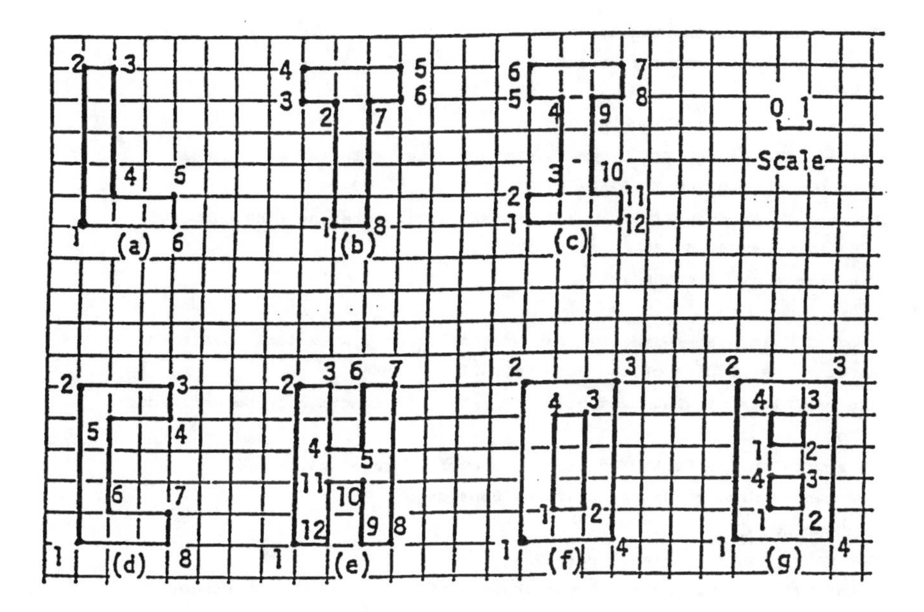

FIGURE 3. Commonly used cross sections in engineering and scientific applications.

```
Enter Y(1) through Y(13), press <Enter> after entering each number :
? 0? 1? 1? 4? 4? 5? 5? 4? 4? 1? 1? 0? 0
The result is  0.90000E+01

Program NuIntGra - Numerical Integration using Trapezoidal or Simpson Rule *
Enter a description of the problem : Fig. 3(d)
Are ordinates tabular data or is an integrand function to be defined?
  Enter ½ : 1
This version is limited to using trapezoidal rule only.
How many closed contours are to be considered? 1
Contour No. 1
How many points are involved?  9
Enter X(1) through X( 9), press <Enter> after entering each number :
? 0? 0? 3? 3? 1? 1? 3? 3? 0
Enter Y(1) through Y( 9), press <Enter> after entering each number :
? 0? 5? 5? 4? 4? 1? 1? 0? 0
The result is  0.90000E+01

Program NuIntGra - Numerical Integration using Trapezoidal or Simpson Rule *
Enter a description of the problem : Fig. 3(e)
Are ordinates tabular data or is an integrand function to be defined?
  Enter ½ : 1
```

```
This version is limited to using trapezoidal rule only.
How many closed contours are to be considered? 1
Contour No. 1
How many points are involved?  13
Enter X(1) through X(13), press <Enter> after entering each number :
? 0? 0? 1? 1? 0? 0? 3? 3? 2? 2? 3? 3? 0
Enter Y(1) through Y(13), press <Enter> after entering each number :
? 0? 1? 1? 4? 4? 5? 5? 4? 4? 1? 1? 0? 0
The result is  0.90000E+01

Program NuIntGra - Numerical Integration using Trapezoidal or Simpson Rule *
Enter a description of the problem : Fig. 3(d)
Are ordinates tabular data or is an integrand function to be defined?
  Enter ½ : 1
This version is limited to using trapezoidal rule only.
How many closed contours are to be considered? 1
Contour No. 1
How many points are involved?  9
Enter X(1) through X( 9), press <Enter> after entering each number :
? 0? 0? 3? 3? 1? 1? 3? 3? 0
Enter Y(1) through Y( 9), press <Enter> after entering each number :
? 0? 5? 5? 4? 4? 1? 1? 0? 0
The result is  0.90000E+01

Program NuIntGra - Numerical Integration using Trapezoidal or Simpson Rule *
Enter a description of the problem : Fig. 3(e)
Are ordinates tabular data or is an integrand function to be defined?
  Enter ½ : 1
This version is limited to using trapezoidal rule only.
How many closed contours are to be considered? 1
Contour No. 1
How many points are involved?  13
Enter X(1) through X(13), press <Enter> after entering each number :
? 0? 0? 1? 1? 2? 2? 3? 3? 2? 2? 1? 1? 0
Enter Y(1) through Y(13), press <Enter> after entering each number :
? 0? 5? 5? 3? 3? 5? 5? 0? 0? 2? 2? 0? 0
The result is  0.11000E+02
Program NuIntGra - Numerical Integration using Trapezoidal or Simpson Rule *
Enter a description of the problem : Fig. 3(f)
Are ordinates tabular data or is an integrand function to be defined?
  Enter ½ : 1
This version is limited to using trapezoidal rule only.
How many closed contours are to be considered? 2
Contour No. 1
How many points are involved?  5
Enter X(1) through X( 5), press <Enter> after entering each number :
? 0? 0? 3? 3? 0
```

```
Enter Y(1) through Y( 5), press <Enter> after entering each number :
? 0? 5? 5? 0? 0
Contour No. 2
How many points are involved?  5
Enter X(1) through X( 5), press <Enter> after entering each number :
? 1? 2? 2? 1? 1
Enter Y(1) through Y( 5), press <Enter> after entering each number :
? 1? 1? 4? 4? 1
The result is  0.12000E+02
```

```
Program NuIntGra - Numerical Integration using Trapezoidal or Simpson Rule *
Enter a description of the problem : Fig. 3(g)
Are ordinates tabular data or is an integrand function to be defined?
  Enter ½ : 1
This version is limited to using trapezoidal rule only.
How many closed contours are to be considered?  3
Contour No. 1
How many points are involved?  5
Enter X(1) through X( 5), press <Enter> after entering each number :
? 0? 0? 3? 3? 0
Enter Y(1) through Y( 5), press <Enter> after entering each number :
? 0? 5? 5? 0? 0
Contour No. 2
How many points are involved?  5
Enter X(1) through X( 5), press <Enter> after entering each number :
? 1? 2? 2? 1? 1
Enter Y(1) through Y( 5), press <Enter> after entering each number :
? 3? 3? 4? 4? 3
Contour No. 3
How many points are involved?  5
Enter X(1) through X( 5), press <Enter> after entering each number :
? 1? 2? 2? 1? 1
Enter Y(1) through Y( 5), press <Enter> after entering each number :
? 1? 1? 2? 2? 1
The result is  0.13000E+02
```

By use of a Function subprogram F(X) which defines the upper branch of a circle of radius equal to 1 and having its center located at X = 1 as listed below, program **NuIntGra** also has been applied for calculating the value of π. The screen display of this interactive run is also listed below after the Function F(X).

QuickBASIC Version

```
FUNCTION F(X)
F = SQR(1(X-1)^2)
END FUNCTION
```

FORTRAN Version

```
FUNCTION F(X)
F = SQRT(1.-(X-1)**2)
RETURN
END
```

```
Program NuIntGra - Numerical Integration using Trapezoidal or Simpson Rule *
Enter a description of the problem : PI calculation
Are ordinates tabular data or is an integrand function to be defined?
  Enter ½ : 2
You must have defined the integrand function by editing and compiling
  subprogram function F(X).
If you have not done so, terminate the program by press <Ctrl Break>.
Using Trapezoidal/Simpson Rule?  Enter ½, respectively : 2
Enter the lower limit of X : 0
Enter the upper limit of X : 2
Enter the number of points for integration : 21
The result is  0.15635E+01
```

More discussion on the accuracy of π will be given later after we have introduced both the trapezoidal and Simpson's rules which have already been incorporated in the program NuIntgra.

Trapezoidal Rule

Returning to Figure 2, we notice that if in approximating the curve by a series of linear segments the points selected on the curve are *equally spaced* in X, Equation 2 can be considerably simplified. In that case, we have:

$$\Delta X = X_2 - X_1 = X_3 - X_2 = \ldots = X_N - X_{N-1} \tag{5}$$

and Equation 2 can be written as:

$$A_T = \Delta X \sum_{i=1}^{N-1} \left(Y_i + Y_{i+1} \right) / 2 \tag{6}$$

Or, in a different form for easy interpretation, it may also be written as:

$$A_T = \Delta X \left(Y_1 + 2 \sum_{i=2}^{N-1} Y_i + Y_N \right) \Big/ 2 \tag{7}$$

All the in-between heights, Y_i for i ranging from 2 to N–1 that is the next to the last, are appearing twice because they are shared by two adjacent strips whereas the first and last heights, Y_1 and Y_N, only appear once in Equation 7. ΔX in Equation 7 is the common width of all strips used in summing the area.

Equation 7 is the well known *Trapezoidal Rule* for numerical integration. In a general case, it can be applied for approximate evaluation of an integral by the formula:

$$\int_{X_L}^{X_R} f(X)dX = \Delta X\left(f_1 + 2\sum_{i=2}^{N-1} f_i + f_N\right) \tag{8}$$

where the increment in X, ΔX, is simply:

$$\Delta X = \left(X_L - X_R\right)/(N-1) \tag{9}$$

when N points are selected on the interval of integration from X_L to X_R. It should be understood that in Equation 8 f_i is the value of the integrand function f(X) calculated at $X = X_i$. That is:

$$f_i \equiv f\left(X_i\right) \tag{10}$$

where for i = 1,2,…,N

$$X_i = (i-1)\Delta X + X_L \tag{11}$$

$$X_1 = X_L \quad \text{and} \quad X_N = X_R \tag{12,13}$$

Program **NuIntGra** allows the user to define the integrand function f(X) by specifying a supporting Subprogram FUNCTION F(X) and to interactively input the integration limits, X_L and X_R, along with the total number of points, N, to determine the value of an integral based on Equation 8. As an example, we illustrate below the estimation of a semi-circular area specifying $X_L = 0$, $X_R = 2$, and N = 21 and defining the integrand function f(X) in FUNCTION F(X) with a statement

F = SQRT(1.-(X–1.)*(X–1.))

This statement describes that the center of the circle is located at (1,0) and radius is equal to 1. When N is increased from 21 to 101 with an increment of 20, the following table shows that the accuracy of trapezoidal rule is steadily increased when the estimated value of the semi-circle is approaching the exact value of π/2.

N	21	41	61	81	101
Area	1.552	1.564	1.567	1.568	1.569
Error	1.21%	0.45%	0.26%	0.19%	0.13%

Simpson's Rule

The Trapezoidal Rule approximates the integrand function f(X) in Equation 8 by a series of connected straight-line segments as illustrated in Figure 2. These straight-line segments can be expressed as *linear* functions of X. A straight line which passes through two points (X_i,Y_i) and (X_{i+1},Y_{i+1}) may be described by the equation:

$$Y = m_i X + b_i \tag{14}$$

where m_i is the slope and b_i is the intercept at the Y-axis of the ith straight line.

Upon substituting the coordinates (X_i,Y_i) and (X_{i+1},Y_{i+1}) into Equation 14, we obtain two equations

$$Y_i = m_i X_i + b_i \quad \text{and} \quad Y_{i+1} = m_i X_{i+1} + b_i$$

The slope m_i can be easily solved by subtracting the first equation from the second equation to be

$$m_i = (Y_{i+1} - Y_i)/(X_{i+1} - X_i) \tag{15}$$

Substituting m_i into the Y_i equation, we obtain the intercept to be:

$$b_i = Y_i - (Y_{i+1} - Y_i)X_i/(X_{i+1} - X_i) \tag{16}$$

For the convenience of further development as well as for ease in computer programming, it is better to write the equation describing the straight line as:

$$Y = a_0 + a_1 X = \sum_{i=0}^{1} a_i X^i \tag{17}$$

That is to replace m_i and b_i in Equation 14 by a_1 and a_0, respectively. From Equations 15 and 16, we therefore can have:

$$a_0 = (Y_{i+1} - Y_i)/(X_{i+1} - X_i) \tag{18}$$

and

$$a_1 = Y_i - (Y_{i+1} - Y_i)X_i/(X_{i+1} - X_i) \tag{19}$$

A logical extension of Equation 17 is to express Y as a second-order *polynomial* of X, namely:

$$Y = a_0 + a_1 X + a_2 X^2 = \sum_{i=0}^{2} a_i X^i \tag{20}$$

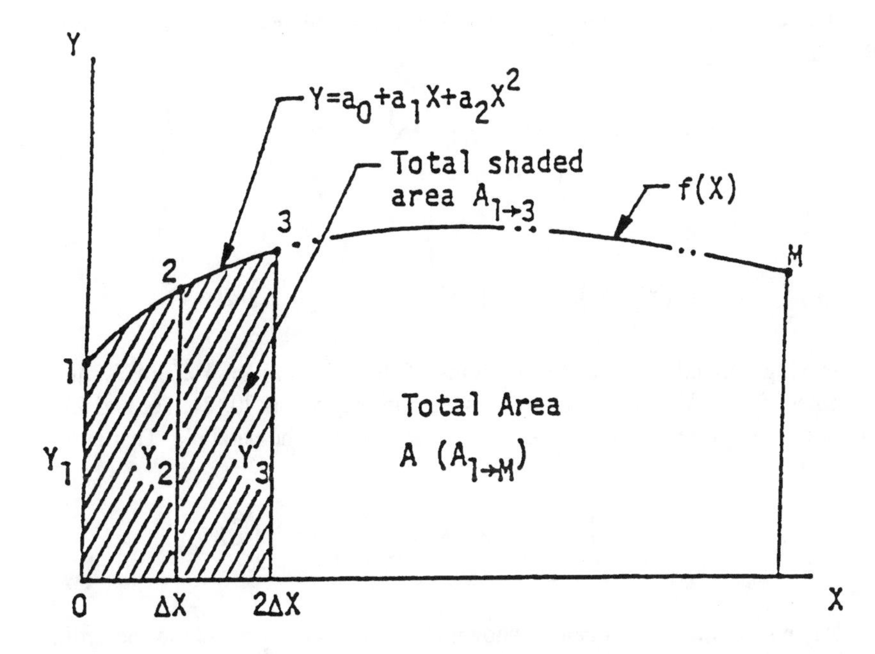

FIGURE 4. Three points are required on the f(X) in order to determine the three coefficients a_0, a_1, and a_2.

It is a *quadratic* equation describing a *parabola*. If we select Equation 20 to approximate a segment of f(X) for numerical evaluation of the integral in Equation 8, three points are required on the f(X) curve as illustrated in Figure 4 in order to determine the three coefficients a_0, a_1, and a_2. For simplicity in derivation, let the three points be $(0,Y_1)$, $(\Delta X,Y_2)$, and $(2\Delta X,Y_3)$. Upon substitution into Equation 20, we obtain:

$$a_0 = Y_1 \tag{21}$$

$$a_0 + (\Delta X)a_1 + (\Delta X)^2 a_2 = Y_2 \tag{22}$$

and

$$a_0 + (2\Delta X)a_1 + (2\Delta X)^2 a_2 = Y_3 \tag{23}$$

As a_0 is already determined in Equation 21, it can be eliminated from Equations 22 and 23 to yield:

$$a_1 + (\Delta X)a_2 = (Y_2 - Y_1)/\Delta X \tag{24}$$

and

$$a_1 + (2\Delta X)a_2 = (Y_3 - Y_1)/2\Delta X \tag{25}$$

The solutions of the above equations can be obtained by application of Cramer's Rule as:

$$a_1 = 2(Y_2 - Y_1)/\Delta X - (Y_3 - Y_1)/2\Delta X = (-3Y_1 + 4Y_2 - Y_3)/2\Delta X \tag{26}$$

and

$$a_2 = [(Y_2 - Y_1)/\Delta X - a_1]/\Delta X = (Y_1 - 2Y_2 + Y_3)/2(\Delta X)^2 \tag{27}$$

Having derived a_0, a_1, and a_2 in terms of the ordinates Y_1, Y_2, and Y_3, and the increment in X, ΔX, we are ready to substitute Equations 20, 21, 26, and 27 into Equation 8 to compute the area $A_{1ø3}$ in Figure 4 under the parabola. That is:

$$A_{1\to 3} = \int_0^{2\Delta X} \sum_{i=0}^{2} a_i X^i dX = \sum_{i=0}^{2} \left[\frac{a_i}{i+1} X^{i+1} \right]_0^2 \Delta X$$

After simplification, it can be shown that the $\text{area}_{1ø3}$ is related to the ordinates Y_1, Y_2, and Y_3, and the increment ΔX by the equation:

$$A_{1\to 3} = \frac{\Delta X}{3}(Y_1 + 4Y_2 + Y_3) \tag{28}$$

When the above-described procedure is applied for numerical integration by approximating the curve of the integrand Y(X) as a series of connected parabolas, we can expand Equation 28 to cover the limits of integration to obtain:

$$A = \sum_{\substack{i=1,3 \\ (\text{odd})}}^{M-2} A_{i\to i+2} = \frac{\Delta X}{3} \sum_{\substack{i=1,3 \\ (\text{odd})}}^{M-2} (Y_i + 4Y_{i+1} + Y_{i+2}) \tag{29}$$

Notice that the limits of integration are treated by having M stations which must be an odd integer, and the stepsize $\Delta X = (X_U - X_L)/(M-1)$. These stations are divided into groups of three stations. Equation 28 has been successively employed for evaluating the adjacent areas $A_{1ø3}$, $A_{3ø5}$, ..., $A_{M-2øM}$ in order to arrive at Equation 29. Equation 29 can also be written in the form of:

$$A = \frac{\Delta X}{3} \left(Y_1 + 4 \sum_{\substack{i=2,4 \\ (\text{even})}}^{M-1} Y_i + 2 \sum_{\substack{i=3,5 \\ (\text{odd})}}^{M-2} Y_i + Y_M \right) \tag{30}$$

which is the well-known *Simpson's rule*. Program **NuIntGra** has the option of using Simpson's rule for numerical integration. It can be shown that when this program is applied for the integration of the semi-circular area under the curve $Y(X) = (1-X^2)^{.5}$, the Simpson's rule using different M stations will result in

M	21	41	61	81	101
Area	1.564	1.568	1.569	1.570	1.570
Error	0.45%	0.19%	0.13%	0.05%	0.05%

Presented below are the program listings of **NuIntGra** in both **QuickBASIC** and **FORTRAN** versions.

QuickBASIC Version

```
        PROGRAM NuIntGra.QB - Numerical Integration using Trapezoidal or Simpson's Rule
        DECLARE FUNCTION F! (X!)
        CLS : CLEAR : DIM FY(1001), MD(9), XD(101, 9), YD(101, 9)
        PRINT "Program NuIntGra - Numerical Integration using Trapezoidal or Simpson Rule *"
        INPUT "Enter a description of the problem : ", Tiltle$
        PRINT "Are ordinates tabular data or is an integrand function to be defined?"
        INPUT "   Enter ½ : ", IN
        IF IN = 1 THEN 6
        PRINT "You must have defined the integrand function by editing and compiling"
        PRINT "   subprogram function F(X)."
        PRINT "If you have not done so, terminate the program by press <Ctrl Break>."
        GOTO 12
6       PRINT "This version is limited to using trapezoidal rule only."
        INPUT "How many closed contours are to be considered?  ", NC
        FOR IC = 1 TO NC
            PRINT USING "Contour No. #"; IC
            INPUT "How many points are involved?  ", MD(IC)
            PRINT "Enter X(1) through X(";
            PRINT USING "##"; MD(IC);
            PRINT "), press <Enter> after entering each number : "
            FOR I = 1 TO MD(IC)
                INPUT ; XD(I, IC)
                NEXT I
            PRINT
            PRINT "Enter Y(1) through Y("; : PRINT USING "##"; MD(IC);
            PRINT "), press <Enter> after entering each number : "
            FOR I = 1 TO MD(IC)
                INPUT ; YD(I, IC)
                NEXT I
            PRINT
            NEXT IC
        GOTO 50
12      INPUT "Using Trapezoidal/Simpson Rule?  Enter ½, respectively : ", K
        INPUT "Enter the lower limit of X : ", XL
        INPUT "Enter the upper limit of X : ", XR
        INPUT "Enter the number of points for integration : ", M
        DX = (XR - XL) / (M - 1)
                    FOR I = 1 TO M:       FY(I) = F(XL + (I - 1) * DX): NEXT I
        ESUMYI = 0
                    FOR I = 2 TO M - 1 STEP 2: ESUMYI = ESUMYI + FY(I): NEXT I
        OSUMYI = 0
                    FOR I = 3 TO M - 2 STEP 2: OSUMYI = OSUMYI + FY(I): NEXT I
        A = DX / (K + 1) * (FY(1) + 2 * K * ESUMYI + 2 * OSUMYI + FY(M))
37      PRINT USING "The result is  #.#####^^^^"; A: GOTO 90
50      A = 0: FOR IC = 1 TO NC
                   FOR I = 1 TO MD(IC) - 1
                      A = A + .5 * (XD(I + 1, IC) - XD(I, IC)) * (YD(I + 1, IC) + YD(I, IC))
                       NEXT I
                   NEXT IC
        GOTO 37
90      END
```

FORTRAN VERSION

```
C     PROGRAM NuIntgra - Numerical Integration using Trapezoidal or Simpson's Rule
      DIMENSION FY(1001),MD(9),XD(101,9),YD(101,9)
      Character*80 Title
      WRITE (*,*) 'Enter a description of the problem :'
      READ  (*,1) Title
    1 FORMAT(A80)
      WRITE (*,2)
    2 FORMAT(' * Program NuIntgra - Numerical Integration using',
     *       ' Trapezoidal or Simpson Rule *')
      WRITE (*,4)
    4 FORMAT(' Are ordinates tabular data or is a integrand',
     *       ' function to be defined?'/'  Enter ½ : ')
      READ  (*,*) IN
      IF (IN.EQ.1) GO TO 6
      WRITE (*,5)
    5 FORMAT(' You must have defined the integrand function by',
     *       ' editing and compiling subprogram function F(X).'
     *      /' If you have not done so, terminate the program',
     *       ' by press <Ctrl Break>.')
      GO TO 12
    6 WRITE (*,7)
    7 FORMAT(' This version is limited to using trapezoidal rule only.')
      WRITE (*,*) 'How many closed contours are to be considered?'
      READ  (*,*) NC
      DO 11 IC=1,NC
      WRITE (*,8) IC
    8 FORMAT(' Contour #',I1,' :')
      WRITE (*,*) 'How many points are involved? '
      READ  (*,*) MD(IC)
      WRITE (*,9) MD(IC)
    9 FORMAT(' Enter X(1) through X(',I2,') :')
      READ  (*,*) (XD(I,IC),I=1,MD(IC))
      WRITE (*,10) MD(IC)
   10 FORMAT(' Enter Y(1) through Y(',I2,') :')
   11 READ  (*,*) (YD(I,IC),I=1,MD(IC))
      GOTO 50
   12 WRITE (*,13)
   13 FORMAT(' Using Trapezoidal/Simpson Rule?  Enter ½'
     *       ' respectively : ')
      READ  (*,*) K
      WRITE (*,14)
   14 FORMAT(' Enter the lower limit of X : ')
      READ (*,*) XL
      WRITE (*,15)
   15 FORMAT(' Enter the upper limit of X : ')
      READ (*,*) XR
      WRITE (*,16)
   16 FORMAT(' Enter the number of points for integration : ')
      READ (*,*) M
      DX=(XR-XL)/(M-1)
      DO 25 I=1,M
   25 FY(I)=F(XL+(I-1)*DX)
      ESUMYI=0
      DO 34 I=2,M-1,2
   34 ESUMYI=ESUMYI+FY(I)
      OSUMYI=0
      DO 35 I=3,M-2,2
   35 OSUMYI=OSUMYI+FY(I)
      A=DX/(K+1)*(FY(1)+2*K*ESUMYI+2*OSUMYI+FY(M))
   37 WRITE (*,40) A
   40 FORMAT(' The result is : ',E12.5)
      GOTO 90
   50 A=0
      DO 60 IC=1,NC
      DO 60 I=1,MD(IC)-1
   60 A=A+.5*(XD(I+1,IC)-XD(I,IC))*(YD(I+1,IC)+YD(I,IC))
      GOTO 37
   90 STOP
      END
```

MATLAB Application

MATLAB has a file **quad.m** which can perform Simpson's Rule. To evaluate the area of a semi-circle by application of Simpson's Rule using **quad.m**, we first prepare the integrand function as a m file as follows:

If this file **integrnd.m** is stored on a disk which has been inserted in the disk drive A, **quad.m** is to be applied as follows:

```
>> Area = quad('A:integrnd',0,2)
```

Notice that **quad** has three arguments. The first argument is the m file in which the integrand function is defined whereas the second and third arguments specify the limits of integration. Since the center of the semi-circle is located at x = 1, the limits of integration are x = 0 and x = 2. The display resulted from the execution of the above **MATLAB** statement is:

```
Recursion level limit reached in quad.  Singularity likely.
Recursion level limit reached in quad.  Singularity likely.
Recursion level limit reached in quad.  Singularity likely.
Recursion level limit reached in quad.  Singularity likely.
Recursion level limit reached in quad.  Singularity likely.
Recursion level limit reached in quad.  Singularity likely.
Recursion level limit reached in quad.  Singularity likely.
Recursion level limit reached in quad.  Singularity likely.
Recursion level limit reached in quad.  Singularity likely.
Recursion level limit reached in quad.  Singularity likely.
Recursion level limit reached in quad.  Singularity likely.
Recursion level limit reached in quad.  Singularity likely.
Recursion level limit reached in quad.  Singularity likely.
Recursion level limit reached in quad.  Singularity likely.
Recursion level limit reached in quad.  Singularity likely.
Area =
         1.5708
```

Notice that warning messages have been printed but the numerical result is not affected.

MATHEMATICA APPLICATION

Mathematica numerically integrate a function f(x) over the interval x = a to x = b by use of the function NIntegrate. The following example demonstrates the computation of one quarter of a circle having a radius equal to 2:

In[1]: = NIntegrate[Sqrt[4x^2], {x, 0, 2}]

Out[1] = 3.14159

5.3 PROGRAM VOLUME — NUMERICAL APPROXIMATION OF DOUBLE INTEGRATION

Program **Volume** is designed for numerical calculation of double integration involving an integrand function of two variables. For convenience of graphical interpretation, the two variables x and y are usually chosen and the integrand function is denoted as z(x,y). If the double integration is to be carried for the region $x_L \le x \le x_U$ and $y_L \le y \le y_U$, the value to be calculated is the volume bounded by the z surface, z = 0 plane, and the four bounding planes $x = x_L$, $x = x_U$, $y = y_L$, and $y = y_U$ where the sub-scripts L and U are used to indicate the lower and upper bounds, respectively. The rectangular region $x_L \le x \le x_U$ and $y_L \le y \le y_U$ on the z = 0 plane is called the *base area*. The volume is there-fore a column which rises above the base area and bounded by the z(x,y) surface, assuming that z is always positive. **Mathematica**lly, the volume can be expressed as:

$$V = \int_{y_L}^{y_U} \int_{x_L}^{x_U} z(x,y) \tag{1}$$

If we are interested in finding the volume of sphere of radius equal to R, the bounds can be selected as $x_L = y_L = 0$ and $x_U = y_U = R$, and let $z = (R^2x^2y^2)^{.5}$. Equation 1 can then be applied to find the one-eighth of spherical volume. In fact, the result can be obtained analytically for this z(x,y) function. We are here, however, interested in a computational method for the case when the integrand function z(x,y) is too complex to allow analytical solution.

The trapezoidal rule for single integration discussed in the program NuIntgra can be extended to double integration by observing from Figure 1 that the total volume V can be estimated as a sum of all *columns* erected within the space bounded by the z surface and the base area. In Figure 5, the integrand functions used are:

$$z(x,y) = \left(R^2 - x^2 - y^2\right)^{0.5}, \quad \text{for} \quad x^2 + y^2 \le R^2 \tag{2}$$

and

$$z(x,y) = 0, \qquad \text{for} \quad x^2 + y^2 > R^2 \tag{3}$$

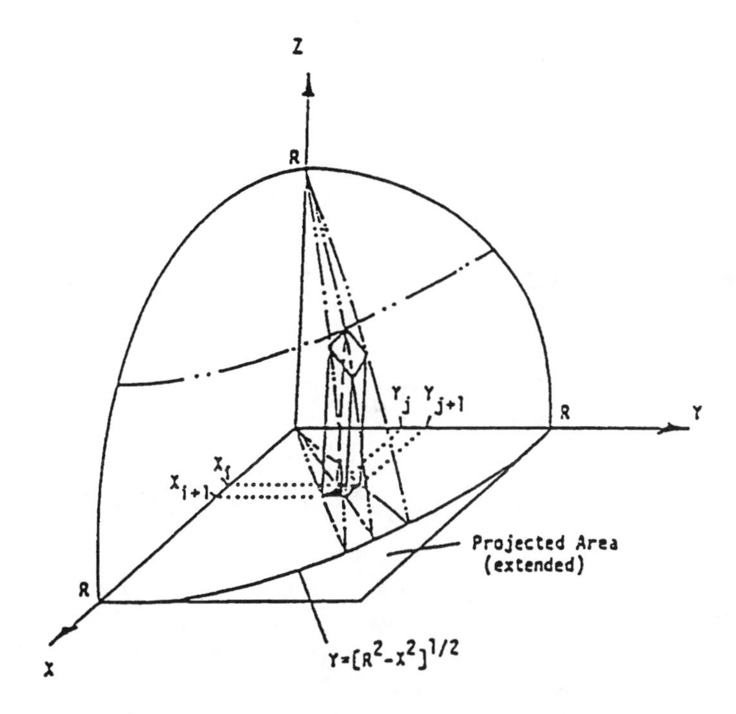

FIGURE 5. In this figure , the integrand functions used are Equations 2 and 3.

Notice that Equation 3 is an added extension of Equation 2 because if we use Equation 1 and the upper limits are $x_U = y_U = R$, a point outside of the boundary $x^2 + y^2 = R^2$ on the base area $0 \le x \le R$ and $0 \le y \le R$ is selected for evaluating z(x,y), the right-hand side of Equation 2 is an imaginary number. Adding Equation 3 will remedy this situation.

If we partition the base area into a gridwork by using uniform increments Δx andΔy along the x- and y-directions, respectively. If there are M and N equally spaced stations along the x- and y-direction, respectively, then the increments can be calculated by the equations:

$$\Delta x = (x_U - x_L)/(M-1) \quad = R/(M-1) \tag{4}$$

and

$$\Delta y = (y_U - y_L)/(N-1) \quad = R/(N-1) \tag{5}$$

At a typical grid-point on the base area, (x_i,y_j), there are three neighboring points (x_i,y_{j+1}), (x_{i+1}, y_j), and (x_{i+1},y_{j+1}). The z values at these four points can be averaged for calculation of the volume, V_{ij}, of this column by the equation:

$$V_{i,j} = \frac{1}{4}\left(z_{i,j} + z_{i,j+1} + z_{i+1,j} + z_{i+1,j+1}\right)\Delta x \Delta y \tag{6}$$

where:

$$z_{i,j} \equiv z\left(x_i, y_j\right) \tag{7}$$

The total volume is then the sum of all $V_{i,j}$ for i ranging from 1 to M and j ranging from 1 to N. Or,

$$\begin{aligned} V &= \int_0^R \int_0^R \left(x^2 + y^2\right)^{.5} dxdy \\ &= \frac{\Delta x \Delta y}{4} \sum_i \sum_j \left(z_{i,j} + z_{i,j+1} + z_{i+1,j} + z_{i+1,j+1}\right) \end{aligned} \tag{8}$$

The two summations in Equation 8 are loosely stated. Actually, the heights calculated at all MxN grid-points on the base area used in Equation 7 can be separated into three groups: (1) those heights at the corners whose coordinates are (0,0), (0,R), (R,0), and (R,R), are needed only once, (2) those heights on the edges of the base area, excluding those at the corners, are needed twice because they are shared by two adjacent columns, and (3) all heights at interior grid-points are needed four times in Equation 8 because they are shared by four adjacent columns. That is to say, in terms of the subscripts I and j the weighting coefficients, $w_{i,j}$, for $z_{i,j}$ can be summarized as follows:

```
wi,j  = 1  for  (i,j) = (1,1),(1,N),(M,1),(M,N),
      = 4  for  i = 2,3,…,M-1 and j = 2,3,…,N-1
      = 2  for  other i and j combinations
```

Subsequently, Equation 8 can be written as:

$$V = \frac{\Delta x \Delta y}{4} \sum_{i=1}^{M} \sum_{j=1}^{N} w_{i,j} z_{i,j} \tag{9}$$

A more precise way to express V in terms $z_{i,j}$ is to introduce a weighting coefficient vector for Trapezoidal rule, $\{w_t\}$. Since we have averaged the four heights of each contributing column, that is *linearly* connecting the four heights. That is, the trapezoidal rule has been applied twice, one in x-direction and another in y-direction. When M and N stations are employed in x- and y-directions, respectively. we may therefore define two weighting coefficient vectors

$$\{w_t\}_x = [1\ 2\ 2 \ldots 2\ 2\ 2]^T \tag{10}$$

and

$$\{w_t\}_y = [1\ 2\ 2 \ldots 2\ 2\ 1]^T \tag{11}$$

It should be noted that the subscripts x and y are added to indicate their association with the x- and y-axes, respectively, and that the orders of these two vectors are M and N, respectively, and that the beginning and ending components in both vectors are equal to one and the other components are equal to two. Having defined $\{w_t\}_x$ and $\{w_t\}_y$, it is now easy to show that Equation 9 can be simply written as:

$$V = \frac{\Delta x}{2} \{w_t\}_x^T [Z] \{w_t\}_y \frac{\Delta y}{2} \tag{12}$$

where [Z] is a matrix of order M by N having $z_{i,j}$ as its elements. Since $\{w_t\}_x$ is a vector of order M by one, its transpose is of order one by M and $\{w_t\}_y$ is of order N by one, the matrix multiplication of the three matrices can be carried out and the result does agree with the requirement on $w_{i,j}$ spelled out in Equation 9.

Use of weighting coefficient vectors has the advantage of extending the numerical evaluation of double integrals from trapezoidal rule to Simpson's rule where three adjacent heights are parabolically fitted (referring to program **NuIntgra** for more details). To illustrate this point, let us first introduce a weighting coefficient vector for Simpson's rule as:

$$\{w_s\} = [1\ 4\ 2 \ldots \text{repeat of } 4 \text{ and } 2 \ldots 4\ 1]^T \tag{13}$$

If we wish to integrate by application of Simpson's rule in both x- and y-directions and using M and N (both must be odd) stations, the formula for the volume is simply:

$$V = \frac{\Delta x}{3} \{w_s\}_x^T [Z] \{w_s\}_y \frac{\Delta y}{3} \tag{14}$$

If for some reason one wants to integrate using Simpson's rule along x-direction by adopting M (odd) stations, and using trapezoidal rule along y-direction by adopting N (no restriction whether odd or even) stations, then:

$$V = \frac{\Delta x}{3} \{w_s\}_x^T [Z] \{w_t\}_y \frac{\Delta y}{2} \tag{15}$$

Let us present a numerical example to further clarify the above concept of numerical volume integration. Consider the problem of estimating the volume between the surface $z(x,y) = 2x + 3y^2 + 4$ and the plane $z = 0$ for $0 \le x \le 2$ and $1 \le y \le 2$ by application of trapezoidal rule along the x direction using an increment of 0.5, and Simpson's rule along y direction using also an increment of 0.5. The increments of o.5 in both x- and y-directions make M = 5 and N = 3. First, we calculate the elements of [Z] which is of order 5 by 3:

$z_{1,1}$=2x0 +3x1^2+4= 7, $z_{1,2}$=2x0 +3x1,5^2+4=10.75, $z_{1,3}$=2x0 +3x2^2+4=16,

$z_{2,1}$=2x0.5+3x1^2+4= 8, $z_{2,2}$=2x0.5+3x1,5^2+4=11.75, $z_{2,3}$=2x0.5+3x2^2+4=17,

$z_{3,1}$=2x1+ 3x1^2+4= 9, $z_{3,2}$=2x1 +3x1,5^2+4=12.75, $z_{3,3}$=2x1 +3x2^2+4=18,

$z_{4,1}$=2x1.5+3x1^2+4=10, $z_{4,2}$=2x1.5+3x1,5^2+4=13.75, $z_{4,3}$=2x1.5+3x2^2+4=19,

$z_{5,1}$=2x2 +3x1^2+4=11, $z_{5,2}$=2x2 +3x1,5^2+4=14.75, $z_{5,3}$=2x2 +3x2^2+4=20

Next, the volume is calculated to be:

$$V = (0.5/2)[1\ 2\ 2\ 2\ 1]\begin{bmatrix} 7 & 10.75 & 16 \\ 8 & 11.75 & 17 \\ 9 & 12.75 & 18 \\ 10 & 13.75 & 19 \\ 11 & 14.75 & 20 \end{bmatrix}\begin{bmatrix} 1 \\ 4 \\ 1 \end{bmatrix}(0.5/3)$$

$$= 0.25[7+2(8+9+10)+11+10.75+2(11.75+12.75+13.75) +14.75+16+2(17+18+19)+20][1\ 4\ 1]^T(0.5)/3)$$

$$= 0.25[72\ 102\ 144][1\ 4\ 1]^T(0.5/3)=0.25x624x0.5/3 = 26$$

Program **Volume** has been developed for interactive specification of the integrand function z(x,y), the integration limits x_L, x_U, y_L, and y_U, the method(s) of integration (i.e., , trapezoidal or Simpson's rule) and number of stations in both x- and y-direction. The integrand function z(x,y) needs to be individually compiled. Both **QuickBASIC** and **FORTRAN** versions are made available. Listings are provided below along with sample applications.

FORTRAN Version

```
C     Program Volume - Calculates volume, double integration of z(x,y) for
C                      x=[xL,xU] and y=[yL,yU] using trapezoidal or
C                      Simpson's rule with M stations along x and N
C                      stations along y.   (Both M and N <=51.)
      DIMENSION WX(51),WY(51),ZV(51,51)
      Character*1 AS
```

```
      WRITE (*,1)
    1 FORMAT(' Program Volume - double integration of z(x,y) over the ',
     *       ' region x=[xL,xU] and y=[yL,yU]'//)
      WRITE (*,2)
    2 FORMAT(' Have you specified the integrand function z(x,y)',
     *       ' by editing a new Subprogram '/'  FUNCTION Z(...)?',
     *       ' Enter Y/N :')
      READ  (*,4) AS
    4 FORMAT(A1)
      IF (AS.NE.'Y') GOTO 90
      WRITE (*,5)
    5 FORMAT(' Enter the limits of integration, xL, xU, yL, and yU :')
      READ  (*,*) xL,xU,yL,yU
      WRITE (*,7)
    7 FORMAT(' Enter ½ for using trapezoidal/Simpson rule',
     *       ' along x-direction')
      READ  (*,*) KX
      WRITE (*,8)
    8 FORMAT(' Enter number of stations along x-direction : ')
      READ  (*,*) M
      WRITE (*,10)
   10 FORMAT(' Enter ½ for using trapezoidal/Simpson rule',
     *       ' along y-direction')
      READ  (*,*) KY
      WRITE (*,12)
   12 FORMAT(' Enter number of stations along y-direction : ')
      READ  (*,*) N
      M1=M-1
      DX=(XU-XL)/M1
      N1=N-1
      DY=(YU-YL)/N1
      WX(1)=1.
      WX(M)=1.
      DO 15 I=2,M1
      WX(I)=2.
      IF (I-(I/2)*2.EQ.0.AND.KX.EQ.2) WX(I)=WX(I)+2.
   15 CONTINUE
      WY(1)=1.
      WY(N)=1.
      DO 18 I=2,N1
      WY(I)=2.
      IF (I-(I/2)*2.EQ.0.AND.KY.EQ.2) WY(I)=WY(I)+2.
   18 CONTINUE
      DO 20 I=1,M
      X=XL+(I-1)*DX
      DO 20 J=1,N
      Y=YL+(J-1)*DY
   20 ZV(I,J)=Z(X,Y)
      VOLUME=0.
      DO 30 J=1,N
      SUM=0.
      DO 25 I=1,M
   25 SUM=SUM+WX(I)*ZV(I,J)
   30 VOLUME=VOLUME+SUM*WY(J)
      VOLUME=VOLUME*DX*DY/(KX+1)/(KY+1)
      WRITE (*,80) VOLUME
   80 FORMAT(5X,'VOLUME=',E12.5)
   90 STOP
      END

      FUNCTION Z(X,Y)
      RSQ=X**2+Y**2
      IF (RSQ.GE.4) GO TO 10
      Z=SQRT(4.-RSQ)
      RETURN
   10 Z=0.
      RETURN
      END
```

Sample Application

The FUNCTION Z(X,Y) listed above is for finding the volume under the surface $z(x,y) = (x^2 + y^{2-4})^{.5}$ over the region $0 \le x \le 2$ and $0 \le y \le 2$. The exact solution is volume = 4.1889. For a sample run of the program **Volume** using trapezoidal rule and 21 stations along both x- and y-directions, the screen display of interactive communication through keyboard input and the calculated result is:

```
Program Volume - double integration of z(x,y) over the region x=[xL,xU] and
                 y=[yL,yU]
Have you specified the integrand function z(x,y) by editing a new subprogram
 FUNCTION Z(...)? Enter Y/N :
Y
Enter the limits of integration, xL, xU, yL, and yU :
0,2,0,2
Enter ½ for using trapezoidal/Simpson rule along x-direction
1
Enter number of station along x-direction :
21
Enter ½ for using trapezoidal/Simpson rule along y-direction
1
Enter number of station along x-direction :
21
    VOLUME=  .41836E+01
Stop - Program terminated.
```

QuickBASIC Version

```
'  Program Volume - Calculates volume, double integration of z(x,y) for
'                   x=[xL,xU] and y=[yL,yU] using trapezoidal or
'                   Simpson's rule with M stations along x and N
'                   stations along y.  (Both M and N <=51.)
            DECLARE FUNCTION Z (X, y)
CLEAR : CLS : DIM WX(51), WY(51), ZV(51, 51)
PRINT "Program Volume - double integration of z(x,y) over the region x=[xL,xU] and"
PRINT "                 y=[yL,yU]."
PRINT "Have you specified the integrand function z(x,y) by editing a new Subprogram"
INPUT "  FUNCTION Z(...)?  Enter Y/N : ", A$
      IF (A$ <> "Y") THEN 90
INPUT "Enter the limits of integration, xL, xU, yL, and yU : ", XL, XU, yL, yU
INPUT "Enter ½ for using trapezoidal/Simpson rule along x-direction : ", KX
INPUT "Enter number of stations along x-direction : ", M
INPUT "Enter ½ for using trapezoidal/Simpson rule along y-direction : ", KY
INPUT "Enter number of stations along y-direction : ", N
      M1 = M - 1: DX = (XU - XL) / M1: N1 = N - 1: DY = (yU-yL)/N1: WX(1) = 1!: WX(M) = 1!
      FOR I = 2 TO M1: WX(I) = 2!
          IF (INT(I) - INT(I / 2) * 2! = 0) THEN 8 ELSE GOTO 15
8         IF (KX = 2) THEN WX(I) = WX(I) + 2!
15        NEXT I: WY(1) = 1!: WY(N) = 1!
      FOR I = 2 TO N1: WY(I) = 2!
          IF (INT(I) - INT(I / 2) * 2 = 0) THEN 16 ELSE GOTO 18
16        IF (KY = 2) THEN WY(I) = WY(I) + 2!
18        NEXT I
      FOR I = 1 TO M: X = XL + (I - 1) * DX
          FOR J = 1 TO N: y = yL + (J - 1) * DY: ZV(I, J) = Z(X, y): NEXT J: NEXT I
VOLUME = 0: FOR J = 1 TO N: SUM = 0!
                FOR I = 1 TO M: SUM = SUM + WX(I) * ZV(I, J): NEXT I
                VOLUME= VOLUME + SUM * WY(J): NEXT J
                VOLUME= VOLUME * DX * DY / (KX + 1) / (KY + 1)
   PRINT USING "Volume = #.#####^^^^"; VOLUME
90 END

   FUNCTION Z (X, y)
   RSQ = X ^ 2 + y ^ 2: IF (RSQ > 4) GOTO 10
   Z = SQR(4 - RSQ): GOTO 95
10 Z = 0!
95 END FUNCTION
```

Sample Applications

The same calculation of one-eighth of a sphere of radius equal to 2 as in the **FORTRAN** version is run but here using Simpson's rule. The screen display is:

```
Program Volume - double integration of z(x,y) over the region x=[xL,xU] and
                 y=[yL,yU]
Have you specified the integrand function z(x,y) by editing a new subprogram
 FUNCTION Z(...)? Enter Y/N : Y
Enter the limits of integration, xL, xU, yL, and yU : 0,2,0,2
Enter ½ for using trapezoidal/Simpson rule along x-direction : 2
Enter number of station along x-direction : 21
Enter ½ for using trapezoidal/Simpson rule along y-direction : 2
Enter number of station along x-direction : 21
Volume = 0.41893E+01
```

We have presented earlier the manual calculation of the double integration for $z(x,y) = 2x + 3y^2 + 4$, program **Volume** can be applied to have the results displayed on the monitor screen as below. The answer is exactly the same as from manual computation.

```
Program Volume - double integration of z(x,y) over the region x=[xL,xU] and
                 y=[yL,yU]
Have you specified the integrand function z(x,y) by editing a new subprogram
 FUNCTION Z(...)? Enter Y/N :
Y
Enter the limits of integration, xL, xU, yL, and yU :
0,2,1,2
Enter ½ for using trapezoidal/Simpson rule along x-direction
1
Enter number of station along x-direction :
5
Enter ½ for using trapezoidal/Simpson rule along y-direction
2
Enter number of station along x-direction :
3
     VOLUME=  .26000E+02
stop - Program terminated.
```

MATLAB Application

A Volume.m file can be created and added to **MATLAB** m files to calculate a double integral when the integrand function is specified by another m file. For integrating a function Z(X,Y) over the region $X_1 \le X \le X_2$ and $Y_1 \le Y \le Y_2$ by either Trapezoidal or Simpson's rules (designated as rule 1 or 2) and with N_X and N_Y stations along the X and Y directions, respectively, this file may be written as followed:

```
function V=Volume(funcfn,X1,X2,RuleX,NX,Y1,Y2,RuleY,NY)
        % RuleX=1, Trapezoidal Rule; =2, Simpson's Rule in X direction.
        % Same for RuleY but for Y direction.
        X=ones(1,NX); Y=ones(NY,1); Z=zeros(NX,NY);
        for I=2:NX-1; X(I)=2; end
        for I=2:NY-1; Y(I)=2; end
        if RuleX == 2
           for I=2:2:NX-1; X(I)=4; end
           end
        if RuleY == 2
           for I=2:2:NY-1; Y(I)=4; end
           end
        DX=(Xend-X0)./(NX-1); DY=(Yend-Y0)./(NY-1);
        for I=1:NX; XV=X0+(I-1).*DX;
            for j=1:NY; YV=Y0+(j-1).*DY;
                Z(i,j)=feval(funfcn,XV,YV);
                end
            end
        V=X*Z*Y.*DX.*DY./(RuleX+1)./(RuleY+1);
```

For each problem, the integrand function Z(X,Y) needs to be prepared as a m file. In case that a hemisphere of radius 2 and centered at X = 0 and Y = 0, we may write:

```
function Z=FuncZ(X,Y)
         T=X.^2+Y.^2;
         if T<4; Z=sqrt(4-T);
            else Z=0;
            end
```

In case of $Z(X,Y) = 2X + 3Y^2 + 4$, we may write a new file as:

```
function Z=FuncZnew(X,Y)
         Z=2.*X+3.*Y.^2+4;
```

Once the files **Volume.m**, **FuncZ.m**, and **FuncZnew.m**, the following **MATLAB** executions can be achieved:

```
>> V=Volume('A:FuncZ',0,2,2,21,0,2,2,21)

 V =

     4.1893

>> V=Volume('A:FuncZnew',0,2,1,5,1,2,2,3)

 V =

     26
```

Notice the first and second integrations of the hemisphere use Trapezoidal and Simpson's rule in both X and Y, respectively. Both use 21 stations in X and Y directions. The third integration of $Z = 2X + 3Y^2 + 4$ over the region $0 \leq X \leq 2$ and $1 \leq Y \leq 2$ is carried out using Trapezoidal rule along X direction with 5 stations and Simpson's rule along Y direction with 3 stations.

MATLAB has a mesh plot capability of generating three-dimensional *hidden-line* surface. For example, when the function FuncZ is used to generate a hemispherical surface of radius equal to 2 described by a square matrix [Z], a plot shown in Figure 6 can be obtained by entering **MATLAB** commands as follows:

```
>> for I=1:31, X(I)=-2+(I-1)*4/30;
       for j=1:31, Y(j)=-2+(j-1)*4/30;
           Z(i,j)=feval('A:FuncZ',X(I),Y(j));
       end
   end
>> mesh(Z)
```

We observe from Figure 6 that the hidden-line feature is apparent but the hemisphere appears like a semiellipsoid. This is due to the aspect ratios of the display monitor and/or of the printer. mesh is the option of specifying different scale factors for the X-, Y-, and Z-axes. To make the three-dimensional surface to appear as a perfect hemisphere, user has to experiment with different scale factors for the three axes. This is left as a homework problem. Also, mesh has option for displaying the surface by viewing it from different angles, user is again urged to try generation of different 3D hidden-line views.

Mathematica Applications

Mathematica has a three-dimensional plot function called **Plot3d** which can be applied for drawing the hemispherical surface. Figure 7 is the result of entering the statement:

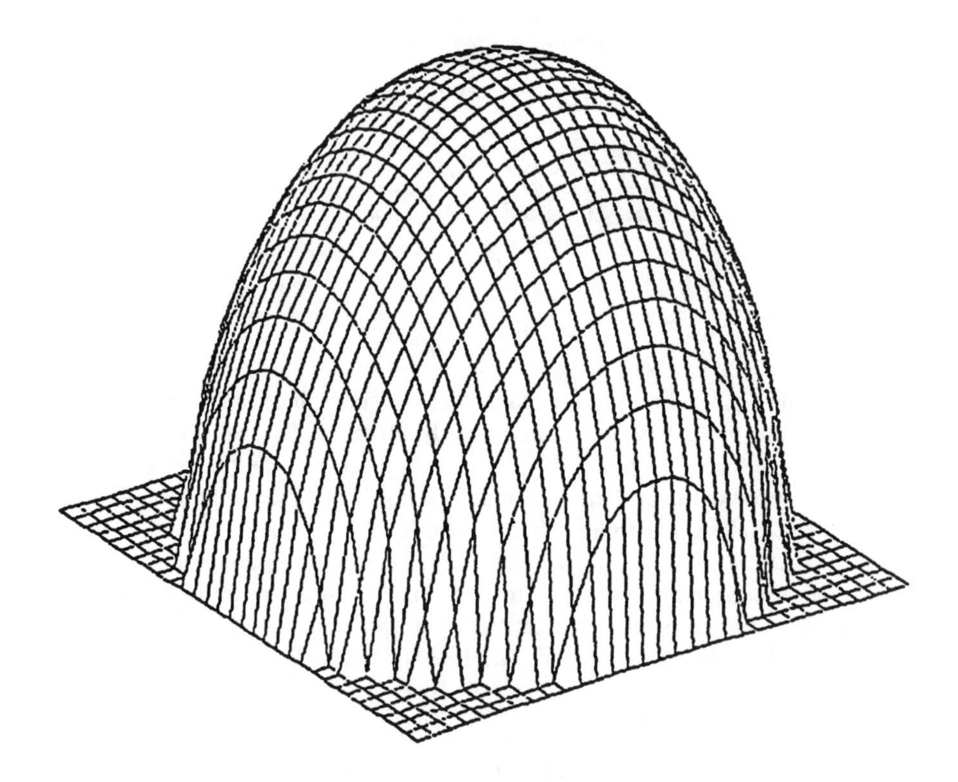

FIGURE 6.

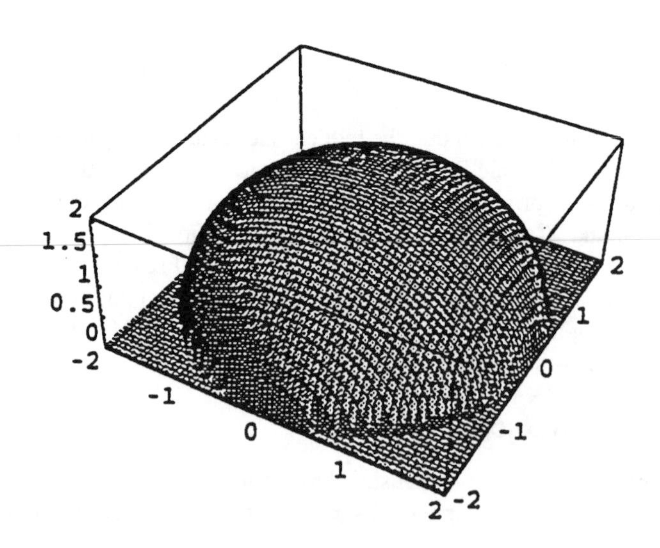

FIGURE 7.

Input]: = Sphere = Plot3D[If[4X^2Y^2>0, Sqrt[4X^2Y^2],0,
{X,–2,2},{Y,–2,2},PlotPoints->{60,60}]

The **If** command tests the first expression inside the brackets, it the condition is true then the statement which follows is implemented and other the last statement inside the bracket is implemented. In this case, the surface only rises over the base circle of radius equal to 2. The **PlotPoints** command specifies how many gird points along X- and Y-directions should be taken to plot the surface. The default number of point is 15 in both directions. The greater the number of grid points, the smoother the surface looks.

The same result can be obtained by first *defining* a surface function, say sf, and then apply **Plot3d** for drawing the surface using sf as follows:

Input]: = sf[X_,Y_] = If[4X^2Y^2>0, Sqrt[4X^2Y^2], 0]

Input[2]: = Plot3D[sf[X,Y],{X,–2,2},{Y,–2,2},PlotPoints->{60,60}]

5.4 PROBLEMS

NuIntGra

1. Having learned how to apply Trapezoidal Rule for numerical integration, how would you find the area under the line $y(x) = 1 + 2x$ and between $x = 1$ and $x = 2$? Do it not by direct integration, but numerically. What should be the stepsize for x in order to ensure an accurate result?
2. Having learned how to apply Simpson's Rule for numerical integration, how would you find the area under the parabolic curve $y(x) = 1 + 2x + 3x^2$ and between $x = 1$ and $x = 2$? Do it not by direct integration but numerically! What should be the stepsize for x in order to ensure an accurate result?
3. If Trapezoidal Rule, instead of Simpson's Rule, is applied for Problem 2, find out how small should be the stepsize for x in order to achieve the same result accurate to the fifth significant digit.
4. Could Simpson's Rule be applied for Problem 1? Would the result be different? If the result is the same, explain why.
5. Given five points (1,1), (2,3), (3,2), (4,5), and (5,4), use a stepsize of $\Delta x = 1$ to compute $\int ydx$ by application of Simpson's and Trapezoidal rules.
6. Use the trapezoidal and Simpson's rules to find the area within the ellipse described by the equation $(x/a)^2 + (y/b)^2 = 1$. Compare the numerical results with the exact solution of πab.
7. Implement the integration of the function $f(x) = 3e^{-2x}\sin x$ over the interval from $x = 0$ to $x = 1$ (in radian) by applying both the Trapezoidal and Simpson's rules and using an increment of $\Delta x = 0.25$.
8. Find the exact solution of Problem 7 by referring to an integration formula for f(x) from any calculus book. Decrease the increment of x (i.e., ,

increase the number of points at which the integrand function is computed) to try to achieve this analytical result using both Trapezoidal and Simpson's Rules.

9. Apply the function **Quad.m** of **MATLAB** to solve Problem 1.
10. Apply the function **Quad.m** of **MATLAB** to solve Problem 2.
11. Apply the function **Quad.m** of **MATLAB** to solve Problem 6.
12. Apply the function **Quad.m** of **MATLAB** to solve Problem 7.
13. Apply **MATLAB** to *spline curve-fit* the five points given in Problem 5 and then integrate.
14. Apply the function **NIntegrate** of **Mathematica** to solve Problem 1.
15. Apply the function **NIntegrate** of **Mathematica** to solve Problem 2.
16. Apply the function **NIntegrate** of **Mathematica** to solve Problem 6.
17. Apply the function **NIntegrate** of **Mathematica** to solve Problem 7.
18. Problem 13 but apply **Mathematica** instead.

VOLUME

1. Apply trapezoidal rule for integration along the x direction and Simpson's rule along the y direction to calculate the volume under the surface $z(x,y) = 3x + 2y^2 + 1$ over the rectangular region $0 \le x \le 2$ and $0 \le y \le 4$ using increments $\Delta x = \Delta y = 1$.
2. Rework Problem 1 except trapezoidal rule is applied for both x and y directions.
3. Find by numerical integration of the ellipsoidal volume based on the double integral $\int\int 3[1-(x/5)^2-(y/4)^2]^{1/2}dxdy$ and for x values ranging from 2 to 4 and y values ranging from 1 to 2. Three stations (for using Simpson's rule) for the x integration and two stations (for using trapezoidal rule) for the y integration are to be adopted.
4. Find the volume between the $z = 0$ plane and the spherical surface $z(x,y) = [4x^2 - y^2]^{1/2}$ for x and y both ranging from 0 to 2 by applying the Simpson's rule for both x and y integrations. *Three* stations are to be taken along the x direction and *five* stations along the y direction for the specified numerical integration.
5. Specify a FUNCTION Z(x,y) for program **Volume** so that the volume enclosed by the ellipsoid $(x/a)^2 + (y/b)^2 + (z/c)^2 = 1$ can be estimated by numerical integration and compare to the exact solution of $4\pi abc/3$.[2]
6. In Figure 8, the shape and dimensions of a pyramid are described by the coordinates of the five points (X_i,Y_i) for I = 1,2,…,5. For application of numerical integration to determine its volume by either trapezoidal or Simpson's rule, we have to partition the projected plane $P_2P_3P_4P_5$ into a gridwork. At each interception point of the gridwork, (X,Y), the height Z(X,Y) needs to be calculated which requires knowing the equations describing the planes $P_1P_2P_3$, $P_1P_3P_4$, $P_1P_4P_5$and $P_1P_5P_2$. The equation of a plane can be written in the form of $2(X-a) + m(Y-b) + n(Z-c) = 1$ where (a,b,c) is a point on the plane and (2,m,n) are the directional cosines

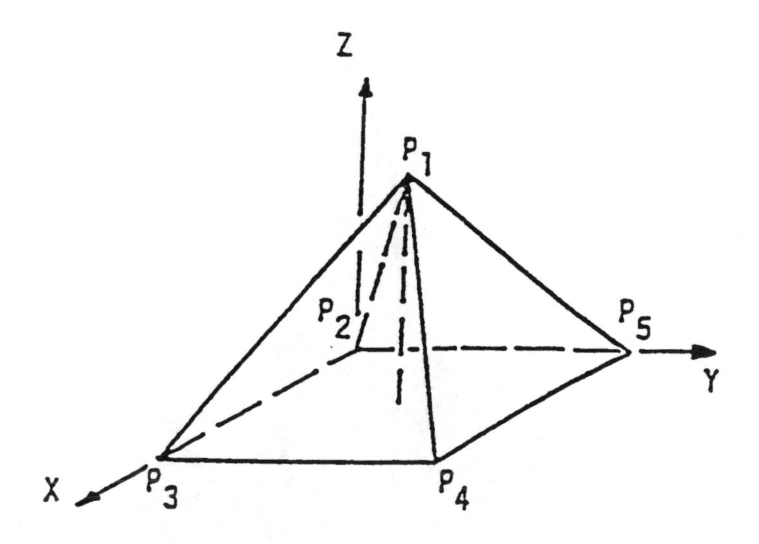

FIGURE 8. Problem 6.

of the unit normal vector of the plane.[3] Apply the equation of plane and assign proper values for the coordinates (X_i,Y_i) describing the pyramid, and then proceed to write a FUNCTION Z(X,Y) to determine its volume by using program **Volume**.

7. Find the volume under the surface $z = 3x^2-4y + 15$ over the base area of $0\le x\le 2$ and $1\le y\le 2$ by applying Simpson's Rule along the x-direction using an increment of $\Delta x = 1$, and Trapezoidal Rule along the y-direction using an increment of $\Delta y = 0.25$.
8. How do you find the volume under the *plane* $z = 2x-0.5y$ and above the rectangular area bounded by $x = 0$, $x = 1$, $y = 0$, and $y = 2$ *numerically* and not by actually integrating the z function? Explain which method and stepsizes in x and y directions you will use, give the numerical result and discuss how accurate it is.
9. Use the function FuncZnew which defines the equation $Z = 2X + 3Y^2 + 4$ and plot the Z surface for $0\le X\le 2$ and $1\le Y\le 2$ by applying **mesh** of **MATLAB**. Experiment with different increments of X and Y.
10. Modify the use of mesh by defining a vector $\{S\} = [S_X\ S_Y\ S_Z\}$ containing the values of scaling factors for the three coordinate axes and then enter mesh(Z,S) to try to improve the appearance of a hemisphere, better than the one shown in Figure 2. Referring to Figure 2, the lowest point is the original and the X-axis is directed to the right (width), Y-axis is directed to the left (depth), and Z-axis is pointing upward (height). Since the hemisphere has a radius equal to 2 and by actually measuring the width, depth, and height to be in the approximate ratios of 2 7/8": 2 7/16": 2 3/4". Based on these values, slowly adjust the values for S_X, S_Y and S_Z.

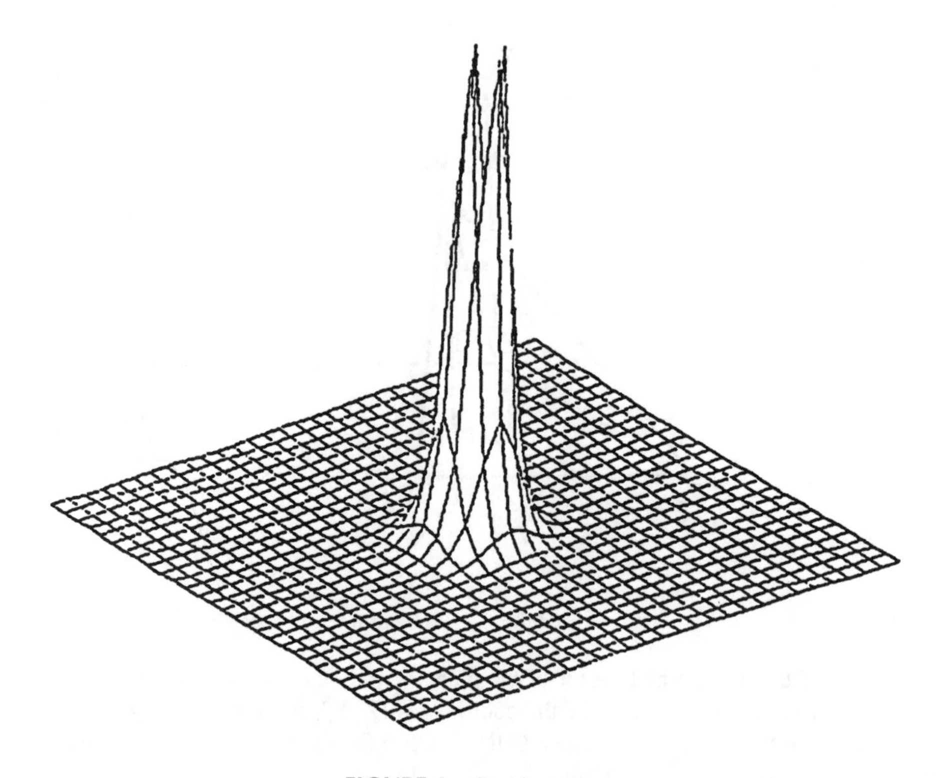

FIGURE 9. Problem 11.

11. Figure 9 is obtained by using **mesh** to plot the surface $Z = 1.5Re^{-2R}$ and $R = (X^2 + Y^2)^{1/2}$ for $-15 \le X,Y \le 15$ with increment of 1 in both X and Y directions. Try to generate this surface by interactively entering **MATLAB** commands. Apply the m file volume and modify the function FuncZnew to accommodate this new integrand function to calculate the volume of this surface above the 30x30 base area.
12. Apply **Mathematica** to solve Problem 6.
13. Apply **Mathematica** to solve Problem 7.
14. Apply **Mathematica** to solve Problem 9.
15. Apply **Mathematica** to solve Problem 11.

5.5 REFERENCES

1. M. Abramowitz and I. A. Stegum, editors, *Handbook of Mathematical Functions with Formulas, Graphs and Mathematical Tables*, National Bureau of Standards Applied Mathematics Series 55, Washington, DC, 1964.
2. R. C. Weast, *Standard Mathematical Tables*, the Chemical Rubber Co. (now CRC Press LLC), Cleveland, OH, 13th edition, 1964.)
3. H. Flanders, R. R. Korfhage, and J. J. Price, *A First Course in Calculus with Analytic Geometry*, Academic Press, New York, 1973.

6 Ordinary Differential Equations — Initial and Boundary Value Problems

6.1 INTRODUCTION

An example of historical interest in solving an unknown function which is governed by an ordinary differential equation and an initial condition is the case of finding y(x) from:

$$\frac{dy}{dx} = y \quad \text{and} \quad y(x=0) = 1 \tag{1}$$

As we all know, $y(x) = e^x$. In fact, the *exponential function* e^x is defined by an infinite series:

$$e^x = 1 + \frac{x^1}{1!} + \frac{x^2}{2!} + \ldots = \sum_{i=0}^{\infty} \frac{x^i}{i!} \tag{2}$$

To prove that Equation 2 indeed is the solution for y satisfying Equation 1, here we apply an iterative procedure of successive integration using a counter k. First, we integrate both sides of Equation 1 with respect to x:

$$\int_0^x \frac{dy}{dx} dx = \int_0^x y \, dx \tag{3}$$

After substituting the initial condition of $y(x = 0) = 1$, we obtain:

$$y(x) = 1 + \int_0^x y \, dx \tag{4}$$

So we are expected to find an unknown y(x) which is to be obtained by integrating it? Numerically, we can do it by assuming a y(x) initially ($k = 1$) equal to 1, investigate how Equation 4 would help us to obtain the next ($k = 2$), guessed y(x), and hope eventually the iterative process would lead us to a solution. The iterative equation, therefore, is for $k = 1,2,\ldots$

$$y^{(k+1)}(x) = 1 + \int_0^x y^{(k)} dx \qquad (5)$$

The results are $y^{(1)} = 1$, $y^{(2)} = 1 + x$, $y^{(3)} = 1 + x + (x^2/2!)$, and eventually the final answer is the infinite series given by Equation 2.

What really need to be discussed in this chapter is not to obtain an analytical expression for y(x) by solving Equation 1 and rather to compute the numerical values of y(x) when x is equally incremented. That is, for a selected x increment, Δx (or, stepsize h), to find $y_i \equiv y(x_i)$ for i = 1,2,... until x reaches a terminating value of x_e and $x_i = (i-1)\,\Delta x$. A simplest method to find y_2 is to approximate the derivative of y(x) by using the forward difference at x_1. That is, according to the notation used in Chapter 4, we can have:

$$\left.\frac{dy}{dx}\right|_{x=x_1} \doteq \frac{\Delta y_1}{\Delta x} = \frac{y_2 - y_1}{\Delta x} = y_1 \qquad (6)$$

Or, $y_2 = (1 + \Delta x)y_1$. In fact, this result can be extended to any x_i to obtain $y_{i+1} = 1 + (x)y_i$. For the general case when the right-hand side of Equation 1 is equal to a prescribed function f(x), we arrive at the Euler's formula $y_{i+1} = y_i + f(x_i)\,\Delta x$. Euler's formula is easy to apply but is inaccurate. For example, using a $\Delta x = 0.1$ in solving Equation 1 with the Euler's formula, it leads to $y_2 = (1 + 0.1)y_1 = 1.1$, $y_3 = (1 + 0.1)y_2 = 1.21$ when the exact values are $y_2 = e^{0.1} = 1.1052$ and $y_3 = e^{0.2} = 1.2214$, respectively. The computational errors accumulate very rapidly.

In this chapter, we shall introduce the most commonly adopted method of Runge-Kutta for solution of the initial-value problems governed by ordinary differential equation(s). For the *fourth-order Runge-Kutta method*, the error per each computational step is of order h^5 where h is the stepsize. Converting the higher-order ordinary differential equation(s) into the standardized form using the *state variables* will be illustrated and computer programs will be developed for numerical solution of the problem.

Engineering problems which are governed by ordinary differential equations and also some associated conditions at certain boundaries will be also be discussed. Numerical methods of solution based on the Runge-Kutta procedure and the finite-difference approximation will both be explained.

6.2 PROGRAM RUNGEKUT — APPLICATION OF THE RUNGE-KUTTA METHOD FOR SOLVING THE INITIAL-VALUE PROBLEMS

Program **RungeKut** is designed for solving the initial-value problems governed by ordinary differential equations using the fourth-order Runge-Kutta method. There are numerous physical problems which are mathematically governed by a set of ordinary differential equations (**ODE**) involving many unknown functions. These unknown functions are all dependent of a variable t. Supplementing to this set of

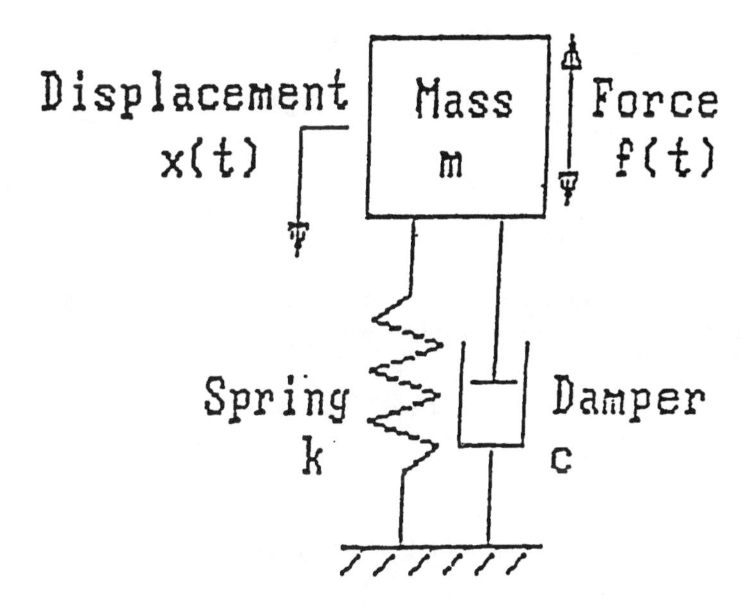

FIGURE 1. The often cited vibration problem shown requires the changes of the elevation x and velocity v to be calculated.

ordinary differential equations are the *initial conditions* of the dependent functions when t is equal to zero. For example, the often cited vibration[1] problem shown in Figure 1 requires the changes of the elevation x and velocity v to be calculated using the equations:

$$m\left(d^2 x/dt^2\right)+c(dx/dt)+kx=f(t) \tag{1}$$

and

$$dx/dt=v \tag{2}$$

where m is the mass, c is the damping coefficient, k is the spring constant, t is the time, and f(t) is a disturbing force applied to the mass. When the physical parameters m, c, and k, and the history of the applied force f(t) are specified, the complete histories of the mass' elevation x and velocity v can be calculated analytically, or, numerically if the initial elevation $x(t = 0)$ and $v(t = 0)$ are known. If m, c, and k remain unchanged throughout the period of investigation and f(t) is a commonly encountered function, Equation 1 can be solved analytically.[1] Otherwise, a numerical method has to be applied to obtain approximate solution of Equation 1.

Many numerical methods are available for solving such initial-value problems governed by ordinary differential equations. Most of the numerical methods require that the governing differential equation be rearranged into a standard form of:

$$\begin{aligned} dx_1/dt &= F_1(x_1, x_2, \ldots, x_n, t;\ \text{parameters}) \\ dx_2/dt &= F_2(x_1, x_2, \ldots, x_n, t;\ \text{parameters}) \\ &\quad \cdots\cdots\cdots \\ dx_n/dt &= F_n(x_1, x_2, \ldots, x_n, t;\ \text{parameters}) \end{aligned} \tag{3}$$

For example, the variables x and v in Equations 1 and 2 are to be renamed x_1 and x_2, respectively. Equation 1 is to be rewritten as:

$$m(dv/dt) + cv + kx = f(t)$$

and then as:

$$dv/dt = [f(t) - cv - kx]/m$$

and finally as:

$$dx_2/dt = F_2(x_1, x_2, t; m, c, k)$$

Meanwhile. Equation 2 is rewritten as:

$$dx_1/dt = F_1(x_1, x_2, t; m, c, k)$$

Or, more systematically the problem is described by the equations:

$$\begin{aligned} dx_1/dt &= F_1(x_1, x_2, t; m, c, k) = x_2 \\ dx_2/dt &= F_2(x_1, x_2, t; m, c, k) = [f(t) - kx_1 - cx_2]/m \end{aligned} \tag{4}$$

and having the initial conditions $x_1(t = 0)$ and $x_2(t = 0)$ prescribed.

Runge-Kutta method is a commonly used method for numerical solution of a system of first-order ordinary differential equations expressed in the general form of (3). It is to be introduced and illustrated with a number of practical applications.

Runge-Kutta Method (Fourth-Order)

Consider the problem of finding x and y values at t>0 when they are governed by the equations:

$$(dx/dt) - 4x + (dy/dt) - 6y = -2 \tag{5}$$

and

$$2(dx/dt) + x + 3(dy/dt) + 2y = 0 \tag{6}$$

when initially their values are x(t = 0) = 7 and y(t = 0) = –4. The analytical solutions are obtainable[2] and they are:

$$x = 5e^{t} + 2 \quad \text{and} \quad y = -3e^{t} - 1 \tag{7}$$

To solve the problem numerically, Equations 5 and 6 need to be decoupled and expressed in the form of Equation 3. Cramer's rule can be applied by treating dx/dt and dy/dt as two unknowns and x and y as parameters, the converted standard form after changing x to x_1 and y to x_2 is:

$$dx_1/dt = F_1(x_1, x_2, t; \text{constants}), \quad x_1(0) = 7 \tag{8}$$

$$dx_2/dt = F_2(x_1, x_2, t; \text{constants}), \quad x_2(0) = -4 \tag{9}$$

where:

$$F_1(x_1, x_2, t; \text{constants}) = -6 + 13x_1 + 20x_2 \tag{10}$$

and

$$F_2(x_1, x_2, t; \text{constants}) = 4 - 9x_1 - 14x_2 \tag{11}$$

Numerical solution of x_1 and x_2 for t>0 is to use a selected time increment Δt (often referred to as the stepsize h for the independent variable t). Denote t_0 as the initial instant t = 0 and t_{j+1} as the instant after j increments of time, that is, t_{j+1} = (j + 1)h. If the values for x_1 and x_2 at t_j, denoted as $x_{1,j}$ and $x_{2,j}$ respectively, are already known, the fourth-order Runge-Kutta method is to use the following formulas to calculate x_1 and x_2 at t_{j+1}, denoted as $x_{1,j+1}$ and $x_{2,j+1}$:

$$x_{i,j+1} = x_{i,j} + (p_{i,1} + 2p_{i,2} + 2p_{i,3} + p_{i,4})/6 \tag{12}$$

for i = 1,2. The p's in Equation 12 are the Runge-Kutta parameters to be calculated using the functions F_1 and F_2 by adjusting the values of the variables x_1 and x_2 at t_j. The formulas for calculating these p's are, for i = 1,2

$$p_{i,1} = hF_i(t_j, x_{1,j}, x_{2,j}) \tag{13}$$

$$p_{i,2} = hF_i\left[t_j + (h/2), x_{1,j} + (p_{1,1}/2), x_{2,j} + (p_{2,1}/2)\right] \quad (14)$$

$$p_{i,3} = hF_i\left[t_j + (h/2), x_{1,j} + (p_{1,2}/2), x_{2,j} + (p_{2,2}/2)\right] \quad (15)$$

$$p_{i,4} = hF_i\left(t_j + h, x_{1,j} + p_{1,3}, x_{2,j} + p_{2,3}\right) \quad (16)$$

Equations 12 to 16 are to be used to generate x_1 and x_2 values at t_j for $j = 1,2,3,\ldots$ which can be tabulated as:

t	0	h	2h	3h	.	.	.	jh	.	.	.	t_e
x_1	7	?	?	?	.	.	.	$x_{1,j}$	.	.	.	?
x_2	-4	?	?	?	.	.	.	$x_{2,j}$	.	.	.	?

where t_e is the ending value of t at which the computation is to be terminated. The first pair of values to be filled into the above table is for x_1 and x_2 at $t = h$ ($j = 1$). Based first on Equations 13 to 16 and then Equation 12, the actual computations for $h = 0.1$ and at t_1 go as follows:

$p_{1,1} = hF_1(t_0,x_{1,0},x_{2,0}) = 0.1F_1(0,7,-4) = 0.1(-6 + 91\text{-}80) = 0.5$
$p_{2,1} = hF_2(t_0,x_{1,0},x_{2,0}) = 0.1F_2(0,7,-4) = 0.1(4\text{-}63 + 56) = -0.3$
$p_{1,2} = hF_1(t_0 + 0.05,x_{1,0} + 0.25,x_{2,0}-0.15) = 0.1F_1(.05,7.25,-4.15)$
$= 0.1(-6 + 13x7.25\text{-}20x4.15) = 0.525$
$p_{2,2} = hF_2(t_0 + 0.05,x_{1,0} + 0.25,x_{2,0}-0.15) = 0.1F_2(.05,7.25,-4.15)$
$= 0.1(4\text{-}9x7.25 + 14x4.15) = -0.315$
$p_{1,3} = hF_1(t_0 + 0.05,x_{1,0} + 0.2625,x_{2,0}-0.1575)$
$= .1F_1(.05,7.2625,-4.1575)$
$= 0.1(-6 + 13x7.2625\text{-}20x4.1575) = .52625$
$p_{2,3} = hF_2(t_0 + 0.05,x_{1,0} + 0.2625,x_{2,0}-0.1575)$
$= .1F_2(.05,7.2625,-4.1575) = 0.1(4-9x7.2625 + 14x4.1575)$
$= -0.31575$
$p_{1,4} = hF_1(t_0 + 0.1,x_{1,0} + 0.52625,x_{2,0}-0.31575)$
$= 0.1F_1(0.1,7.52625,-4.31575)$
$= 0.1(-6 + 13x7.52625-20x4.31575) = 0.552625$
$p_{2,4} = hF_2(t_0 + 0.1,x_{1,0} + 0.52625,x_{2,0}-0.31575)$
$= .1F_2(.1,7.52625,-4.31575) = .1(4-9x7.52625 + 14x4.31575) = -0.331575$
$x_{1,1} = 7 + (0.5 + 2x0.525 + 2x0.52625 + 0.552625)/6$
$= 7 + (3.155125)/6 = 7.5258541 \quad (17)$
$x_{2,1} = -4 + (-0.3-2x0.315-2x0.31575-0.331575)/6$
$= -4 + (-1.893075)/6 = -4.3155125 \quad (18)$

The exact solution calculated by using Equation 7 are:

$$x_{1,1} = 7.5258355 \quad \text{and} \quad x_{2,1} = -4.3155013 \quad (19)$$

The errors are 0.000247% and 0.000260% for x_1 and x_2, respectively. Per-step error for the fourth-order Runge-Kutta method is difficult to estimate because the method is derived by matching terms in Equation 12 with Taylor-series expansions of x_1 and x_2 about t_i through and including the h^4 terms. But approximately, the per-step error is of order h^5. For better accuracy, the fifth-order Runge-Kutta method should be applied. For general use, the *classic* fourth-order Runge-Kutta method is, however, easier to develop a computer program which is to be discussed next.

Subroutine RKN

A subroutine called **RKN** has been written for applying the fourth-order Runge-Kutta method to solve the initial-value problems governed by a set of first-order ordinary differential equations. It has been coded according to the procedure described in the preceding section. That is, the equations must be in the form of Equation 3 by having the first derivatives of the dependent variables (x_1 through x_N) all on the left sides of the equations and the right sides be called F_1 through F_N. These functions are to be defined in a Function subprogram F.

The **FORTRAN** version of Subroutine **RKN** is listed below. There are seven arguments for this subroutine, the first four are input arguments where the last is an output argument. The fifth argument P keeps the Runge-Kutta parameters generated in this subroutine. The sixth argument XT is needed for adjusting the input argument XIN. These two arguments, P and XT, are included for handling the general case of N variables. Listing them as arguments makes possible to specify them as matrix and vector of *adjustable* sizes.

FORTRAN Version

```
      SUBROUTINE RKN(XIN,N,DT,T,P,XT,XOUT)
C     SOLVING N FIRST-ORDER O.D.E. D(XIN)/DT=F(XIN,T)
C     KNOWING XIN AT T, THIS SUB. FINDS XOUT AT T+DT
C     P ARE THE RUNGE-KUTTA 4TH-ORDER PARAMETERS.
      DIMENSION P(N,4),XIN(N),XOUT(N),XT(N)
      DO 5 I=1,N
    5 P(I,1)=DT*F(XIN,T,I,N)
      DO 15 J=2,3
      DO 10 I=1,N
   10 XT(I)=XIN(I)+P(I,J-1)/2.
      DO 15 I=1,N
   15 P(I,J)=DT*F(XT,T+DT/2.,I,N)
      DO 20 I=1,N
   20 XT(I)=XIN(I)+P(I,3)
      DO 25 I=1,N
      P(I,4)=DT*F(XT,T+DT,I,N)
   25 XOUT(I)=XIN(I)+(P(I,1)+2.*(P(I,2)+P(I,3))+P(I,4))/6.
      RETURN
      END
```

QuickBASIC Version

```
      SUB RKN (XIN(), N, DT, T, P(), XT(), XOUT())
'     SOLVING N FIRST-ORDER O.D.E. D(XIN)/DT=F(XIN,T)
'     KNOWING XIN AT T, THIS SUB. FINDS XOUT AT T+DT
'     P ARE THE RUNGE-KUTTA 4TH-ORDER PARAMETERS.
      FOR I = 1 TO N: P(I,1) = DT * F(XIN(), T, I, N): NEXT I
      FOR J = 2 TO 3
         FOR I=1 TO N: XT(I)=XIN(I)+P(I,J-1)/2        : NEXT I
         FOR I=1 TO N
            P(I,J)=DT*F(XT(),T+DT/2,I,N): NEXT I    : NEXT J
      FOR I = 1 TO N: XT(I)=XIN(I)+P(I,3) : NEXT I
      FOR I = 1 TO N: P(I,4) = DT*F(XT(),T+DT,I,N)    : NEXT I
      FOR I = 1 TO N
         XOUT(I)=XIN(I)+(P(I,1)+2*(P(I,2)+P(I,3))+P(I,4))/6
         NEXT I
      END SUB
```

Program RungeKut

Program **RungeKut** which calls the subroutine **RKN** is to be run interactively by specifying the inputs through the keyboard. Displayed messages on screen instruct user how to input the necessary data and describe the problem in proper sequence. From the provided listing of the program **RungeKut**, user will find that the following inputs and editing need to be executed in the sequence specified:

(1) Input the number of variables, N, involved.
(2) Define the N functions on the right sides of Equation 3, F_1 through F_N, by editing the **DEF** statements starting from statement 161. Presently only 9 functions can be handled by the program **RungeKut**, but the user should be able to expand the program to accommodate any N value which is greater than 9 by renumbering the program and adding more DEF statements.
(3) Type RUN 161 to run the program.
(4) Reenter the N value.
(5) Enter the beginning (not necessary equal to zero!) and ending values of the independent variable, denoted as t_0 and t_e (T0 and TEND in the program **RungeKut**), respectively. It is over this range, the values of the N dependent variables are to be calculated.
(6) Enter the stepsize, h (DT in the program **RungeKut**), with which the independent variable is to be incremented.
(7) Enter the N initial values.

QuickBASIC Version

```
' Program RungeKut.QB - Solving N 1st-order ordinary differential equations
'                      initial-value problem using Runge-Kutta 4th-order
'                      method.  dx(I)/dt=F(i;t;x1,x2,...,Xn), N<=100
'                      F(X,T,I,N) must be defined in a subprogram.
      DECLARE SUB RKN (X(), N, DT, T, P(), XT(), XN())
      DECLARE FUNCTION F (X(), T, I, N)
      CLEAR : CLS : DIM P(100, 4), X(100), XN(100), XT(100)
      INPUT " Input number of equation : ", N
      INPUT " Enter the initial value of the independent variable : ", T
      INPUT " Enter the increment of the independent variable : ", DT
      INPUT " Enter the ending value of the independent variable : ", TEND
      PRINT " Enter the initial value of the dependent variables : "
      FOR I = 1 TO N
         PRINT "X("; I; ") = "; : INPUT "", X(I): NEXT I
      PRINT "     T            X(1)          X(2)       "
      PRINT USING "##.####^^^^"; T;
      FOR I = 1 TO N: PRINT USING " ##.####^^^^"; X(I); : NEXT I: PRINT
6     CALL RKN(X(), N, DT, T, P(), XT(), XN())
10    T = T + DT
      FOR I = 1 TO N: X(I) = XN(I): NEXT I
      PRINT USING "##.####^^^^"; T;
      FOR I = 1 TO N
         PRINT USING " ##.####^^^^"; X(I); : NEXT I: PRINT
      IF (T >= TEND) THEN 50 ELSE 6
50    PRINT : PRINT "    Last set of R-K Parameters are :"
      FOR I = 1 TO N: PRINT "    ";
         FOR J = 1 TO 4: PRINT USING "  ##.###^^^^"; P(I, J);
                        NEXT J: PRINT : NEXT I
      END
```

FORTRAN Version

```
C  ** Program RungeKut - Solving N 1st-order ordinary differential equations
C                        initial-value problem using Runge-Kutta 4th-order
C                        method.  dx(I)/dt=F(i;t;x1,x2,...,Xn), N<=100
C                        F(X,T,I,N) must be defined in a subprogram.
      DIMENSION P(100,4),X(100),XN(100),XT(100)
      WRITE (*,10)
   10 FORMAT(1X,'Input number of equation :')
      READ (*,*) N
      WRITE (*,12)
```

```
12 FORMAT(1X,'Enter the initial value of the independent variable :')
   READ (*,*) T
   WRITE (*,14)
14 FORMAT(1X,'Enter the increment of the independent variable : ')
   READ (*,*) DT
   WRITE (*,16)
16 FORMAT(1X,'Enter the ending value of the independent variable : ')
   READ (*,*) TEND
   WRITE (*,18)
18 FORMAT(1X,'Enter the initial value of the dependent variables : ')
   DO 20 I=1,N
20 READ (*,*) X(I)
   WRITE (*,22)
22 FORMAT(9X,'T',5X,'X(1) through X(N)')
   WRITE (*,24) T,(X(I),I=1,N)
24 FORMAT(F10.3,5E14.5)
26 CALL RKN(X,N,DT,T,P,XT,XN)
   T=T+DT
   DO 30 I=1,N
30 X(I)=XN(I)
   WRITE (*,24) T,(X(I),I=1,N)
   IF (T.GE.TEND) GOTO 50
   GOTO 26
50 WRITE (*,52)
52 FORMAT(1X,'R-K Parameters are :')
   DO 54 I=1,N
54 WRITE (*,56) (P(I,J),J=1,4)
56 FORMAT(4E15.5)
   END
```

Function F

According to Equations 12 to 15, the Runge-Kutta parameters $p_{i,j}$ (matrix P in the program RungeKut) are calculated using two FOR-NEXT loops — an I loop covering N sets of variables and a J loop covering the four parameters in each set. As an illustrative example, let us apply program **RungeKut** for the problem defined by Equations 8 to 11. We create a supporting function program F as follows:

QuickBASIC Version

```
  FUNCTION F(X(), T, I, N)
  IF I=1 THEN F=-6+13*X(1)+20*X(2): GOTO 5
              F= 4-9*X(1)-14*X(2)
5 END FUNCTION
```

FORTRAN Version

```
      FUNCTION F(X,T,I,N)
      DIMENSION X(N)
      GO TO (10,20),I
   10 F=-6+13*X(1)+20*X(2)
      RETURN
   20 F=4-9*X(1)-14*X(2)
      RETURN
      END
```

The computation can then commence by entering through keyboard the beginning and ending values of t, $t_0 = 0$ and $t_e = 3$, respectively, the stepsize h = 0.1, and the initial values $x_{1,0} = 7$ and $x_{2,0} = -4$. The complete sequence of question-and-answer steps in running the program RungeKut for the problem described by Equations 8 to 11 is manifested by a copy of the screen display:

```
 Input number of equation : 2
 Enter the initial value of the independent variable : 0
 Enter the increment of the independent variable : 0.1
 Enter the ending value of the independent varaible : 3
 Enter the initial value of the dependent variables:
X( 1 ) = 7
X( 2 ) = -4
      T            X(1)          X(2)
 0.0000E+00  7.0000E+00 -4.0000E+00
 1.0000E-01  7.5259E+00 -4.3155E+00
 2.0000E-01  8.1070E+00 -4.6642E+00
 3.0000E-01  8.7493E+00 -5.0496E+00
 4.0000E-01  9.4591E+00 -5.4755E+00
 5.0000E-01  1.0244E+01 -5.9462E+00
 6.0000E-01  1.1111E+01 -6,4664E+00
 7.0000E-01  1.2069E+01 -7.0413E+00
 8.0000E-01  1.3128E+01 -7.6766E+00
 9.0000E-01  1.4298E+01 -8.3788E+00
 1.0000E+00  1.5591E+01 -9.1548E+00
 1.1000E+00  1.7021E+01 -1.0012E+01
 1.2000E+00  1.8601E+01 -1.0960E+01
 1.3000E+00  2.0345E+01 -1.2008E+01
 1.4000E+00  2.2276E+01 -1.3166E+01
 1.5000E+00  2.4408E+01 -1.4445E+01
```

```
1.6000E+00  2.6765E+01 -1.5859E+01
1.7000E+00  2.9370E+01 -1.7422E+01
1.8000E+00  3.2248E+01 -1.9149E+01
1.9000E+00  3.5429E+01 -2.1058E+01
2.0000E+00  3.8945E+01 -2.3167E+01
2.1000E+00  4.2831E+01 -2.5498E+01
2.2000E+00  4.7125E+01 -2.8075E+01
2.3000E+00  5.1871E+01 -3.0922E+01
2.4000E+00  5.7116E+01 -3.4069E+01
2.5000E+00  6.2912E+01 -3.7547E+01
2.6000E+00  6.9319E+01 -4.1391E+01
2.7000E+00  7.6398E+01 -4.5639E+01
2.8000E+00  8.4223E+01 -5.0334E+01
2.9000E+00  9.2871E+01 -5.5522E+01
3.0000E+00  1.0243E+02 -6.1256E+01
3.1000E+00  1.1299E+02 -6.7594E+01

    Last set of R-K Parameters are :
       1.004E+01    1.054E+01    1.057E+01    1.110E+01
      -6.026E+00   -6.327E+00   -6.342E+00   -6.660E+00
```

Sample Applications of the program RungeKut

As a first example, consider the problem of a beam shown in Figure 2 which is built into the wall so that it is not allowed to displace or rotate at the left end. For consideration of the general case when the beam is loaded by (1) a uniformly distributed load of 1 N/cm over the leftmost quarter-length, x between 0 and 10 cm, (2) a linearly varying distributed load of 0 at x = 10 cm and 2 N/cm at x = 20 cm, (3) a moment of 3 N-cm at x = 30 cm, and (4) a concentrated load of 4 N at the free end of the beam, x = 40 cm. It is of concern to the structural engineers to know how the beam will be deformed. The equation for finding the deflected curve of the beam, usually denoted as y(x), is:[3]

$$EI\left(d^2y/dx^2\right) = M(x) \tag{21}$$

where EI is the beam stiffness and M(x) is the variation of bending moment along the beam. It can be shown that the moment distribution for the loading shown in Figure 4 can be described by the equations:

$$M = \begin{cases} -1121/3 + 24x - x^2/2, & \text{for } 0 \,\%\, x \,\%\, 10 \text{ cm} \\ -871/3 + 4x + x^2 - x^3/30, & \text{for } 10 \,\%\, x \,\%\, 20 \text{ cm} \\ 4x - 157, & \text{for } 20 < x \,\%\, 30 \text{ cm} \\ 4x - 160, & \text{for } 30 < x \,\%\, 40 \text{ cm} \end{cases} \tag{22}$$

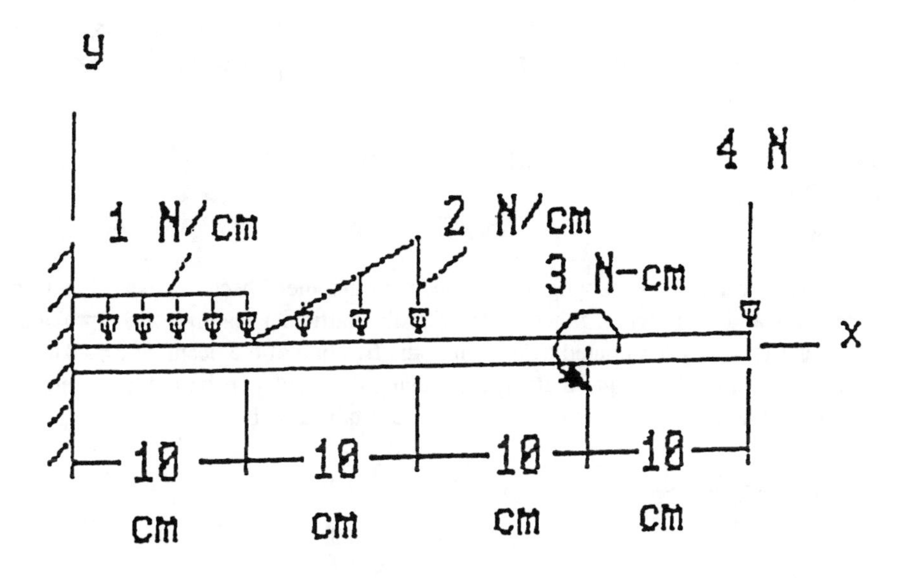

FIGURE 2. The problem of a beam, which is built into the wall so that it is not allowed to displace or rotate at the left end.

To convert the problem into the form of Equation 3, we replace y, dy/dx, and x by x_1, x_2, and t, respectively. Knowing Equation 21 and that the initial conditions are y = 0 and dy/dx = 0 at x = 0, we can obtain from Equation 21 the following system of first-order ordinary differential equations:

$$dx_1/dx = F_1(x, x_1, x_2) = x_2, \qquad x_1(x = 0) = 0 \tag{23}$$

and

$$dx_2/dx = F_2(x, x_1, x_2) = M/EI, \qquad x_2(x = 0) = 0 \tag{24}$$

For Equation 24, the moment distribution has already been described by Equation 21 whereas the beam stiffness EI is to be set equal to $2x10^5$ N/cm^2 in numerical calculation of the deflection of the beam using program **RungeKut**.

To apply program RungeKut for solving the deflection equation, y(x) which has been renamed as $x_1(t)$, we need to define in the **QuickBASIC** program RungeKut two functions:

```
DEF FNF1(X) = X(2)
DEF FNF2(X) = BM(X)/(2*10^5)
```

Notice that the bending moment M is represented by BM in **QuickBASIC** programming. In view of Equation 21, F_2 in Equation 24 has to be defined by modifying the subprogram function, here we illustrate it with a **FORTRAN** version:

```
FUNCTION BM(X)
IF ((X.LE.0.).AND.(X.LT.10.)) BM = -1121./3 + 24*X-X**2/2
IF  ((X.GE.10.).AND.(X.LT.20.))  BM =  -871./3  +  4*T  +
T**2T**3/30
IF ((X.GE.20.).AND.(X.LT.30.)) BM = 4*X-157
IF ((X.GE.30.).AND.(X.LT.40.)) BM = 4*X-160
RETURN
END
```

In fact, each problem will require such arrangement because these function statements and subprogram function describe the particular features of the problem being analyzed. The computed deflection at the free end of the beam, y(x = 40 cm), by application of the program **RungeKut** using different stepsizes, and the errors in % of the analytical solution (= –0.68917) are tabulated below:

Stepsize h, cm	y at x = 40 cm,	Error, %
5	-0.79961	16.0
2	-0.73499	6.65
1	-0.71236	3.36
.5	-0.70083	1.69
.25	-0.69501	.847
.1	-0.69151	.340
.05	-0.69033	.168

This problem can also be arranged into a set of four first-order ordinary differential equations and can be solved by using the expressions for the distributed loads directly. This approach saves the reader from deriving the expressions for the bending moment. Readers interested in such an approach should solve Problem 4.

A Nonlinear Oscillation Problem Solved by RungeKut

The numerical solution using the Runge-Kutta method can be further demonstrated by solving a nonlinear problem of two connected masses m_1 and m_2 as shown in Figure 3. A cable of constant length passing frictionless rings is used. Initially, both masses are held at rest at positions shown. When they are released, their instantaneous positions can be denoted as y(t) and z(t), respectively. The instantaneous angle and the length of the inclined portion of the cable can be expressed in term of y as:

$$\theta = \tan^{-1}[(y+h)/b] \quad \text{and} \quad L = [(y+h)^2 + b^2]^{0.5} \qquad (25,26)$$

If we denote the cable tension as T, then the Newton's second law applied to the two masses leads to $m_1g-2T\sin\theta = m_1d^2y/dt^2$ and $T-m_2g = m_2d^2z/dt^2$. By eliminating T, we obtain:

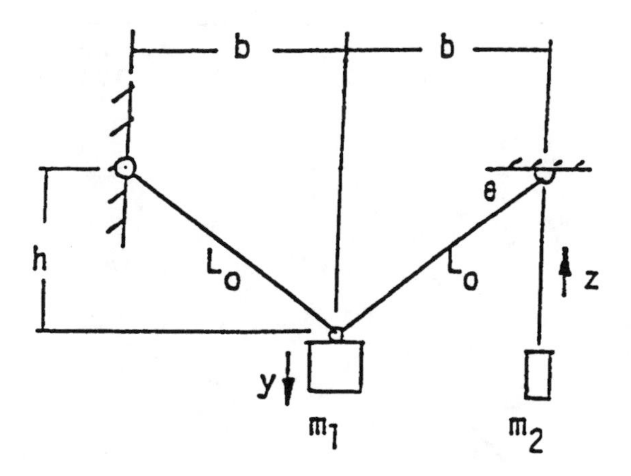

FIGURE 3. The numerical solution using the Runge-Kutta method can be further demonstrated by solving a nonlinear problem of two connected masses m_1 and m_2.

$$\frac{d^2y}{dt^2}+\frac{2m_2}{m_1}\sin\theta\frac{d^2z}{dt^2}=g\left(1-\frac{2m_2}{m_1}\sin\theta\right) \tag{27}$$

The displacements y and z are restricted by the condition that the cable length must remain unchanged. That is $z = 2\{[(y + h)^2 + b^2]^{1/2}[h^2 + b^2]^{1/2}\}$. This relationship can be differentiated with respect to t to obtain another equation relating d^2y/dt^2 and d^2z/dt^2 which is:

$$\frac{d^2z}{dt^2}=\frac{-2}{L^2}(y+h)\frac{dLdy}{dtdt}+\frac{2}{L}\left[\left(\frac{dy}{dt}\right)^2+(y+h)\frac{d^2y}{dt^2}\right] \tag{28}$$

Equation 28 can be substituted into Equation 27 to obtain:

$$\frac{d^2y}{dt^2}=\frac{\sin\theta}{L(2y+h+L)}\left\{gL^2-\frac{2m_2}{m_1}\left[gL^2-2(y+h)\frac{dLdy}{dtdt}+2L\left(\frac{dy}{dt}\right)^2\right]\right\} \tag{29}$$

By letting $x_1 = y$, $x_2 = dy/dt$, $x_3 = z$, and $x_4 = dz/dt$, then according to Equation 3 we have $F_1 = x_2$, F_2 to be constructed using the right-hand side of Equation 29, $F_3 = x_4$, and F_4 to be constructed using the right-side of Equation 28. It can be shown that the final form of the system of four first-order differential equations are:

$$\frac{dx_1}{dt}=x_2,\qquad \frac{dx_2}{dt}=\frac{gL^2(1-2r_m\sin\theta)+4r_mv^2\sin\theta}{L(1+4r\sin^2\theta)},$$
$$\frac{dx_3}{dt}=x_4,\qquad \text{and}\quad \frac{dx_4}{dt}=\frac{-2v^2}{L}+2\frac{dx_2}{dt}\sin\theta \tag{30}$$

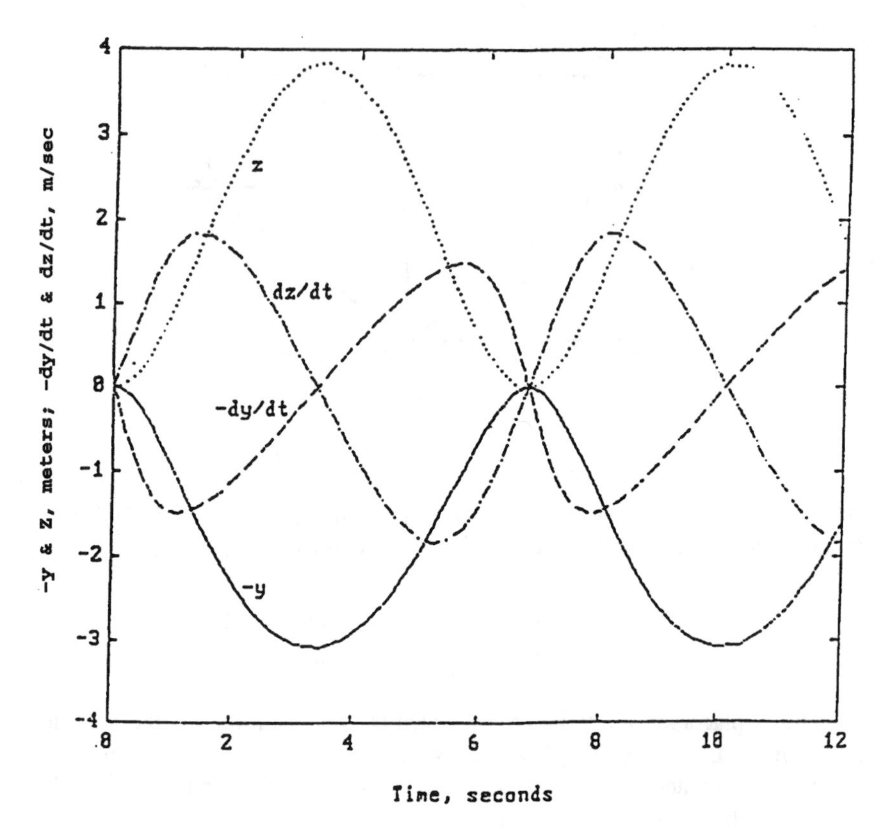

FIGURE 4. A numerical case where b = 3.6 m, h = .15 m, $m_2/m_1 = 0.8$, and initial conditions y = z = dy/dt = dz/dt = 0 has been investigated and the results for -y, z, -dy/dt, and dz/dt have been plotted for 0≤t≤12 seconds.

where:

$$r_m = \frac{m_2}{m_1} \quad \text{and} \quad v^2 = x_2\left(\frac{dL}{dt}\sin\theta - x_2\right) \tag{31,32}$$

A numerical case where b = 3.6 m, h = .15 m, $m_2/m_1 = 0.8$, and initial conditions y = z = dy/dt = dz/dt = 0 has been investigated and the results for -y, z, -dy/dt, and dz/dt have been plotted in Figure 4 for 0≤t≤12 seconds. The reason that negative values of y and dy/dt are used is that the mass m_1 is moving downward as positive. It can be observed from Figure 4 that y varies between 0 and 3.0704 m, and z varies between 0 and 3.8358 m. The oscillation has a period approximately equal to 2x6.6831 = 13.3662 seconds, so the frequency is about 0.47 rad/sec. The masses reach their maximum speeds, $|dy/dt|_{max} = 1.4811$ and $|dz/dt|_{max} = 1.8336$ in m/sec, when $y = .5y_{max} = 1.5352$ and $z = .5z_{max} = 1.9179$ m, respectively. Details for the oscillation for the studied period are listed below:

t	-y	-dy/dt	z	dz/dt
0	0	0	0	0
0.1200	-0.0183	-0.3042	0.0142	0.2365
0.2203	-0.0611	-0.5455	0.0478	0.4340
0.3225	-0.1285	-0.7695	0.1024	0.6343
0.4289	-0.2215	-0.9718	0.1809	0.8385
0.5395	-0.3389	-1.1443	0.2849	1.0411
0.6542	-0.4784	-1.2818	0.4156	1.2347
0.7783	-0.6444	-1.3859	0.5805	1.4185
0.9146	-0.8384	-1.4535	0.7856	1.5835
1.0644	-1.0588	-1.4816	1.0335	1.7167
1.2249	-1.2961	-1.4702	1.3167	1.8636
1.4002	-1.5501	-1.4215	1.6368	1.8377
1.5942	-1.8180	-1.3360	1.9917	1.8108
1.8108	-2.0945	-1.2132	2.3745	1.7144
2.0581	-2.3746	-1.0488	2.7778	1.5367
2.3474	-2.6476	-0.8348	3.1840	1.2608
2.6849	-2.8844	-0.5663	3.5460	0.8754
3.0632	-3.0395	-0.2517	3.7874	0.3944
3.4510	-3.0734	0.0768	3.8407	-0.1207
3.8449	-2.9776	0.4087	3.6907	-0.6370
4.2378	-2.7537	0.7287	3.3452	-1.1125
4.6224	-2.4165	1.0201	2.8393	-1.5020
4.9961	-1.9879	1.2648	2.2251	-1.7601
5.2483	-1.6520	1.3929	1.7697	-1.8351
5.5033	-1.2854	1.4715	1.3035	-1.8009
5.1730	-0.9752	1.4760	0.9377	-1.6725
5.8940	-0.7124	1.4152	0.6510	-1.4820
6.0557	-0.4927	1.2930	0.4294	-1.2523
6.2049	-0.3124	1.1110	0.2609	-0.9991
6.3373	-0.1793	0.8900	0.1449	-0.7525
6.4583	-0.0863	0.6415	0.0680	-0.5173
6.5749	-0.0271	0.3686	0.0210	-0.2879
6.6831	-0.0018	0.0968	0.0014	-0.0745
6.7854	-0.0053	-0.1647	0.0041	0.1271
6.8857	-0.0344	-0.4140	0.0267	0.3246
6.9865	-0.0882	-0.6480	0.0695	0.5231
7.0904	-0.1669	-0.8629	0.1343	0.7251
7.1993	-0.2716	-1.0538	0.2246	0.9306
7.3115	-0.3990	-1.2107	0.3403	1.1297
7.4299	-0.5501	-1.3332	0.4856	1.3195
7.5592	-0.7285	-1.4212	0.6679	1.4961
7.7002	-0.9360	-1.4713	0.8934	1.6491
7.8563	-1.1640	-1.4811	1.1570	1.7621
8.0231	-1.4090	-1.4527	1.4569	1.8261
8.2063	-1.6698	-1.3874	1.7932	1.8336
8.4101	-1.9425	-1.2851	2.1620	1.7763
8.6380	-2.2199	-1.1451	2.5533	1.6457
8.9035	-2.4997	-0.9590	2.9624	1.4256
9.2138	-2.7607	-0.7210	3.3560	1.1015
9.5745	-2.9680	-0.4273	3.6761	0.6655
9.9570	-3.0704	-0.1056	3.8358	0.1659
10.3483	-3.0467	0.2261	3.7987	-0.3546
10.7427	-2.8924	0.5547	3.5583	-0.8581
11.1326	-2.6151	0.8640	3.1351	-1.3006
11,5121	-2.2343	1.1368	2.5740	-1.6368
11.8752	-1.7810	1.3497	1.9416	-1.8184
12.0000	-1.6090	1.4055	1.7133	-1.8373

MATLAB Application

MATLAB has a file **called ode45.m** which implement the fourth- and fifth-order Runge-Kutta integration. Here, we demonstrate how the sample problem used in the **FORTRAN** and **QuickBASIC** versions can also be solved by use of the m file. The forcing functions given in Equations 10 and 11 are first prepared as follows:

```
function XDOT = FunF(T,X)
        XDOT = zeros(2,1);
     XDOT(1) = -6+13.*X(1)+20.*X(2);
     XDOT(2) =  4- 9.*X(1)-14.*X(2);
```

If this file **FunF.m** is stored on a disk which has been inserted in disk drive A, **ode45.m** is to be applied with appropriate initial conditions and a time interval of investigation as follows:

```
>> T0 = 0; Tend = 3; X0 = [7;–4]; [T,X] = ode45('A:Funf',T0,Tend,X0); plot(T,X)
```

Notice that **ode45** has four arguments. The first argument is the m file in which the forcing functions are defined. The second and third argument the initial and final values of time, respectively. The fourth argument is a vector containing the initial values of the dependent variables. The resulting display, after rearranging T and X side-by-side for saving space instead of one after the other, is:

```
T =                    X =
0                       7.0000    -4.0000
0.0300                  7.1523    -4.0914
0.2350                  8.3244    -4.7947
0.4424                  9.7826    -5.6695
0.6478                 11.5563    -6.7338
0.8511                 13.7117    -8.0270
1.0529                 16.3304    -9.5982
1.2534                 19.5113   -11.5068
1.4528                 23.3747   -13.8248
1.6512                 28.0667   -16.6400
1.8489                 33.7646   -20.0588
2.0460                 40.6839   -24.2103
2.2425                 49.0860   -29.2516
2.4387                 59.2887   -35.3732
2.6344                 71.6776   -42.8066
2.8299                 86.7210   -51.8326
3.0000                102.4276   -61.2566
```

The plots of X(1) and X(2) vs. T using the solid and broken lines, respectively, are shown in Figure 5.

As another example, **MATLAB** is applied to obtain the displacement and velocity histories of a vibration system, Figure 1. First, a m file **FunMCK.m** is created to describe this system as:

```
function XDOT = FunMCK(T,X)
         XDOT = zeros(2,1);
      XDOT(1) = X(2);
      XDOT(2) = .5*(1-3*X(2)-4*X(1));
```

Notice that the first and second variables X(1) and X(2) are displacement (x) and velocity (v = dx/dt), respectively, and the mass (m), damping constant (c), and spring constant (k) are taken as 2 N-sec^2/cm, 3 N-sec/cm, and 4 N/cm, respectively. For a system which is initially at rest and disturbed by a constant force of F(t) = 1

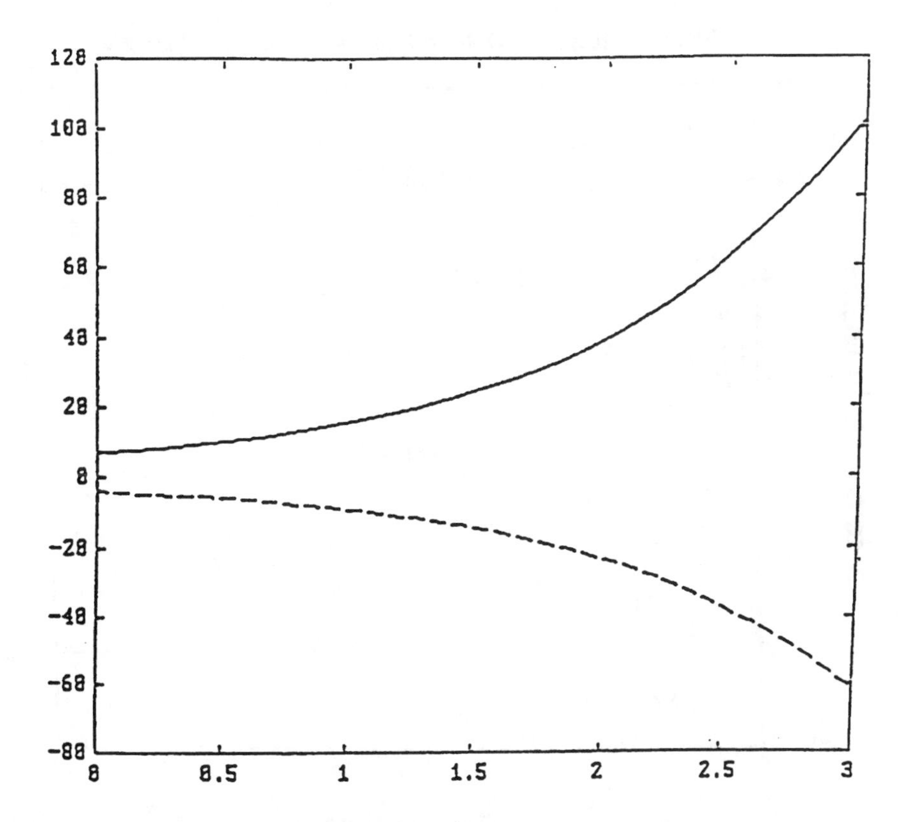

FIGURE 5. The plots of X(1) and X(2) vs. T using the solid and broken lines, respectively.

N, the **MATLAB** solution for $0 \le t \le 25$ seconds shown in Figure 6 is obtained by entering the commands:

```
>> T0 = 0; Tend = 25; X0 = [0;0]; X = [0;0]; [T,X] = ode45('a:FunMCK',T0,Tend,X0)
```

We can observe from Figure 6 that the mass has a overshoot (referring to Figure 2 in the program **NewRaphG**) of about 0.28 cm at approximately t = 2.5 seconds and finally settles to a static deflection of 0.25 cm, and that the maximum ascending velocity is about 0.18 cm/sec.

As another example of dynamic analysis in the field of fluid mechanics, Figure 8 shows the flow of a fluid between two connected tanks. The valve settings control the amount of flows, q_i for i = 1,2,3. The levels of the tanks h_1 and h_2 change in time depending on these settings and also on the supply rates Q_1 and Q_2 and the discharge rate q_3. Expressing the valve settings in terms of the resistances R_i for i = 1,2,3, the conservation of masses requires that the flow rates be computed with the formulas:

$$A_1 \frac{dh_1}{dt} = Q_1 - \frac{1}{R_1}(h_1 - h_3) \quad \text{and} \quad A_2 \frac{dh_2}{dt} = Q_2 - \frac{1}{R_2}(h_2 - h_3) \qquad (33,34)$$

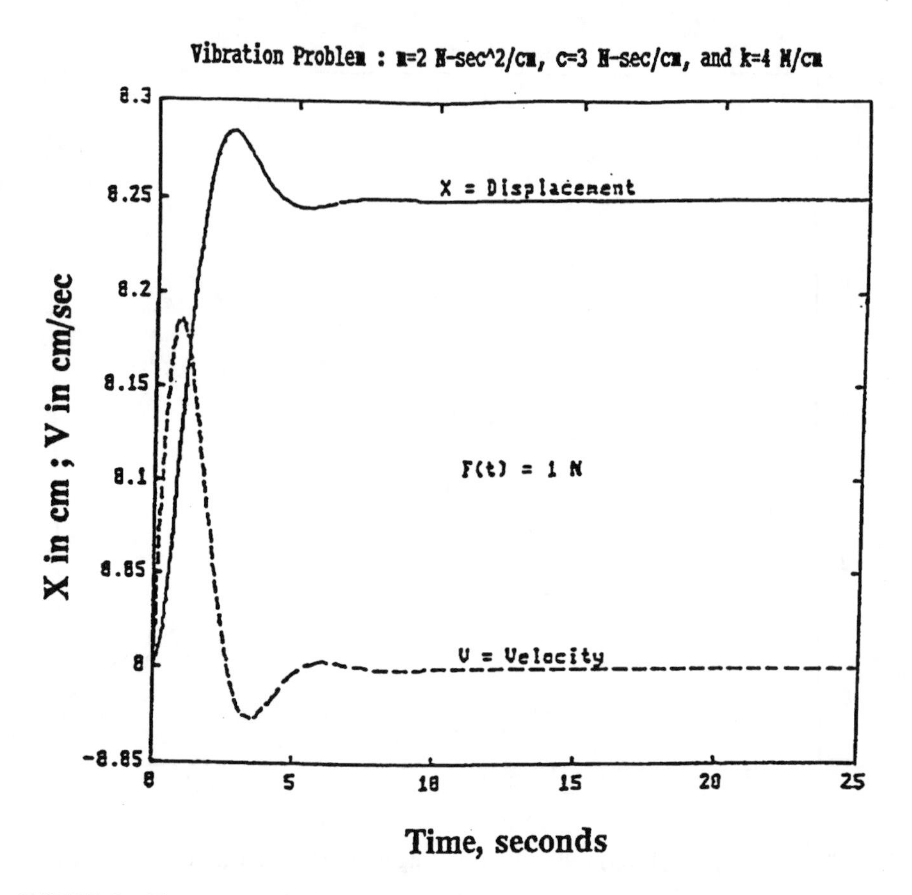

FIGURE 6. For a system initially at rest and disturbed by a constant force of F(t) = 1 N, the **MATLAB** solution for 0≤t≤25 seconds shown here is obtained by entering the commands shown below.

where A's are the cross-sectional areas of the tanks, and h_3 is the pressure head at the junction indicated in Figure 7 and is related to the discharge rate q_3 by the equation:

$$q_3 = q_1 + q_2 = h_3/R_3 = (h_1 - h_3)/R_1 + (h_2 - h_3)/R_2 \tag{35}$$

Or, h_3 can be written in terms of R's and h_1 and h_2 as:

$$h_3 = R_3(R_2h_1 + R_1h_2)/(R_1R_2 + R_1R_3 + R_2R_3) \tag{36}$$

By eliminating h_3 terms from Equations 33 and 34, we obtain two differential equations in h_1 and h_2 to be:

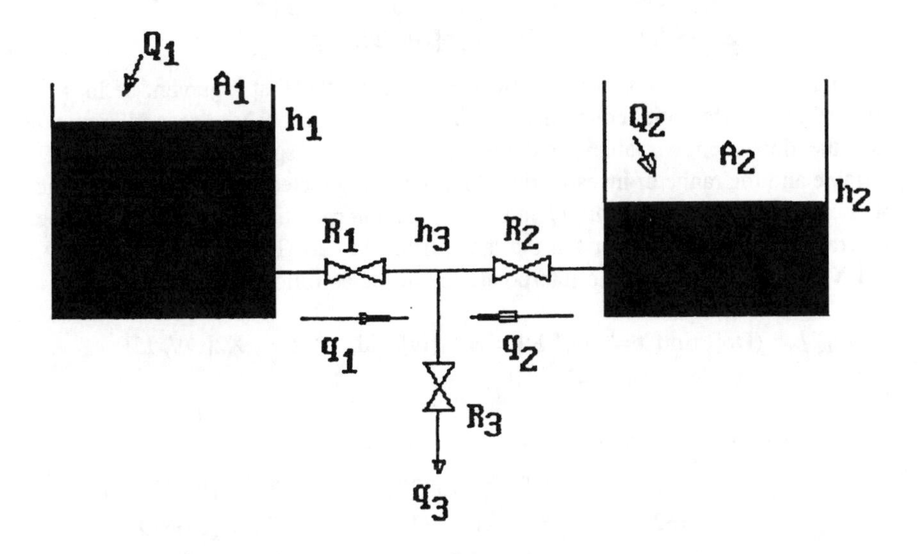

FIGURE 7. The flow of a fluid between two connected tanks.

$$A_1\frac{dh_1}{dt} = Q_1 - a_1h_1 + a_3h_2 \quad \text{and} \quad A_2\frac{dh_2}{dt} = Q_2 - a_2h_2 + a_3h_1 \qquad (37,38)$$

where:

$$a_1 = (R_2 + R_3)/\Delta \quad \text{and} \quad a_2 = (R_1 + R_3)/\Delta \qquad (39,40)$$

and

$$\Delta = R_1R_2 + R_2R_3 + R_3R_1 \qquad (41)$$

By assigning values for the parameters involved in the above problem, Runge-Kutta method can again be applied effectively for computing the fluid levels in both tanks.[4]

Mathematica Applications

Mathematica solves a set of ordinary differential equations based on the Runge-Kutta method by a function called **NDSolve**. The following run illustrates its interactive application using the same example in the **MATLAB** presentation of Figure 7:

```
In[1]: = Id = (NDSolve[{X1'[t] = = X2[t], X2'[t] = = .5*(1–3*X2[t]–4*X1[t]),
                 X1[0] = = 0, X2[0] = = 0},
                 {X1,X2}, {t,0,25}])
```

```
Out[1] = {{X1 -> InterpolatingFunction[{0., 25.}, <>],
          X2 -> InterpolatingFunction[{0., 25.}, <>]}}
```

In[1] shows that NDSolve has three arguments: the first argument defines the two ordinary differential equations and the initial conditions, the second argument lists the dependent variables, and the third argument specifies the independent variable and the range of investigation. Id is a name selected for easy later reference of the results generated. *Out[1]* indicates that interpolation functions have been generated for X1 and X2 for t in the range from 0 to 25. To print the values of X1 and X2, Id can be referred to interpolate the needs as follows:

```
In[2]: = (Do[Print["t =", tv," X1 =", X1[tv]/. Id, " X2 =", X2[tv]/. Id],
                {tv, 0, 25, 1}])
```

```
Out[2] = t = 0     X1 = {0.}           X2 = {0.}
         t = 1     X1 = {0.13827}      X2 = {0.183528}
         t = 2     X1 = {0.267432}     X2 = {0.0629972}
         t = 3     X1 = {0.280915}     X2 = {–0.0193286}
         t = 4     X1 = {0.256722}     X2 = {–0.0206924}
         t = 5     X1 = {0.245409}     X2 = {–0.00278874}
         t = 6     X1 = {0.246925}     X2 = {0.00365961}
         t = 7     X1 = {0.249969}     X2 = {0.00187841}
         t = 8     X1 = {0.250675}     X2 = {–0.000172054}
         t = 9     X1 = {0.250238}     X2 = {–0.000477574}
         t = 10    X1 = {0.249931}     X2 = {–0.000125015}
         t = 11    X1 = {0.249923}     X2 = {0.0000639464}
         t = 12    X1 = {0.24999}      X2 = {0.0000489215}
         t = 13    X1 = {0.250014}     X2 = {4.83211 10–7}
         t = 14    X1 = {0.250006}     X2 = {–0.0000104109}
         t = 15    X1 = {0.249999}     X2 = {–3.5685 10–6}
         t = 16    X1 = {0.249998}     X2 = {1.38473 10–6}
         t = 17    X1 = {0.25}         X2 = {1.35708 10–6}
         t = 18    X1 = {0.25}         X2 = {1.39173 10–7}
         t = 19    X1 = {0.25}         X2 = {–4.40766 10–7}
         t = 20    X1 = {0.25}         X2 = {–4.61875 10–7}
         t = 21    X1 = {0.25}         X2 = {–1.93084 10–8}
         t = 22    X1 = {0.25}         X2 = {2.47763 10–8}
         t = 23    X1 = {0.25}         X2 = {9.81364 10–8}
         t = 24    X1 = {0.25}         X2 = {6.7364 10–8}
         t = 25    X1 = {0.25}         X2 = {3.57288 10–8}
```

These results are in agreement with the plotted curves shown in the **MATLAB** application. *In[2]* shows that the replacement operator/. is employed in X1[tv]/.Id which requires all t appearing in the resulting interpolating function for X1(t) created in the Id statement to be substituted by the value of tv. The looping **DO** statement instructs that the tv values be changed from a minimum of 0 and a maximum of 25 using an increment of 1. To have a closer look of the overshoot region, In[2] can be modified to yield

In[3]: = (Do[Print["t = ",tv," X1 = ",X1[tv]/. Id," X2 = ",X2[tv]/. Id],
{tv, 2, 3, 0.1}])

Out[3] =	t = 2	X1 = {0.267432}	X2 = {0.0629972}
	t = 2.1	X1 = {0.273097}	X2 = {0.0504262}
	t = 2.2	X1 = {0.277544}	X2 = {0.0386711}
	t = 2.3	X1 = {0.280862}	X2 = {0.0278364}
	t = 2.4	X1 = {0.283145}	X2 = {0.0179948}
	t = 2.5	X1 = {0.284496}	X2 = {0.00919032}
	t = 2.6	X1 = {0.285019}	X2 = {0.00144176}
	t = 2.7	X1 = {0.284819}	X2 = {–0.00525428}
	t = 2.8	X1 = {0.284002}	X2 = {–0.0109202}
	t = 2.9	X1 = {0.282669}	X2 = {–0.0155945}
	t = 3.	X1 = {0.280915}	X2 = {–0.0193286}

This detailed study using a time increment of 0.1 reaffirms that the overshoot occurs at t = 2.6 and X1 has a maximum value equal to 0.28502.

6.3 PROGRAM ODEBVPRK — APPLICATION OF RUNGE-KUTTA METHOD FOR SOLVING BOUNDARY-VALUE PROBLEMS

The program **OdeBvpRK** is designed for numerically solving the linear boundary-value problems governed by the ordinary differential equation by superposition of two solutions obtained by application of the Runge-Kutta fourth-order method. To explain the procedure involved, consider the problem of a loaded beam shown in Figure 8. Mathematically, the deflection y(x) satisfies the well-known flexural equation.[5]

$$\frac{d^2y}{dx^2} = \frac{M}{EI} \tag{1}$$

where M is the internal moment distribution, E is the Young's modulus and I is the moment of inertia of the cross section of the beam. For the general case, M, E and I can be function of x. The *boundary conditions* of this problem are:

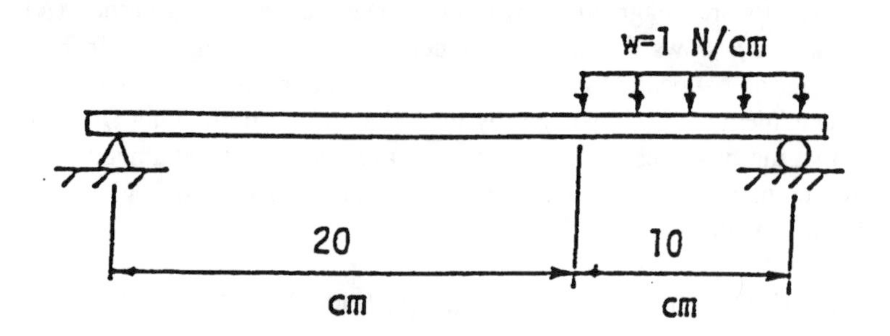

FIGURE 8. The problem of a loaded beam.

$$y(x=0)=0 \quad \text{and} \quad y(x=L)=0 \tag{2}$$

where L is the length of the beam. The problem is to determine the resulting deflection y(x). Knowing y, the moment and shearing force, V, distributions can subsequently be determined based on Equation 1 and V = dM/dx. The final objective is to calculate the stress distributions in the loaded beam using the M and V results.

The Runge-Kutta method for solving an initial problem can be applied here if in additional to the initial condition given in (2), y(x = 0) = 0, we also know the slope, Θ, at x = 0. But, we can always make a guess and hope that by making better and better guesses the trial process will eventually lead to one which satisfies the other boundary condition, namely y(x = L) = 0 given in Equation 3. In fact, if the problem is linear, all we need to do is making two guesses and linearly combine these two trial solutions to obtain the solution y(x).

Let us first convert the governing differential Equation 1 into two first-order equations as:

$$\left[\begin{array}{ll} \dfrac{dx_1}{dx}=x_2, & x_1(x=0)=y(x=0)=0 \\ \dfrac{dx_2}{dx}=\dfrac{M}{EI}, & x_2(x=0)=\theta(x=0)=\theta_0 \end{array}\right. \tag{3,4}$$

To apply the fourth-order Runge-Kutta method, we have to first decide on a stepsize h, for example h = L/N we then plan to calculate the deflections at N + 1 locations, x + jh for j = 1,2,…N since j = 0 is the initial location. If we assume a value for Θ_0, say A, the Runge-Kutta process will be able to generate the following table

x	0	h	2h	...	jh	...	Nh
$x_1=y^{(1)}$	0	$y_1^{(1)}$	$y_2^{(1)}$	...	$y_j^{(1)}$	...	$y_N^{(1)}$
$x_2=\theta^{(1)}$	A	$\theta_1^{(1)}$	$\theta_2^{(1)}$	...	$\theta_j^{(1)}$	...	$\theta_N^{(1)}$

(5)

If $y_N^{(1)} = 0$, then the value A selected for $\Theta(x = 0)$ is correct and the y and Θ values listed in Equation 5 are the results for the selected stepsize. If $y_N^{(1)}$ is not equal to zero, then the value incorrectly selected, we have to make a second try by letting $\Theta(x = 0) = B$ to obtain a second table by application of the Runge-Kutta method. Let the second table be denoted as:

x	0	h	2h	...	jh	...	Nh
$x_1=y^{(2)}$	0	$y_1^{(2)}$	$y_2^{(2)}$	...	$y_j^{(2)}$	...	$y_N^{(2)}$
$x_2=\theta^{(2)}$	B	$\theta_1^{(2)}$	$\theta_2^{(2)}$	...	$\theta_j^{(2)}$	...	$\theta_N^{(2)}$

(6)

Again, if $y_N^{(2)}= 0$, then the value B selected for $\theta(x = 0)$ is correct and the y and θ values listed in Equation 6 are the results for the selected stepsize. Otherwise, if the problem is linear, the solutions can be obtained by linearly combining the two trial results as, for j = 1,2,...,N:

$$Y_j = \alpha y_j^{(1)} + \beta y_j^{(2)} \quad \text{and} \quad \theta_j = \alpha\theta_j^{(1)} + \beta\theta_j^{(2)} \tag{7,8}$$

where the weighting coefficients and are to be determined by solving the equations:

$$\alpha + \beta = 1 \quad \text{and} \quad \alpha y_N^{(1)} + \beta y_N^{(2)} = 0 \tag{9,10}$$

Equation 10 is derived from the boundary condition $y(x = Nh = L) = 0$ and based on Equation 7. Equation 9 needs more explanation because it cannot be derived if $y(x = 0) = 0$. Let us assume that for the general case, $y(x = 0) = \delta$. Then Equation 7 gives $\alpha\delta + \beta\delta = \delta$ which leads to Equation 9. Using Cramer's rule, we can easily obtain:

$$\alpha = y_N^{(2)}/D \quad \text{and} \quad \beta = -y_N^{(1)}/D \tag{11,12}$$

and

$$D = y_N^{(2} - y_N^{(1)} \tag{13}$$

Numerical Examples

Let us consider the problem of a loaded beam as shown in Figure 8. The crosssection of the beam has a width of 1 cm and a height of 2 cm which results in a moment of inertia $I = 2/3\ \text{cm}^4$. The reactions at the left and right supports can be computed to be 5/3 N and 25/3 N, respectively. Based on these data, it can be shown that the equations for the internal bending moments are:

$$M(x) = \begin{cases} 5x/3, & \text{for } 0 \le x \le 20 \text{ cm} \\ -200 + \dfrac{65}{3}x - \dfrac{1}{2}x^2, & \text{for } 20 < x \le 30 \text{ cm} \end{cases} \tag{14}$$

If the beam has a Young's modulus of elasticity $E = 2x10^7$ N/cm^2, we may decide on a stepsize of h = 1 cm and proceed to prepare a computer program using the fourth-order Runge-Kutta method to generate two trial solutions and then linearly combining to arrive at the desired distributions of the deflected shape y(x) and slope (x). The **FORTRAN** version of this program called **OdeBvpRK** to be presented later has produced the following display on screen:

```
              Trial  1

     X             Y            THETA

   0.00      0.0000E+00    100.0000E-03

   0.30    300.0000E-04    100.0000E-03

   0.60    600.0000E-04    100.0000E-03

   0.90    900.0000E-04    100.0001E-03

        . . . . . . . . . . . . . .

        . . . . . . . . . . . . . .

  29.70    297.0517E-02    100.0437E-03

  30.00    300.0530E-02    100.0437E-03


              Trial  2

     X             Y            THETA

   0.00      0.0000E+00    200.0000E-03

   0.30    600.0000E-04    200.0000E-03

   0.60    120.0000E-04    200.0000E-03

   0.90    180.0000E-04    200.0000E-03

        . . . . . . . . . . . . . .

        . . . . . . . . . . . . . .

  29.70    594.0519E-02    200.0437E-03

  30.00    600.0532E-02    200.0437E-03


          The solutions are:

     X             Y            Theta

   0.00       0.000E+00   -175.9827E-07

   0.30   -527.8736E-08   -175.8337E-01
```

```
 0.30  -527.8736E-08  -175.8337E-01
 0.60  -105.4257E-07  -175.6847E-07
 1.20  -210.5534E-07  -175.0886E-07
       . . . . . . . . . . . . . .
       . . . . . . . . . . . . . .
15.90  -196.4569E-06  -177.3238E-08
16.20  -197.1722E-06  -116.2291E-08
16.50  -197.4106E-06  -551.3430E-09
16.80  -197.1722E-06   745.0581E-10
17.10  -197.4106E-06   700.3546E-09
       . . . . . . . . . . . . . .
       . . . . . . . . . . . . . .
28.80  -305.1758E-07   256.8960E-07
29.10  -224.1135E-07   258.8332E-07
29.40  -143.0511E-07   260.1743E-07
29.70  -715.2557E-08   260.9193E-07
30.00   476.8372E-09   261.2174E-07
```

FORTRAN Version

```
C     PROGRAM OdeBvpRK - Ordinary differential equation, boundary-value problem using two
C                        Runge-Kutta trial solutions . Up to 1000 steps.
      DIMENSION XT(2,1001,2),XS(2,1001),P(2,4),TEMP(2)
      REAL INERTIA,L
      COMMON E,INERTIA
C     2 unknowns and 100 steps
      DATA DX,L,NV,NS,NSTATION/0.3,30.,2,100,101/
      E=2.E7
      INERTIA=.66667
      DO 10 NTRY=1,2
      WRITE (*,5) NTRY
    5 FORMAT(1X,'Trial #',I2/9X,'X',14X,'Y',10X,'THETA')
      XT(1,1,NTRY)=0.
      XT(2,1,NTRY)=0.1*NTRY
      XV=0
      WRITE (*,8) XV, (XT(IP,1,NTRY),IP=1,NV)
    8 FORMAT(F10.2,2E15.4)
      DO 10 KS=1,NS
      CALL RKN(XT(1,KS,NTRY),NV,DX,XV,P,TEMP,XT(1,KS+1,NTRY))
      XV=XV+DX
   10 WRITE (*,8) XV, (XT(IP,KS+1,NTRY),IP=1,NV)
      D=XT(1,NS+1,2)-XT(1,NS+1,1)
      ALPHA=XT(1,NS+1,2)/D
      BETA=-XT(1,NS+1,1)/D
      WRITE (*,12)
   12 FORMAT(1X,'The solutions are:'/9X,'X',14X,'Y',10X,
     *       'Theta'/)
      XV=-DX
      DO 15 I=1,NSTATION
      XV=XV+DX
```

```
      XS(1,I)=ALPHA*XT(1,I,1)+BETA*XT(1,I,2)
      XS(2,I)=ALPHA*XT(2,I,1)+BETA*XT(2,I,2)
   15 WRITE (*,8) XV,(XS(IP,I),IP=1,2)
      STOP
      END


      FUNCTION F(X,T,I,N)
      COMMON E,INERTIA
      REAL INERTIA,M
      DIMENSION X(N)
      GO TO (10,20),I
   10 F=X(2)
      RETURN
   20 IF (T.GT.20.) GO TO 25
      M=5*T/3
   22 F=M/E/INERTIA
      RETURN
   25 M=-200+65*T/3-T**2/2
      GO TO 22
      END
      SUBROUTINE RKN(XIN,N,DT,T,P,XT,XOUT)
C     SOLVING N FIRST-ORDER O.D.E. D(XIN)/DT=F(XIN,T) KNOWING XIN AT T, THIS SUB. FINDS XOUT
C      AT T+DT
C     P ARE THE RUNGE-KUTTA 4TH-ORDER PARAMETERS.
      DIMENSION P(N,4),XIN(N),XOUT(N),XT(N)
      DO 5 I=1,N
    5 P(I,1)=DT*F(XIN,T,I,N)
      DO 15 J=2,3
      DO 10 I=1,N
   10 XT(I)=XIN(I)+P(I,J-1)/2.
      DO 15 I=1,N
   15 P(I,J)=DT*F(XT,T+DT/2.,I,N)
      DO 20 I=1,N
   20 XT(I)=XIN(I)+P(I,3)
      DO 25 I=1,N
      P(I,4)=DT*F(XT,T+DT,I,N)
   25 XOUT(I)=XIN(I)+(P(I,1)+2.*(P(I,2)+P(I,3))+P(I,4))/6.
      RETURN
      END
```

The Subprogram FUNCTION F which defines the initial-value problem is coded in accordance with Equation 14. The two trial initial slopes are selected as equal to 0.1 and 0.2. The trial results are kept in the three-dimensional variable XT, in which the deflection $y^{(k)}(j)$ for the kth try at station $x = x_j = jh$ is stored in XT(1,j,k) whereas the slope there is stored in XT(2,j,k) for $j = 1,2,\ldots,30$ and $k = 1,2$. Such a three-subscripts arrangement facilitates the calling of the subroutine **RKN** because XT(1,KS,NTRY) is transmitted as XIN(1) and automatically the next value XT(2,KS, NTRY) as XIN(2), and the computed results XOUT(1) and XOUT(2) are to be stored as XT(1,KS + 1,NTRY) and XT(2,KS + 1,NTRY), respectively. Notice that there are only two dependent variables, NV = 2.

After the weighting coefficients α (ALPHA in the program) and β (BETA) have been calculated, the final distributions of the deflection and slope are saved in first and second rows of the two-dimensional variable X, respectively. It should be emphatically noted that the solutions obtained is only good for the selected stepsize h = 1 cm. Whether it is accurate or not remains to be tested by using finer stepsizes and by repeated application of the Runge-Kutta methods.

It can be shown that the maximum deflection of the loaded beam is equal to –2.019 cm and is obtained when the stepsize is continuously halved and two consecutively calculated values is different less than 0.0001 cm in magnitude. The needed modification of the above listed program to include this change in the stepsize and testing of the difference in the maximum deflection is left as homework for the reader.

QuickBASIC Version

```
'       PROGRAM OdeBvpRK - Ordinary differential equation, boundary-value problem using two
'            Runge-Kutta trial solutions.  Up to 1000 steps.
        DECLARE SUB RKN(XIN(),N,DT,T,P(),TEMP(),XOUT())
        DECLARE FUNCTION F(X(),T,I,N)
        DIM XT(2,1001,2),XS(2,1001),P(2,4),TEMP(2): COMMON SHARED E, INERTIA
'       2 unknowns and 100 steps
        READ E,INERTIA,L,NV,NS: DATA 2.E7,,66667.,30.,2,100
             DX = L / NS: DIM XIN(NV), XOUT(NV)
        FOR NTRY = 1 TO 2
            PRINT USING " Trial ## "; NTRY
            PRINT SPC(9); "X"; SPC(14); "Y"; SPC(10); "THETA"
            XT(1,1,NTRY) = 0: XT(2,1,NTRY) = .1 * NTRY: XV = 0
            PRINT USING "#######.##"; XV;
            FOR IP=1 TO NV: PRINT USING "  ####.####^^^^";XT(IP,1,NTRY);: NEXT IP: PRINT
            FOR KS = 1 TO NS
                FOR I=1 TO NV: XIN(I)=XT(I,KS,NTRY)    : NEXT I
                CALL RKN(XIN(),NV,DX,XV,P(),TEMP(),XOUT())
                FOR I=1 TO NV: XT(I,KS+1,NTRY)=XOUT(I): NEXT I
                XV = XV + DX: PRINT USING "#######.##"; XV;
               FOR IP=1 TO NV: PRINT USING "  ####.####^^^^";XT(IP,KS+1,NTRY);: NEXT IP: PRINT
                NEXT KS
            NEXT NTRY
            D=XT(1,NS+1,2)-XT(1,NS+1,1): ALPHA=XT(1,NS+1,2)/D: BETA=-XT(1,NS+1,1)/D
        PRINT " The solutions are:"
        PRINT SPC(9);"X";SPC(14);"Y";SPC(10);"Theta": PRINT: XV=-DX: NSTATION=NS+1
        FOR I = 1 TO NSTATION: XV=XV+DX
            XS(1,I)=ALPHA*XT(1,I,1)+BETA*XT(1,I,2): XS(2,I)=ALPHA*XT(2,I,1)+BETA*XT(2,I,2)
            PRINT USING "#######.##";XV;
            FOR IP=1 TO 2: PRINT USING "  ####.####^^^^";XS(IP,I);: NEXT IP
            PRINT: NEXT I: END

        FUNCTION F(X(),T,I,N)
        ON I GOTO 10,20
10 F=X(2): GOTO 50
20 IF T>20 THEN 25
   M=5*T/3
22 F=M/E/INERTIA: GOTO 50
25 M=-200+65*T/3-T^2/2: GOTO 22
50 END FUNCTION

        SUB RKN (XIN(), N, DT, T, P(), XT(), XOUT())
'       SOLVING N FIRST-ORDER O.D.E. D(XIN)/DT=F(XIN,T) KNOWING XIN AT T, THIS SUB. FINDS XOUT
AT T+DT
'       P ARE THE RUNGE-KUTTA 4TH-ORDER PARAMETERS.
        FOR I = 1 TO N: P(I,1) = DT * F(XIN(),T,I,N): NEXT I
        FOR J = 2 TO 3
            FOR I=1 TO N: XT(I)=XIN(I)+P(I,J-1)/2!      : NEXT I
            FOR I=1 TO N: P(I,J)=DT*F(XT(),T+DT/2!,I,N): NEXT I: NEXT J
        FOR I=1 TO N:  XT(I)=XIN(I)+P(I, 3): NEXT I
        FOR I=1 TO N: P(I,4)=DT*F(XT(),T+DT,I,N)
            XOUT(I)=XIN(I)+(P(I,1)+2!*(P(I,2)+P(I,3))+P(I,4))/6!: NEXT I
    END SUB
```

MATLAB Applications

In the program RungeKut, **MATLAB** is used for solving initial value problems by application of its m function **ode45** based on the fourth- and fifth-order Runge-Kutta method. Here, this function can be employed twice to solve a boundary-value problem governed by linear ordinary differential equations. To demonstrate the procedure, the sample problem discussed in **FORTRAN** and **QuickBASIC** versions of the program **OdeBvpRK**, with the aid of function **BVPF.m** listed in the subdirectory <mFiles>, can be solved by interactive **MATLAB** operations as follows:

```
>> format compact
>> % ** First trial solution **
>> X0=0; Xend=30; YT10=[0;0.1]; [X,YT1]=ode45('a:BVPF',X0.Xend,YT10)
   X =
           0
      0.3000
      6.3000
     12.3000
     18.3000
     24.3000
     30.0000
   YT1 =
           0    0.1000
      0.0300    0.1000
      0.6300    0.1000
      1.2300    0.1000
      1.8301    0.1000
      2.4302    0.1000
      3.0005    0.1000
>> % ** Second trial solution **
>>                YT20=[0;0.2]; [X,YT2]=ode45('a:BVPF',X0.Xend,YT20)
   X =
           0
      0.3000
      6.3000
     12.3000
     18.3000
     24.3000
     30.0000
   YT2 =
           0    0.2000
      0.0600    0.2000
      1.2600    0.2000
      2.4600    0.2000
      3.6601    0.2000
      4.8603    0.2000
      6.0005    0.2000
>> % ** Linearly combining the two solutions **
>> D=YT2(7,1)-YT1(1,7); Alpha=YT2(1,7)/D
   Alpha =
      2.0002
>> Beta=-YT1(1,7)/D
   Beta =
     -1.0002
   >> % ** Find the solution for deflection (first column) and slope (second column) **
>> Yanswer=Alpha*YT1+Beta*YT2
   Yanswer =
     1.0e-003 *
           0   -0.0177
     -0.0053   -0.0177
     -0.1064   -0.0152
     -0.1791   -0.0083
     -0.1965    0.0032
     -0.1325    0.0182
           0    0.0261
>> plot(X,Yanswer)
```

Notice that **format compact** enables the display to use fewer spacings; YT1 and YT2 keep the two trial solutions, and **ode45** automatically determines the best stepsize which if used directly will result in a coarse plot as shown in Figure 9. The plot showing solid-line curve for the deflection and broken-line curve for the slope can, however, be refined by linear interpolation of the data (X,Yanswer) and expanding X and the two columns of Yanswer into new data arrays of XSpline, YSpline, and YPSpline, respectively. Toward that end, the m function spline in **MATLAB** is to be applied as follows:

```
>> XSpline=0:0.1:30; YSpline=spline(X,Yanswer,XSpline);
>>                   YPSpline=spline(X,Yanswer(:,2),XSpline);
>> plot(XSpline,YSpline,XSpline,YPSpline)
```

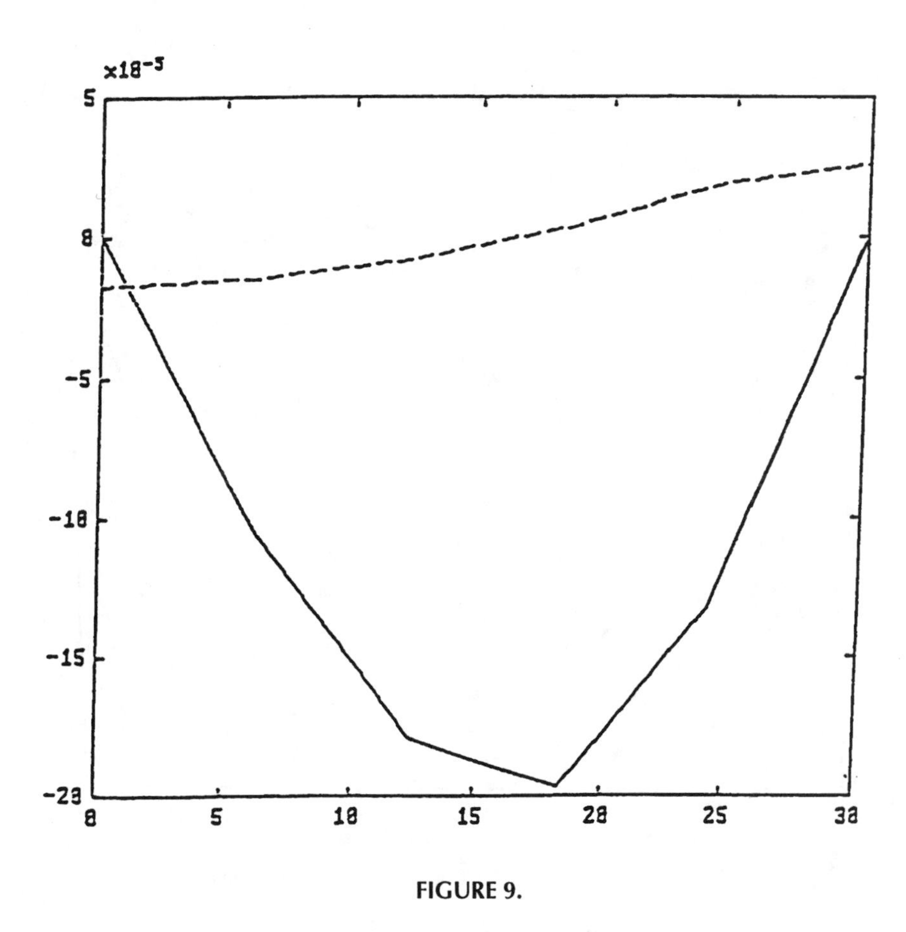

FIGURE 9.

The result of plotting the spline curves is shown in Figure 10.

Mathematica Applications

The Runge-Kutta method, particularly the most popular fourth-order method, can be applied for solution of boundary-value problem governed by ordinary differential equation(s). Here, only the application of this method is elaborated; readers are therefore referred to program **RungeKut** to review the method itself and the development of related programs and subprograms. The boundary-value problem is to be solved by continuously guessing the initial condition(s) which are not provided until all boundary conditions are satisfied if the problem is *nonlinear*. When the problem is linear, then only a finite number of guesses are necessary. A system of two first-order ordinary differential equations which governed the loaded elastic beam problem previously solved in the **MATLAB** application is here adapted to demonstrate the application of the Runge-Kutta method.

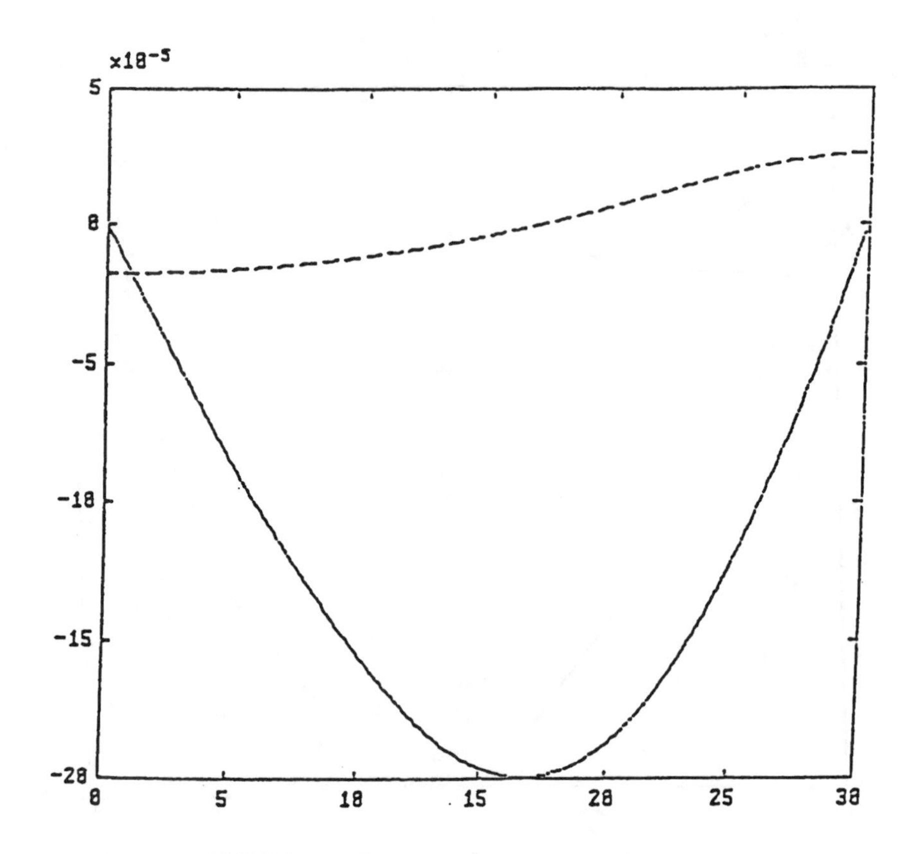

FIGURE 10. The result of plotting the spline curves.

In[1]: = EI = 4.*10^7/3; F[X_] = If[X>20,
(–200 + 65*X/3X^2/2)/EI, 5*X/3/EI]

In[2]: = Id1 = (NDSolve[{Y′[X] = = YP[X], YP'[X] = = F[X], Y[0] = = 0,
YP[0] = = 0.1}, {Y,YP}, {X, 0, 30}])

In[3]: = Y30Trial1 = Y[30]/. Id1

Out[3] = {3.00052}

EI value and F(X) are defined in *In[1]*. *In[2]* specifies the two first-order ordinary differential equations involving the deflection Y and slope YP, describes the correct initial condition Y(X = 0) = 0, gives a guessed slope Y′(X = 0) = 0.1, and decides on the limit of investigation from X = 0 to X = 30. In[3] interpolates the ending Y value by using the data obtained in Id1. A second trial is then to follow as:

In[4]: = Id2 = (NDSolve[{Y′[X] == YP[X], YP'[X] = = F[X],
Y[0] == 0, YP[0] == 0.2},
{Y,YP}, {X, 0, 30}])

In[5]: = Y30Trial2 = Y[30]/. Id2

Out[5] = {6.00052}

Linear combination of the two trial solutions is now possible by calculating a correct Y′(X = 0) which should be equal to the value given in Out[8].

In[6]: = d = Y30Trial2Y30Trial1; a = Y30Trial2/d; b = -Y30Trial1/d;

In[7]: = Print["Alpha = ",a," Beta = ",b]

Out[7] = Alpha = {2.00017} Beta = {–1.00017}

In[8]: = YP0 = 0.1*a + 0.2*b

Out[8] = {–0.0000174888}

Finally, the actual deflection and slope can be obtained by providing the correct set of initial conditions and again applying the Runge-Kutta method.

```
In[9]: = Id = (NDSolve[{Y'[X] = = YP[X], YP'[X] = = F[X],
                Y[0] = = 0, YP[0] = = –0.0000174888},
               {Y,YP}, {X,0,30}])
```

```
In[10]: = (Do[Print["X = ", Xv, " Y = ", Y[Xv]/.Id, " DY/DX = ",
               YP[Xv]/.Id], {Xv,0,30}])
```

Out[10] =

X = 0	Y = {0.}	DY/DX = {–0.0000174888}
X = 1	Y = {–0.0000174645}	DY/DX = {–0.0000174262}
X = 2	Y = {–0.0000348041}	DY/DX = {–0.0000172387}
X = 3	Y = {–0.0000518936}	DY/DX = {–0.0000169262}
X = 4	Y = {–0.0000686077}	DY/DX = {–0.0000164887}
X = 5	Y = {–0.0000848222}	DY/DX = {–0.0000159262}
X = 6	Y = {–0.000100412}	DY/DX = {–0.0000152387}
X = 7	Y = {–0.000115251}	DY/DX = {–0.0000144262}
X = 8	Y = {–0.000129215}	DY/DX = {–0.0000134887}
X = 9	Y = {–0.00014218}	DY/DX = {–0.0000124262}
X = 10	Y = {–0.000154019}	DY/DX = {–0.0000112387}
X = 11	Y = {–0.000164608}	DY/DX = {–0.00000992619}
X = 12	Y = {–0.000173788}	DY/DX = {–0.00000848869}
X = 13	Y = {–0.000181454}	DY/DX = {–0.00000692619}
X = 14	Y = {–0.000187493}	DY/DX = {–0.00000523869}

X = 15 Y = {–0.000191786} DY/DX = {–0.00000342619}
X = 16 Y = {–0.00019422} DY/DX = {–0.00000148869}
X = 17 Y = {–0.00019468} DY/DX = {0.000000573812}
X = 18 Y = {–0.000193037} DY/DX = {0.00000276006}
X = 19 Y = {–0.000189163} DY/DX = {0.00000506839}
X = 20 Y = {–0.000182912} DY/DX = {0.0000074989}
X = 21 Y = {–0.000174247} DY/DX = {0.0000100026}
X = 22 Y = {–0.00016319} DY/DX = {0.0000125002}
X = 23 Y = {–0.00014959} DY/DX = {0.0000149974}
X = 24 Y = {–0.000133573} DY/DX = {0.0000174176}
X = 25 Y = {–0.00011547} DY/DX = {0.0000196352}
X = 26 Y = {–0.0000951574} DY/DX = {0.0000216325}
X = 27 Y = {–0.0000725124} DY/DX = {0.0000233628}
X = 28 Y = {–0.0000483635} DY/DX = {0.0000246913}
X = 29 Y = {–0.0000233292} DY/DX = {0.0000255217}
X = 30 Y = {0.0000023099} DY/DX = {0.0000258107}

6.4 PROGRAM ODEBVPFD — APPLICATION OF FINITE DIFFERENCE METHOD FOR SOLVING BOUNDARY-VALUE PROBLEMS

The program **OdeBvpFD** is designed for numerically solving boundary-value problems governed by the ordinary differential equation which are to be replaced finite-difference equations. To illustrate the procedure involved, let us consider the problem of an annular membrane which is tightened by a uniform tension T and rigidly mounted along its inner and outer boundaries, $R = R_i$ and $R = R_o$, respectively. As shown in Figure 11, it is then inflated by application of a uniform pressure p. The deformation of the membrane, Z(R), when its magnitude is small enough not to affect the tension T, can be determined by solving the ordinary differential equation[6] Z(R) satisfies Equation 1 is for $R_i<R<R_o$ and the boundary conditions.

$$\frac{d^2Z}{dR^2}+\frac{1}{R}\frac{dZ}{dR}=-\frac{p}{T} \tag{1}$$

$$Z(R_i)=0 \quad \text{and} \quad Z(R_o)=0 \tag{2}$$

If the finite-difference approximation is to be applied for solving Equation 1, we will be seeking not for the expression Z(R) but for the numerical values at a selected stations of R in the interval $R_i<R<R_o$, say N. Let these stations be designated as R_k for k = 1 to N and the lateral displacements of the membrane as $Zk \equiv Z(R_k)$. Using the first-order and second-order central differences (see the program DiffTabl),

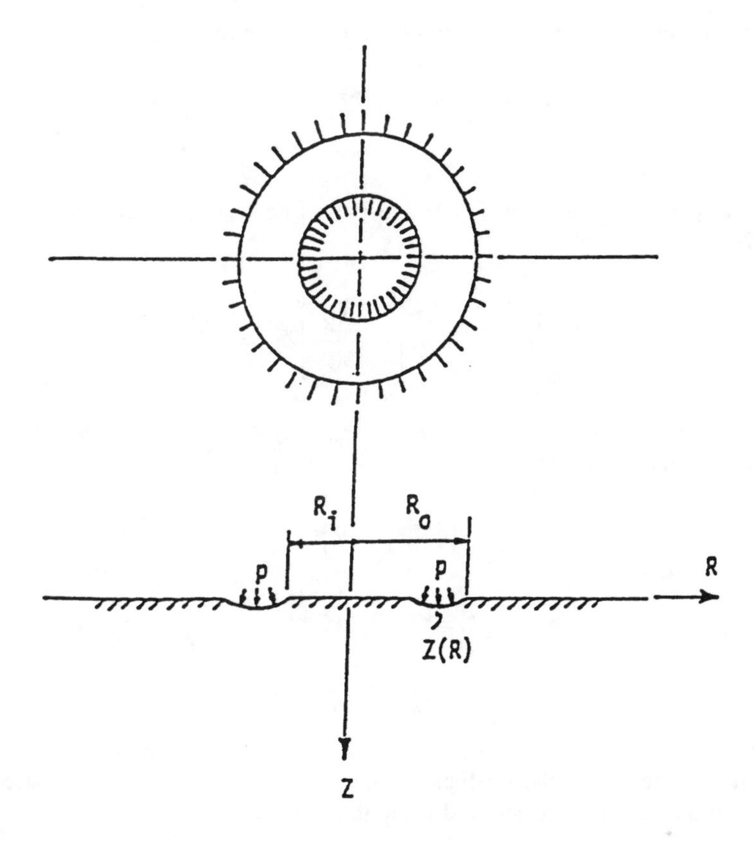

FIGURE 11. An annular membrane tightened by a uniform tension T and rigidly mounted along its inner and outer boundaries, $R = R_i$ and $R = R_o$, respectively, and then inflated by application of a uniform pressure p.

the first and second derivatives of Z(R) appearing in Equation 1 at R_k can be approximated as, respectively:

$$\left.\frac{dZ}{dR}\right|\text{at } R_k \approx \frac{Z_{k+1} - Z_{k-1}}{2\Delta R} \tag{3}$$

and

$$\left.\frac{d^2Z}{dR^2}\right|\text{at } R_k \approx \frac{Z_{k-1} - 2Z_k + Z_{k+1}}{(\Delta R)^2} \tag{4}$$

where ΔR is the increment in R and is related to the decided number of station N by the equation:

$$\Delta R = (R_o - R_i)/(N+1) \tag{5}$$

Substituting Equations 3 and 4 into Equation 1, we obtain for $R = R_k$ that:

$$\frac{Z_{k-1}-2Z_k+Z_{k+1}}{(\Delta R)^2}-\frac{1}{R_k}\frac{Z_{k+1}-Z_{k-1}}{2\Delta R}=-\frac{p}{T}$$

Multiplying both sides by $(\Delta R)^2$ and collecting terms, we can have for k = 1,2,...,N:

$$\left[1+\frac{\Delta R}{2R_k}\right]Z_{k-1}-2Z_k+\left[1-\frac{\Delta R}{2R_k}\right]Z_{k+1}=\frac{-p(\Delta R)^2}{T} \tag{6}$$

The two boundaries are $Z_o \equiv Z(R = R_i)$ and $Z_{N+1} \equiv Z(R = R_o)$, at which $Z = 0$. The above equation thus lead to the matrix equation:

$$[C]\{Z\}=\{R\} \tag{7}$$

where:

$$\{Z\}=\left[Z_1\ \ Z_2 \ldots Z_N\right]^T \tag{8}$$

$$\{R\}=-p(\Delta R)^2[1\ \ 1\ldots 1]^T/T \tag{9}$$

and if the elements for the coefficient matrix [C] are denoted as $c_{k,j}$, based on Equation 6 they are to be calculated using the formulas:

$$c_{k,k}=-2, \qquad \text{for } k=1,2,\ldots N$$

$$\begin{aligned} c_{k,k-1} &= 1-\left(\Delta R/2R_k\right), && \text{for } k=2,3,\ldots,N \\ c_{k,k+1} &= 1+\left(\Delta R/2R_k\right), && \text{for } k=1,2,\ldots,N-1 \\ c_{k,j} &= 0, && \text{elsewhere} \end{aligned} \tag{10}$$

and

$$R_k = k\Delta R \tag{11}$$

After having calculated [C] and {R}, Equation 7 can be solved by calling the subroutine **GAUSS** for {Z}. This solution is only for the selected value of N. N should be continuously increased to test if the maximum Z value would be affected.

The program **OdeBvpFD** has been developed in both **QuickBASIC** and **FORTRAN** versions based on the above described procedure for solving the boundary problems governed by ordinary differential equations. It is designed for the general case where the coefficient matrix [C] and the right-hand side vector {R} of the matrix equation derived from the finite-difference approximation, [C]{Z} = {R}, are

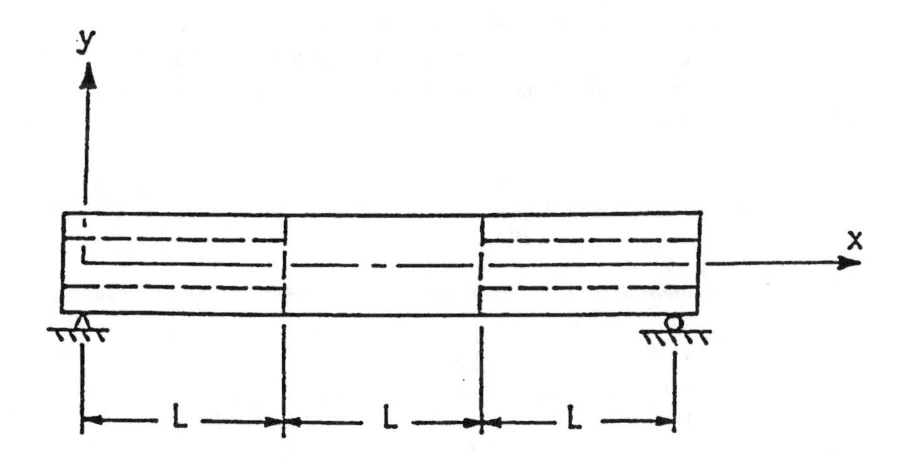

FIGURE 12. As a second example, we consider the simply supported beam of length 3L which has a solid section at its middle portion and hollow sections at both end portions as shown

to be defined by the user. In addition, the user needs to specify the number of station, N, the stepsize, ΔR, and the location of the left boundary, R_i. Here, Z and R are referred as the dependent and independent variables, respectively using the membrane problem only for the convenience of explanation; user could have other notations as dependent and independent variables.

As a second example, we consider the simply supported beam of length 3L which has a solid section at its middle portion and hollow sections at both end portions as shown in Figure 12. It is of interest to know how the beam will deflect under its own weight. This is the case of a beam subjected to uniformly distributed loads. For simplicity, let us assume that the distributed loads to be w_m and w_e, in N/cm^2, for the middle and end portions, respectively. To determine the deflection y(x) for $0<x<3L$, we need the following relevant equations which are available in a standard textbook on mechanics of materials[7]:

$$\frac{dy}{dx} = \theta, \qquad \frac{d\theta}{dx} = \frac{d^2y}{dx^2} = \frac{M}{EI}, \qquad \frac{dM}{dx} = V, \qquad \frac{dV}{dx} = w \tag{12–15}$$

where Θ, M, and V are the slope, bending moment, and shearing force, respectively. E is the Young's modulus and I is the moment of inertia of the beam. The distributed load w is considered as positive when it is applied upward in the direction of the positive y-axis and the moment is considered as positive if it causes the beam to bend concave up. The shearing forces are considered as positive if they are related to M according to Equation 14.

By successive substitutions, Equations 12–15 can yield an equation which relates the deflection y directly to the distributed load w as:

$$\frac{d^2}{dx^2}\left(EI\frac{d^2y}{dx^2}\right) = w \tag{16}$$

This is a fourth-order ordinary differential equation and needs fourth supplementary initial, boundary, or mixed conditions in order to completely solve for the deflection y(x). For the simply supported beam subjected to its own weight, the four boundary conditions are:

$$y = 0 \quad \text{and} \quad M = EI\frac{d^2y}{dx^2} = 0 \quad \text{at} \quad x = 0 \quad \text{and} \quad x = L \tag{17}$$

If EI is not a function of x, the central-difference approximation for Equation 16 at $x = x_k$ is:

$$\left.\frac{d^4y}{dx^4}\right|_{\text{at } x_k} \approx \frac{y_{k-2} - 4y_{k-1} + 6y_k - 4y_{k+1} + y_{k+2}}{(\Delta x)^4} = \left.\frac{w}{EI}\right|_{\text{at } x_k} \tag{18}$$

If N in-between stations are selected for determination of the deflections there, then the two boundaries are at k = 0 and k = N + 1. In view of the subscripts k–2 and k + 2 in Equation 18, k can thus only take the values 2 through N–1. For solving y_1 through y_N, we hence need two more equations which are the two moment conditions in Equation 17. At $x_0 = 0$, the second-order, forward-difference formula can be applied to give

$$M = EI\left.\frac{d^2y}{dx^2}\right|_{\text{at } x_0} \approx \frac{y_2 - 2y_1 + y_0}{(\Delta x)^2} EI \,\Big|_{\text{at } x_0} = 0$$

Since y = 0 at the left support $x = x_0 = 0$, the above equation yields

$$-2y_1 + y_2 = 0 \tag{19}$$

Similarly, at the right support $x = x_{N+1} = 3L$, the second-order, backward-difference formula can be applied to give

$$M = EI\left.\frac{d^2y}{dx^2}\right|_{\text{at } x_{N+1}} \approx \frac{y_{N+1} - 2y_n + y_{N-1}}{(\Delta x)^2} EI\Big|_{\text{at } x_{N+1}} = 0$$

Since y = 0 at the right support $x = x_{N+1} = 3L$, the above equation yields

$$y_{N-1} - 2y_N = 0 \tag{20}$$

Meanwhile, the boundary condition y = 0 at the left support $x = x_0 = 0$ can be substituted into Equation 18 for k = 2 to obtain:

$$-4y_1 + 6y_2 - 4y_3 + y_4 = (\Delta x)^4 \left.\frac{w}{EI}\right|_{\text{at } x_2} \tag{21}$$

Similarly, the boundary condition y = 0 at the right support $x = x_{N+1} = 3L$ can be substituted into Equation 18 for k = N–1 to obtain:

$$y_{N-3} - 4y_{N-2} + 6y_{N-1} - 4y_N = (\Delta x)^4 \frac{w}{EI}\Big|_{\text{at } x_{N-1}} \tag{22}$$

Equation 18 now should have the modified form of, for k = 3,4,…,N–2:

$$y_{k-2} - 4y_{k-1} + 6y_k - 4y_{k+1} + y_{k+2} = (\Delta x)^4 \frac{w}{EI}\Big|_{\text{at } x_k} \tag{23}$$

In matrix form, Equations 19–23 can be written as [C]{Y} = {R} where {Y} = $[y_1\ y_2 \bullet\bullet\bullet\ y_N]^T$ and the elements of the coefficient matrix [C] denoted as $c_{i,j}$ are to be calculated using the formulas:

$$\begin{aligned}
&c_{1,1} = c_{N,N} = -2, && c_{1,2} = c_{N,N-1} = 1\\
&c_{i,i} = 6, && \text{for } i = 2,3,\ldots,N-1\\
&c_{i,i-1} = c_{i,i+1} = -4, && \text{for } i = 2,3,\ldots,N-1\\
&c_{i,i-2} = c_{i,i+2} = 1, && \text{for } i = 3,4,\ldots,N-2\\
&c_{i,j} = 0 && \text{elsewhere}
\end{aligned} \tag{24}$$

For the right-side vector {R} in the matrix equation [C]{Y} = {R}, its elements denoted as r_i are to be calculated using the formula:

$$r_1 = r_N = 0 \quad \text{and} \quad r_i = (\Delta x)^4 \frac{w}{EI}\Big|_{\text{at } x_i} \text{ for } i = 2,3,\ldots,N-1 \tag{25}$$

where $w(x_i)$ means the distributed load at $x = x_i$. Consider the beam shown in Figure 12 with dead loads w_m and w_e. That is:

$$\begin{aligned}
&w(x_i) = w_e, && \text{for} \quad 0 \le x_i < L \quad \text{and} \quad 2L < x_i \le 3L\\
&w(x_i) = w_m, && \text{for} \quad L < x_i < 2L\\
&w(x_i) = (w_m + w_e)/2 && \text{for} \quad x_i = L \quad \text{and} \quad x_i = 2L
\end{aligned} \tag{26}$$

Notice that the average value of w_m and w_e at the junctions of hollow and solid portions of the beam. Expressions for the EI product can be given similar to those for w. In a sample application of the **QuickBASIC** version of the program **OdeBvpFD**, we will give two subprogram functions which are prepared for the beam problem shown in Figure 12 based on Equations 25 and 26.

An alternative approach for solving the simply supported beam by finite-difference approximation of Equation 13 and using the two boundary conditions y = 0 at x = 0 and x = 3L is given as a homework problem. This approach requires the moment equation M(x) be derived prior to the solution of the matrix equation [C]{Y} = {R}.

FORTRAN Version

```
C     PROGRAM OdeBvpFD - Ordinary differential equation, boundary-value problem
C                        solved by finite-difference method.  Up to 99 stations.
      DIMENSION X(99),C(99,99),R(99)
      WRITE (*,2)
    2 FORMAT(1X,'Program OdeBvpFD - Finite-difference solution of ',
     *       'O.D.E. boundary-value problem.'/)
      WRITE (*,4)
    4 FORMAT(1X,'Have you edited the functions CIJ and RI for ',
     *          'generating the elements of [C] and {R} in '
     *          '[C]{Y}={R}?'/
     *       1X,'If not, press <Shift Break>;if yes, press 1.')
      READ (*,*) I1
      WRITE (*,6)
    6 FORMAT(1X,'Enter the number of (in-between, excluding '
     *          'boundaries) stations and stepsize :')
      READ (*,*) N,DX
      WRITE (*,8)
    8 FORMAT(1X,'Enter the first (left boundary) X value : ')
      READ (*,*) X1
C     Calculate [C] and {R}
      DO 10 I=1,N
      X(I)=X1+I*DX
      R(I)=RI(X(I),DX)
      DO 10 J=1,N
   10 C(I,J)=CIJ(X(I),I,J,DX)
      CALL GAUSS(C,N,99,R)
C     {Y} is in {R}.
      WRITE (*,12) (R(K),K=1,N)
   12 FORMAT(1X,'The solution is : '//5(5E16.5/))
      STOP
      END


      FUNCTION RI(X,DX)
      RI=-5*DX**2/100.
      RETURN
      END


      FUNCTION CIJ(X,I,J,DX)
      CIJ=0.
      IF (I.EQ.J    ) CIJ=-2.
      IF (I.EQ.(J-1)) CIJ=1.-DX/2./X
      IF (I.EQ.(J+1)) CIJ=1.+DX/2./X
      RETURN
      END


      SUBROUTINE GAUSS(C,N,M,V)

      (See Program Gauss.)
```

Sample Application

Consider the membrane problem shown in Figure 1. Let the inner radius R_i be equal to 3 in, the stepsize or radial increment ΔR be equal to 9.5 in, and the number of stations N between the two boundaries be selected as equal to 11 (that is, the other boundary is at $R = R_i + (N + 1)\ \Delta R = 3 + 12x0.5 = R_o = 9$ in.) If the tension T is equal to 100 lbs/in and the pressure p is equal to 5 lbs/in^2, a function subprogram RI can then be accordingly prepared as listed in the program OdeBvpFD. An interactive run of this program has resulted in a display on screen as follows:

```
Program OdeBvpFD - Finite-difference solution of O.D.E. boundary-value problems.

Have you edited the functions CIJ and RI for generating the elements of [C] and
{R} in [C]{Y}={R}?
If not, press <Shift Break>; if yes, press 1.
1
Enter the number of (in-between, excluding boundaries) stations and stepsize :
11,0.5
Enter the first (left boundary) X value :
3.
The solution is :

     .53374E-01     .10159E+00     .14271E+00     .17552E+00     .19864E+00
     .21086E+00     .21111E+00     .19837E+00     .17172E+00     .13031E+00
     .73321E-01
Stop - Program terminated.
```

QuickBASIC Version

```
' PROGRAM OdeBvpFD - Ordinary differential equation, boundary-value problem
'                   solved by finite-difference method.  Up to 119 stations.
      DECLARE FUNCTION  RI(X,DX)
      DECLARE FUNCTION CIJ(X,I,J,DX)
      DECLARE SUB  GauJor(C(),N,R(),D)
      COMMON  SHARED N
CLS : CLEAR : DIM X(119),C(119,119),R(119)
      PRINT "Program OdeBvpFD - Finite-difference solution of O.D.E. ";
      PRINT "boundary-value problem. ": PRINT
      PRINT "Have you edited the functions CIJ and RI for generating ";
      PRINT "the elements of [C]"
      PRINT "  and {R} in [C]{Y}={R}?"
      INPUT "If not, press <Shift Break>;if yes, press any key : ",I1: PRINT
      PRINT "Enter the number of (in-between, excluding boundaries) ";
      INPUT "stations and stepsize :  ",N,DX
      INPUT "Enter the first (left boundary) value of the independent variable : ",X1
'     Calculate [C] and {R}
      FOR I = 1 TO N     :    X(I) = X1+I*DX          : R(I)=RI(X(I),DX)
         FOR J = 1 TO N: C(I,J) = CIJ(X(I),I,J,DX): NEXT J: NEXT I
      CALL GauJor(C(),N,R(),D)                                 '{Y} is in {R}.
      PRINT : PRINT "The solution is : ": PRINT
      FOR K = 1 TO N: PRINT USING "    ##.#####^^^^";R(K); : NEXT K
      END

      FUNCTION CIJ(X,I,J,DX)
      CIJ = 0 : IF I = J     THEN CIJ = -2!
      IF I = J - 1 THEN CIJ =  1! - DX / 2! / X
      IF I = J + 1 THEN CIJ =  1! + DX / 2! / X
      END FUNCTION

      SUB GauJor(A(),N,C(),D)
'     Gauss-Jordan Elimination method for solving [A]{X}={C}.  {C} exits as {X}.
      FOR I = 1 TO N: IF A(I,I)=0 THEN 220              ' *** Normalization ***
200      FOR J=I+1 TO N     : A(I,J) = A(I,J)/A(I,I): NEXT J
         C(I) = C(I)/A(I,I): GOTO 250
220      FOR J = I+1 TO N: IF A(J,I)=0 THEN 230              ' *** Pivoting ***
            FOR K = I TO N: T=A(I,K): A(I,K)=A(J,K): A(J,K)=T: NEXT K
            T=C(I): C(I)=C(J): C(J)=T: GOTO 200
230         NEXT J
250      FOR K = 1 TO N: IF K = I THEN 265                ' *** Elimination ***
            IF A(K, I) = 0 THEN 265
            C(K) = C(K) - A(K, I) * C(I)
            FOR J = I + 1 TO N: A(K,J)=A(K,J)-A(K,I)*A(I,J): NEXT J
265         NEXT K
         NEXT I
      D=1: FOR I=1 TO N: D=D*A(I,I): NEXT I
      END SUB

      FUNCTION RI(X,DX)
      RI=-5*DX^2/100
      END FUNCTION
```

Sample Applications

Consider the same membrane problem as for the sample application using the **FORTRAN** version of the program **OdeBvpFD**. The **QuickBASIC** version has a COMMON SHARED statement allowing N to be shared in the subprograms. Also, this version has been expanded to solving up to 119 stations. It gives an interactive run as follows:

```
Program OdeBvpFD - Finite-difference solution of O.D.E. boundary-value problems.

Have you edited the functions CIJ and RI for generating the elements of [C]
  and {R} in [C]{Y}={R}?
If not, press <Shift Break>; if yes, press any key :

Enter the number of (in-between, excluding boundaries) stations and stepsize :
11,0.5
Enter the first (left boundary) X value : 3.
The solution is :

      .53374E-01       .10159E+00       .14271E+00       .17552E+00       .19864E+00
      .21086E+00       .21111E+00       .19837E+00       .17172E+00       .13031E+00
      .73321E-01
```

The simply supported beam under its own weight is considered as a second example. For $w_m = w_e = -2$ and a uniform EI value equal to 1 (that is when the beam has a uniform cross section without the hollow end portions), L = 100, $\Delta x = 30$, and N = 9 stations between the supports, the subprograms CIJ and RI are prepared according to Equations 24 to 26 as follows:

```
        FUNCTION CIJ(X,I,J,DX)
        CIJ = 0
        IF I<> 1    THEN 15
        IF J = 1    THEN CIJ = -2: GOTO 99
        IF J = 2    THEN CIJ = 1!: GOTO 99 ELSE 99
15      IF I<> N    THEN 25
        IF I = J    THEN CIJ = -2: GOTO 99
        IF J = N-1  THEN CIJ = 1!: GOTO 99 ELSE 99
25      IF I = J    THEN CIJ = 6!: GOTO 99
        IF J=(I-1)  THEN CIJ = -4: GOTO 99
        IF J=(I+1)  THEN CIJ = -4: GOTO 99
        IF J=(I-2)  THEN CIJ = 1!: GOTO 99
        IF J=(I+2)  THEN CIJ = 1!
99      END FUNCTION

        FUNCTION RI(X,DX)
          WM =-2 :       WE=-2        : EIM=1    : EIE=1
           L = 100
        IF X = DX         THEN RI = 0                          : GOTO 88
        IF X = N * DX     THEN RI = 0                          : GOTO 88
        IF X < L          THEN RI = DX^4*WE/EIE                : GOTO 88
        IF X = L          THEN RI = DX^4*(WM+WE)/(EIM+EIE): GOTO 88
        IF X>= 2 * L      THEN 22
                               RI = DX^4*WM/EIM                : GOTO 88
22      IF X = 2 * L      THEN RI = DX^4*(WM+WE)/(EIM+EIE): GOTO 88
                               RI = DX^4*WE/EIE
88      END FUNCTION
```

It should be noted that the arguments X and DX for FUNCTION CIJ are not actually involved in any of the statements therein. They are kept so that the statements in the program **OdeBvpFD** involving CIJ can remain as general as possible to accommodate other problems which may need these arguments as linkages. The interactive application of the above two FUNCTION subprograms is demonstrated below.

```
Program OdeBvpFD - Finite-difference solution of O.D.E. boundary-value problems.

Have you edited the functions CIJ and RI for generating the elements of [C]
  and {R} in [C]{Y}={R}?
If not, press <Shift Break>; if yes, press 1.

Enter the number of (in-between, excluding boundaries) stations and stepsize :
9,30
Enter the first (left boundary) X value : 0

The solution is :

   -3.40199E+07     -6.80398E+07     -9.63898E+07     -1.15020E+08     -1.21500E+08
   -1.15020E+08     -9.63898E+07     -6.80398E+07     -3.40199E+07
```

The analytical solution of this beam problem can be easily obtained. From any textbook on mechanics of materials (see footnote *), the maximum deflection for a beam of length 3L and subjected to a uniformly distributed load w is $y_{max} = 5w(3L)^4/384EI$. Since w = –2, L = 100, and EI = 1 are used in the above illustrative run, y_{max} is equal to $5x(-2)x(300)^4/384 = 2.109x10^8$. The computed result of y at x_7 (for N = 9, the 5th station is the mid-length of the beam) by the program **OdeBvpFD** is equal to $-1.21500x10^8$ which has an error about 42%. A homework problem is given for readers to exercise different N and Δx values, which will show that for N equal to 19, 29, 59, 99, and 119, the respective y_{max} values are $-1.63285x10^8$, $-1.78535x10^8$, — $1.93665x10^8$, $-1.92011x10^8$, and $-1.83x10^8$. Notice that the accuracy continue to improve until the roundoff errors begin to affect the solution when N becomes large, for such cases the double precision should be used in solving the matrix equation [C]{Y} = {R} by the Gaussian elimination method.

MATLAB Applications

The two sample problems discussed in **FORTRAN** and **QuickBASIC** version of the program **OdeBvpFD** can be executed interactively by **MATLAB** with the commands entered from keyboard as follows:

Membrane Problem

```
>>N=11; DX=0.5; R=-5/100*DX^2*ones(1,N)

 R =

   Columns 1 through 7

    -0.0125    -0.0125    -0.0125    -0.0125    -0.0125    -0.0125    -0.0125
```

```
>>N=11; DX=0.5; R=-5/100*DX^2*ones(1,N)

 R =

   Columns 1 through 7

    -0.0125   -0.0125   -0.0125   -0.0125   -0.0125   -0.0125   -0.0125

   Columns 8 through 11

    -0.0125   -0.0125   -0.0125   -0.0125

>>for I=1:N, X=3+I*DX;
      for j=1:N
          if I==j, C(i,j)=-2;
             elseif I==j-1, C(i,j)=1-DX/2/X;
             elseif I==j+1, C(i,j)=1+DX/2/X;
             else             C(i,j)=0;
          end
      end
  end, C

 C =

   Columns 1 through 7

    -2.0000    0.9286         0         0         0         0         0
     1.0625   -2.0000    0.9375         0         0         0         0
          0    1.0556   -2.0000    0.9444         0         0         0
          0         0    1.0500   -2.0000         0         0         0
          0         0         0    1.0455   -2.0000         0         0
          0         0         0         0    1.0417   -2.0000         0
          0         0         0         0         0    1.0385   -2.0000
          0         0         0         0         0         0    1.0357
          0         0         0         0         0         0         0
          0         0         0         0         0         0         0
          0         0         0         0         0         0         0

   Columns 8 through 11

          0         0         0         0
          0         0         0         0
          0         0         0         0
          0         0         0         0
          0         0         0         0
          0         0         0         0
     0.9615         0         0         0
    -2.0000    0.9643         0         0
     1.0333   -2.0000    0.9667         0
          0    1.0313   -2.0000    0.0688
          0         0    1.0294   -2.0000

>>V=C\R'; V'
ans =

  Columns 1 through 7

    0.0534    0.1015    0.1427    0.1755    0.1986    0.2109    0.2111

  Columns 8 through 11

    0.1984    0.1717    0.1303    0.0733
```

Beam Deflection Problem

```
>>C=zeros(9,9); C(1,1)=-2; C(1,2)=1; C(9,9)=-2; C(9,8)=1;
>>for I=2:8
     for j=1:9
         if I==j, C(i,j)=6;
            elseif j==I-1, C(i,j)=-4;
            elseif j==I+1, C(i,j)=-4;
            elseif j==I-2, C(i,j)=1;
            elseif j=I+2, C(i,j)=1;
         end
     end
  end, C

C =
     -2     1     0     0     0     0     0     0     0
     -4     6    -4     1     0     0     0     0     0
      1    -4     6    -4     1     0     0     0     0
      0     1    -4     6    -4     1     0     0     0
      0     0     1    -4     6    -4     1     0     0
      0     0     0     1    -4     6    -4     1     0
      0     0     0     0     1    -4     6    -4     1
      0     0     0     0     0     1    -4     6    -4
      0     0     0     0     0     0     0     1    -2

>>R=zeros(1,9); WE=-2; WM=-2; EIE=1; EIM=1; L=100;
  for I=2:8, X=I*DX;
      if X<L, R(I)=DX^4*WE/EIE;
         elseif X==L, R(I)=DX^4*)WE+WM)/(EIE+EIM);
         elseif X>L
                  if X<2*L, R(I)=DX^4*WM/EIM;
                     elseif X==2*L, R(I)=DX^4*(WE+WM)/(EIE+EIM);
                     elseif X>2*L, R(I)=DX^4*WE/EIE;
                  end
      end
  end, R

 R =

   Columns 1 through 6

            0   -1620000   -1620000   -1620000   -1620000   -1620000

   Columns 7 through 9

     -1620000   -1620000          0

>>V=C\R'; V'

 ans =

   1.0e+008 *

   Columns 1 through 7

    -0.3402   -0.6804   -0.9639   -1.2150   -1.1502   -0.9639

   Columns 8 through 9

    -0.6804   -0.3402
```

Mathematica Applications

For solving the membrane problem, **Mathematica** can be applied as follows:

In[1]: = Ns = 11; DX = 0.5; c = Table[0,{Ns},{Ns}];
R = Table[–5*DX^2/100,{Ns}];
Print["R = ",R]

Out[1] = R = {–0.0125, –0.0125, –0.0125, –0.0125, –0.0125, –0.0125,
–0.0125, –0.0125, –0.0125, –0.0125, –0.0125}

In[2]: = (Do[Do[x = 3 + i*DX; If[i = = j, c[[i,j]] = –2;,
If[i = = j–1, c[[i,j]] = 1DX/2/x;,
If[i = = j + 1, c[[i,j]] = 1 + DX/2/x;, Continue]]],
{i,Ns}], {j,Ns}]); Print["Matrix c = ",c]

Out[2] = Matrix c = {{–2, 0.928571, 0, 0, 0, 0, 0, 0, 0, 0, 0},
{1.0625,–2, 0.9375, 0, 0, 0, 0, 0, 0, 0, 0},
{0, 1.05556, –2, 0.944444, 0, 0, 0, 0, 0, 0, 0},
{0, 0, 1.05, –2, 0.95, 0, 0, 0, 0, 0, 0},
{0, 0, 0, 1.04545, –2, 0.954545, 0, 0, 0, 0, 0},
{0, 0, 0, 0, 1.04167, –2, 0.958333, 0, 0, 0, 0},
{0, 0, 0, 0, 0, 1.03846, –2, 0.961538, 0, 0, 0},
{0, 0, 0, 0, 0, 0, 1.03571, –2, 0.964286, 0, 0},
{0, 0, 0, 0, 0, 0, 0, 1.03333, –2, 0.966667, 0},
{0, 0, 0, 0, 0, 0, 0, 0, 1.03125, –2, 0.96875},
{0, 0, 0, 0, 0, 0, 0, 0, 0, 1.02941, –2}}

In[3]: = V = Inverse[c].R

Out[3] = {0.0533741, 0.101498, 0.142705, 0.175525, 0.198641, 0.210864,
0.211107, 0.198368, 0.171723, 0.13031, 0.0733212}

These results are in agreement with those obtained by the **MATLAB** application. The loaded beam problem also can be treated in a similar manner as follows:

In[1]: = c = Table[0,{9},{9}]; c[[1,1]] = –2; c[[1,2]] = 1; c[[9,9]] = –2;
c[[9,8]] = 1;

In[2]: = (Do[Do[If[i = = j,c[[i,j]] = 6;, If[(j = = i–1)||(j = = i + 1), c[[i,j]] = –4;,
If[(j = = i + 2)||(j = = i–2), c[[i,j]] = 1;, Continue]]],
{i,2,8}],{j,9}]); Print["Matrix c = ",c]

Out[2] = Matrix c = {{–2, 1, 0, 0, 0, 0, 0, 0, 0}, {–4, 6, –4, 1, 0, 0, 0, 0, 0},
{1, –4, 6, –4, 1, 0, 0, 0, 0}, {0, 1, –4, 6, –4, 1, 0, 0, 0},
{0, 0, 1, –4, 6, –4, 1, 0, 0}, {0, 0, 0, 1, –4, 6, –4, 1, 0},

```
{0, 0, 0, 0, 1, –4, 6, –4, 1}, {0, 0, 0, 0, 0, 1, –4, 6, –4},
{0, 0, 0, 0, 0, 0, 0, 1, –2}}
```

```
In[3]: = R = Table[0,{9}]; WE = –2; WM = –2; EIE = 1; EIM = 1; L = 100;
         DX = 30;
```

```
In[4]: = (Do[X = i*DX; If[X<L, R[[i]] = DX^4*WE/EIE;,
          If[X = = L, R[[i]] = DX^4*(WE + WM)/(EIE + EIM);,
           If[X>L, If[X<2*L, R[[i]] = DX^4*WM/EIM;,
            If[X = = 2*L, R[[i]] = DX^4*(WE + WM)/(EIE + EIM);,
             IF[X>2*L, R[[i]] = DX^4*WE/EIE;,
    Continue]]]]]],
      {i,2,8}]); Print["R = ",R]
```

```
Out[4] = R = {0, –1620000, –1620000, –1620000, –1620000, –1620000,
                                      –1620000, –1620000, 0}
```

```
In[5]: = V = Inverse[c].R
```

```
Out[5] = {–34020000, –68040000, –96390000, –115020000, –121500000,
          –115020000, –96390000, –68040000, –34020000}
```

Again, the results are in agreement with those obtained by the **MATLAB** application.

6.5 PROBLEMS

RungeKut

1. The differential equation of motion of a spring-and-mass system is $d^2x/dt^2 + \omega^2 x = 0$ where $\omega^2 = k/m$, m is the mass and k is the spring constant. If the weight is 5 lb., g = 32.2 ft/sec^2, k = 1.5 lb/in, and initially the displacement x = 2 in. and velocity dx/dt = 0, use the 4th-order Runge-Kutta method and a step-size of 0.05 sec. to manually calculate the values of x and dx/dt at t = 0.05 sec. Note that the given *second-order* ordinary differential equations should first be converted into two first-order differential equations.
2. Initially, the two functions x(t) and y(t) have values 1 and –1, respectively. That is x(t = 0) = 1 and y(t = 0) = –1. For t>0, they satisfy the differential equations dx/dt = 5x–2y + 2t and dy/dt = x^2–0.25sin2t. Use Runge-Kutta classic fourth-order method and a time increment of 0.01 second to calculate x(t = 0.01) and y(t = 0.01).
3. In the following two equations, the terms dx/dt and dy/dt both appear.

$$2dx/dt - 3x + 5dy/dt - 7y = .5t + 1$$

$$-3dx/dt - x - 4dy/dt + 9y = .2e^{-3t}$$

Separate them by treating them as unknowns and solve them by simple substitution or Cramer's rule. The resulting equations can be expressed in the forms of $dx/dt = F_1(t,x,y)$ and $dy/dt = F_2(t,x,y)$. Carry out by manual computation using Runge-Kutta method to obtain the x and y values when $t = 0.1$ if the time increment is 0.1 and the initial conditions are $x(t = 0) = 1$ and $y(t = 0) = -2$.

4. The distributed loads on the beam shown in Figure 3 can be described as $w = -1$ N/cm, for $0<x<10$ cm; $w = 2.2x$ N/cm, for $10<x<20$ cm; $w = 0$, for $20<x<40$ cm. Meanwhile, the bending moment applied at $x = 30$ cm can be described as $M_a = 0$, for $0<x<30$ cm and $M_a = -3$ N-cm, for $30<x<40$ cm. By introducing new variables t, and x_1 through x_4 so that $t = x$, $x_1 = y$, $x_2 = dy/dx$, $x_3 = d^2y/dx^2$, and $x_4 = d^3y/dx^3$, the problem of finding the deflection y(x) of the beam can be formulated (see the reference cited in footnote) as:

$$dx_1/dt = F_1(t,x_1,x_2,x_3,x_4) = x_2, \qquad x_1(t = 0) = 0$$
$$dx_2/dt = F_2(t,x_1,x_2,x_3,x_4) = (x_3 + M_a)/EI, \qquad x_2(t = 0) = 0$$
$$dx_3/dt = F_3(t,x_1,x_2,x_3,x_4) = x_4, \qquad x_3(t = 0) = -1121/3$$
$$dx_4/dt = F_4(t,x_1,x_2,x_3,x_4) = w, \qquad x_4(t = 0) = 24$$

with M_a and w being the applied bending moment and distributed loads, respectively. The initial conditions specified above are all at the left end of the beam which is built into the wall and for the deflection (x_1), slope (x_2), bending moment (x_3), and shearing force (x_4), respectively. Apply the program **RungeKut** by using $EI = 2x10^5$ N/cm^2 and various stepsizes to tabulate the results and errors similar to that given in the text.

5. Apply the fourth-order Runge-Kutta method to find the values of x_1 and x_2 at the time $t = 0.2$ second using a time increment of 0.1 second based on the following governing equations:

$$\frac{dx_1}{dt} = 4 - 5x_1 + 7x_2 - 2t \quad \text{and} \quad \frac{dx_2}{dt} = -3x_1 + x_1x_2 + 6e^{-t}$$

At $t = 0.1$ second, $x_1 = -1$ and $x_2 = 1$.

6. Use different stepsizes to calculate y values at $x = 0.1$, 0.2, and 0.3 by application of the program **RungeKut** for the initial-value problem $dy/dx = x^2-y$, $y(x = 0) = 1$. The analytical solution is $y = 2-2x + x^2e^{-x}$, by which the exact solutions can be easily computed to be $y(0.1) = 0.90516$, $y(0.2) = 0.82127$, and $y(0.3) = 0.74918$. Determine the stepsize which will lead to a Runge-Kutta numerical solutions of y(0.1), y(0.2), and y(0.3) accurate to five decimal figures.

7. For the loaded beam shown in Figure 13, the deflection y(x) is to be determined by solving Equation 21. Let the stiffness EI be equal to $2x10^7$ N-cm^2 and it can be shown that the bending moment M can be described by the equations:

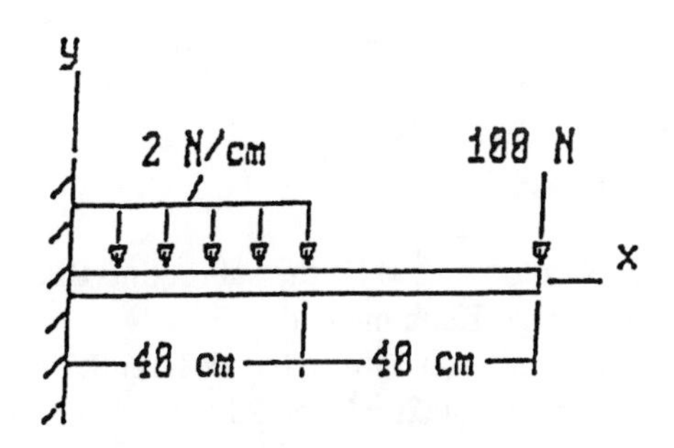

FIGURE 13. Question 7.

$$M = -x^2 + 180x - 9600 \quad \text{for} \quad 0<x<40 \text{ cm}$$

and

$$M = 100x - 8000 \quad \text{for } 40<x<80 \text{ cm}$$

Apply the program **RungeKut** to find y at x = 80 cm by using stepsizes h = 4, 2, 1, 0.5, 0.25, and 0.1 and calculate the error by comparing with the expected value of y(x = 80) = –0.928 cm.

8. Convert the following two differential equations into three first-order differential equations in the forms of $dx_i/dt = F_i(t;x_1,x_2,x_3;\text{constants})$ for i = 1,2,3 so that the subroutine RKN can be readily applied:

$$\frac{d^2u}{dt^2} + 2\frac{du}{dt} + 3\frac{dv}{dt} + 4u - 5v - 0.6\sin 7t = 0$$

$$\frac{dv}{dt} - 0.1\frac{du}{dt} + 20uv - 300e^{-4t} = 0$$

9. Write a subprogram FUNCTION F(X,T,I,N) which includes the statements

COMMON R1,R2,R3,R4

for transmitting the values of R1, R2, R3, and R4 from the main program. These four variables are r_1, r_2, r_3, and r_4, respectively, appearing in the equations:

$$\frac{dA}{dt} = -r_1A$$

$$\frac{dB}{dt} = r_1A - r_2B$$

$$\frac{dC}{dt} = r_2B - r_3C$$

$$\frac{dD}{dt} = r_3C - r_4D$$

$$\frac{dE}{dt} = r_4D$$

This FUNCTION is to be used by the subroutine **RKN** in application of the fourth-order Runge-Kutta method.

10. The functions x(t) and y(t) satisfy the differential equations $d^2x/dt^2 + 3dx/dt + 5dy/dt - 7e^{-9t} + 9\sin 2t = 0$ and $2dx/dt - 4dy/dt + 6x - 8y + 10t - 12 = 0$. Convert the above two equations into the standard form $dx_i/dt = f_i(t; x_1, x_2, x_3; \text{constants})$ for i = 1,2,3 where $x_1 = x$, $x_2 = y$, and $x_3 = dx/dt$. Give the expressions for f_1, f_2, and f_3 in terms of t, x_1, x_2, and x_3 so that the Runge-Kutta method can be applied.
11. Apply the fourth-order Runge-Kutta method to find the values of y and z at x = 0.35 if at x = 0.3, y = 1 and z = 2 respectively and they satisfy thedifferential equations dy/dx = xy + z and dz/dx = yz + x. Use a stepsize of x = 0.5 and show all details of how the Runge-Kutta parameters are calculated.
12. The deflection y of the load beam shown below satisfies the ordinary differential equation $EI(d^2y/dx^2) = M$ where the Young's modulus $E = 2x10^8$ N/m^2, moment of inertia $I = 4.5x10^{-8}$ m^4 and the internal bending moment, in N-m, has been derived in terms of x as M(x) = 200x–30 for $0 \le x \le .1$ m and M(x) = 100x–20 for $.1 \le x \le .2$ m. (1) Using an increment of $\Delta x = 0.01$ m, standardize the above problem into a system of two first-order ordinary differential equations $dx_i/dx = f_i(x; x_1, x_2; \text{constants})$ for i = 1,2 where $x_1 = y$ and $x_2 = dy/dx$ (slope). (2) Write a FUNCTION F(...) needed in SUBROUTINE RKN which we have discussed in class for using the fourth-order Runge-Kutta method, based on the result of Step (1) and also the M(x) equations. (3) Calculate the eight RungeKutta parameters and then the value of y and dy/dx at x = 0.01 m.
13. Convert the following differential equation into a set of two first-order differential equations so that the fourth-order Runge-Kutta method can be applied: $d^2x/dt^2 + 4dx/dt + 3x = 4e^{-t}$. If at t = 0, x = 0 and dx/dt = 2, use a time increment oft = 0.1, compute the x and dx/dt values at Δt = 0.1 based on the fourth-order Runge-Kutta method.
14. The forced swaying motion of a three-story building can be simulated as a system of three lumped masses m_i connected by springs with stiffnesses k_i and subjected to forces $f_i(t)$ for i = 1,2,3 as shown in Figure 14. Here, the dampingcharacteristics are not considered but could be incorporated. Derive the governing differential equations for the and then convert them into a system of 6 first-order differential equations so that the programs **RungeKut** and **ode45** can be applied to find the histories of displacements, $x_i(t)$, and velocities $v_i(t) = dx_i(t)/dt$. Solve a numerical case of m_i = 2i N-sec^2/cm, k_i = 3i N/cm, and $f_i(t) = (2i-3)\sin(2i-1)t$ N, $x_i(t = 0) = 0$ and $v_i(t = 0) = 0$ for $0 < t \le 20$ seconds. Plot all displacement and velocity histories.

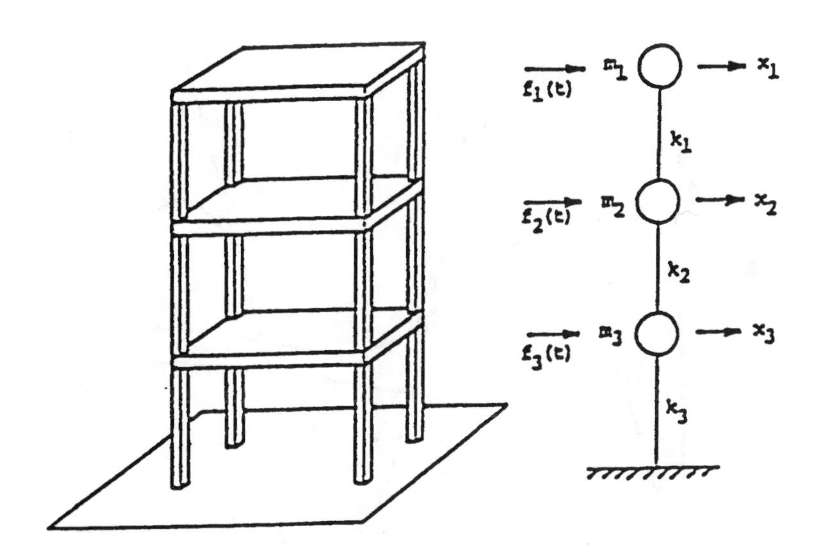

FIGURE 14. Problem 14.

15. Instead of f(t) = 1 in obtaining the system's response of the mechanical vibration problem using the **MATLAB** file FunMCK.m shown in Figure 5, resolve the problem for the case of f(t) = 5sin(0.5t–0.3) by changing FunMCK.m and plot the resulting displacement and velocity.
16. Implement the Runge-Kutta solution of Equation 30 by defining a subprogram function TwoMs in **FORTRAN**, **QuickBASIC**, or, **MATLAB** to obtain the result shown in Figure 5 for the case of b = 3.6 m, h = 1.5 m, m_2/m_1 = 0.8, and initial conditions y = z = dy/dt = dz/dt = 0. And calculate the histories of the cable tension T(t) and angle (t).
17. For the nonlinear oscillation problem of two connected masses shown in Figures 4 and 5, we observe that the oscillation goes on continuously. The motion can be damped by adding a viscous device vertically connected to the mass whose displacement is denoted as z(t). This could be a frictional wall, on which the mass slides vertically. Usually, the retarding force of such a damping device, F_r, could be assumed to be linearly proportional to the velocity of the motion, dz/dt. That is, F_r = cdz/dt where c is constant. Figure 15 is a result of the oscillation when a damping device having c = 1 N-sec/m and m_1 = 1 N-sec^2/m is added to that system. We notice that amplitudes of y(t) and z(t) shown in Figure 5 are steadily decreased. Develop this modified program in **FORTRAN**, **QuickBASIC**, or,**MATLAB** to generate Figure 15.
18. Use **Mathematica**'s function **NDSolve** to solve Problem 7.

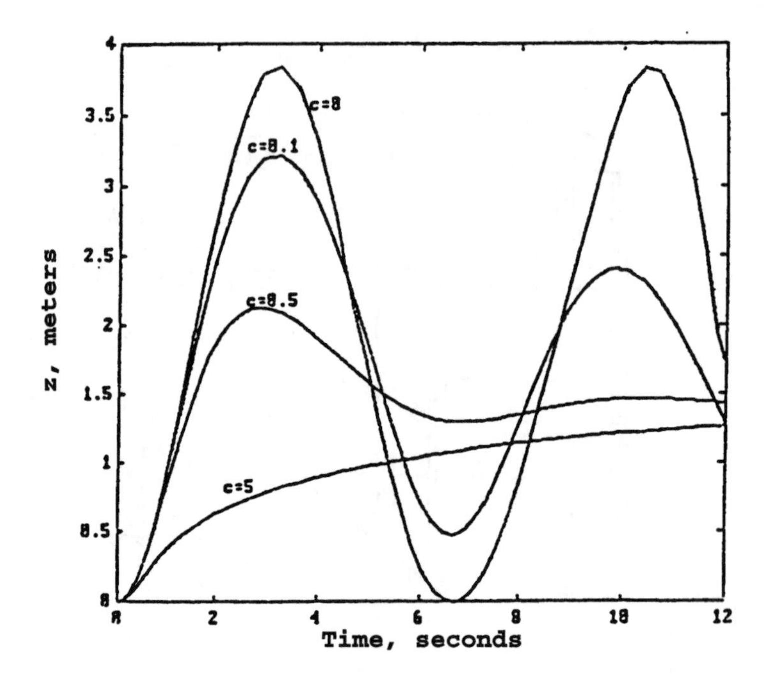

FIGURE 15. Problem 17.

19. Apply **Mathematica** to solve Problem 17 for a time increment $\Delta t = 0.2$ sec and until $t = 12$ seconds.
20. Apply **Mathematica** for solving Problem. 14.

ODEBVPRK

1. The function y(x) satisfies the boundary conditions $y(x = 0) = 2$ and $y(x = 3) = 4$ and the differential equation $5d^2y/dx^2-3dy/dx + y = 13x-15$ for $0<x<3$. Apply the fourth-order Runge-Kutta method to find the y values at $x = 1$ and $x = 2$ based on an increment of x equal to 1.
2. The function y(x) has the boundary values of $y(x = 0) = 1$ and $y(x = 3) = 5$ and for x between 0 and 3, y(x) satisfies the ordinary differential equation:

$$\frac{d^2y}{dx^2} - 3\frac{dy}{dx} + 2y = 2x + 1$$

   Apply the fourth-order Runge-Kutta method to find the y values at $x = 1$ and $x = 2$ based on a stepsize of $\Delta x = 1$.
3. For a membrane (Figure 16) under uniform tension T and fastened at the inner radius R_i and outer radius R_o, the axisymmetric deformation z resulted by the acting uniform pressure p can be shown to satisfy the differential equation:[7]

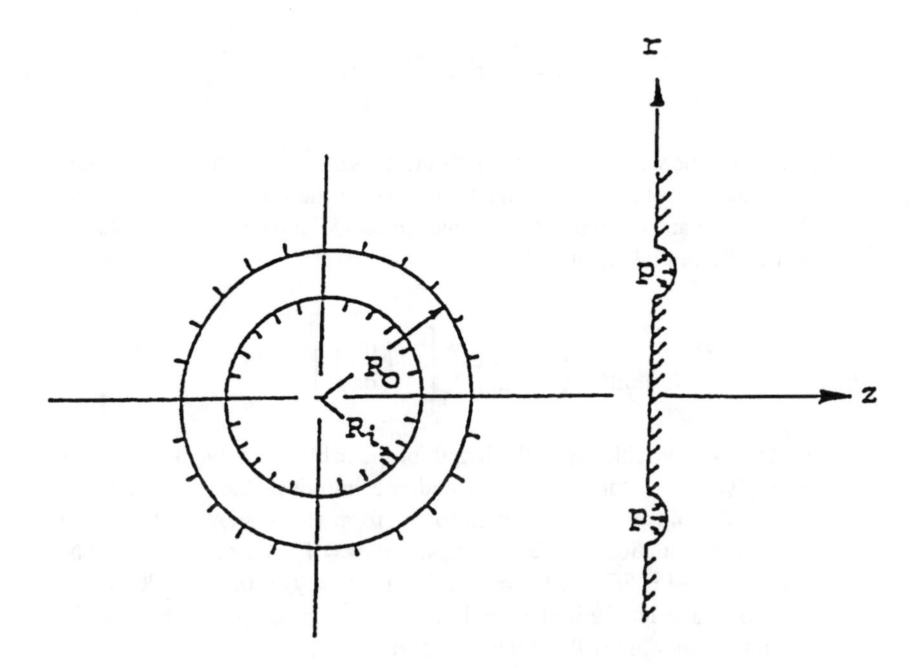

FIGURE 16. Problem 3.

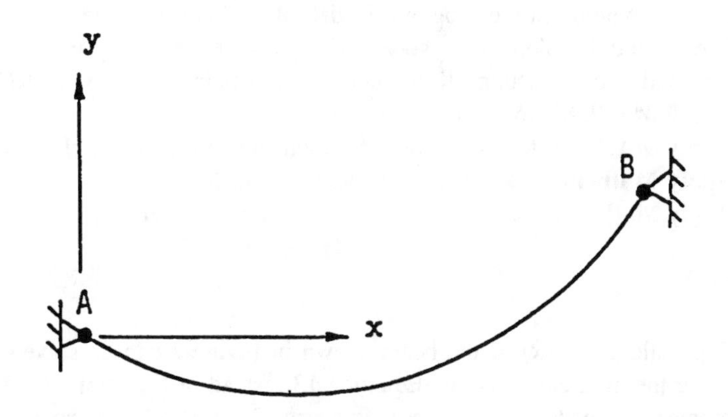

FIGURE 17. Problem 4.

$$\frac{d^2z}{dr^2}+\frac{1}{r}\frac{dz}{dr}=-\frac{p}{T}$$

for $R_i<r<R_o$. The boundary conditions are $z(R_i) = 0$ and $z(R_o) = 0$. Modify the program **OdeBvpRK** to solve this problem.

4. A cable hung at its two ends as shown in Figure 17 by its own weight will have a catenary shape described by the equation:

$$y = \frac{T_x}{w}\left[\cosh\frac{wx}{T_x} - 1\right] \quad \text{(a)}$$

where w is the weight per unit length and T_x is the horizontal, x-component of the tension of the cable. Equation (a) is for the case when both w and T_x are constant throughout the cable. In fact, Equation a is the solution of the differential equation:[9]

$$\frac{d^2y}{d^2x} = \frac{w}{T_x}\frac{ds}{dx} = \frac{w}{T_x}\left[1 + \left(\frac{dy}{dx}\right)^2\right]^{1/2} \quad \text{(b)}$$

where s is a variable along the length of the cable. To solve this problem by applying the Runge-Kutta method, we introduce the *slope* variable, $= \theta$ dy/dx and convert Equation (b) to form the system of first-order differential equations $dy/dx = f_1(x)$ and $d/dx = f_2(x)$ where $f_1(x) =$ and $f_2(x) = w[1 + \theta^2]^{1/2}/T_x$. Let w = 0.12 KN/m, $x_A = y_A = 0$, $x_B = 200$ m, and $y_B = 50$ m and let the initial conditions be y = = 0 at $x = x_A = 0$, iterate T_x value until y_B is within 99.9% of 50 m.

5. Actually, the value of T_x in Problem 4 can be obtained by solving the transcendental equation (a) for $y = y_B = 50$, $x = x_B = 200$, and w = 0.12. Select a method in the program **FindRoot** to find this value.
6. How could Problem 4 be solved if $x_A = -100$ m and $y_A = 25$ m by application of the Runge-Kutta method (w remains equl to 0.12 KN/m)?
7. Apply **MATLAB** to solve Problem 2.
8. Apply **MATLAB** to solve Problem 4 by using an increment of $\Delta x = 2$ m.
9. Apply **Mathematica** to solve Problem 2.
10. Apply **Mathematica** to solve Problem 4 using an incremwnt of $\Delta x = 2$ m.

OdeBvpFD

1. The deflection y(x) of the beam shown in Figure 2 can be solved from using the moment equation, Equation 13 instead of Equation 16, but the moment M needs to be expressed in terms of x. A similar matrix equation [C]{Y} = {R} should be derived using only two boundary conditions y = 0 at x = 0 and x = 3L instead of the four boundary conditions specified in (17). Using the second-order, central-difference formula for d^2y/dx^2, derive the formulas for calculations of the elements of [C] and {R}.
2. Based on the results of Problem 1, proceed to prepare the subprogram functions CIJ and RI and solve for {Y}. Using the data presented in the sample application of the **QuickBASIC** version of program **OdeBvpFD**, compute {Y} and compare the two approaches.
3. Following the illustrative example, run the **QuickBASIC** version of program **OdeBvoFD** for the beam problem shown in Figure 2 but for N equal to 19, 29, 59, 99, and 119.

4. Roundoff errors in the Gaussian elimination steps begin to affect the accuracy of the computed values of the deflection for Problem 3 when N = 119. Change the program **OdeBvpFD** into double precision arithmetics and rerun the case N = 119 and compare the computed y_{max} to that of analytical solution.
5. Make necessary changes in the **FORTRAN** version of program **OdeBvpFD** to solve Problems 3 and 4.
6. For the second sample problem (deflection of beam, Figure 2), change the distributed loads to $w_m = 2$ and $w_e = 1$, and the rigidities to $EI_m = 1$ and $EI_e = 0.5$ to recalculate the maximum deflection y_{max}.
7. Show that for the beam deflection problem shown in Figure 2 when Equation 13 is approximated by use of second-order, central difference and by incorporating the boundary conditions y = 0 at x = 0 and x = 3L, it will lead to the solution of the matrix equation [C]{Y} = {R} where the elements of [C] and {R} denoted as $c_{i,j}$ and r_i, respectively can be calculated by the formulas:

$$\begin{aligned} c_{i,i} &= -2, && \text{for } i = 1,2,\ldots,N \\ c_{i,i+1} &= c_{i+1,i} = 1, && \text{for } i = 1,2,\ldots,N-1 \\ c_{i,j} &= 0, && \text{elsewhere} \end{aligned} \tag{a}$$

and

$$r_i = (\Delta x)^2 \frac{M}{EI}\bigg|_{\text{at } x_i} \tag{b}$$

where N is the number of stations between the two supports and Δx is the stepsize equal to 3L/(N + 1).
8. For Figure 2, if the uniformly distributed loads for the middle and ending portions are designated as w_m and w_e, respectively, derive the expressions for the internal bending moments in the three portions of the beam, $0 \le x \le L, L \le x \le 2L$, and $2L \le x \le 3L$.
9. Prepare subprogram FUNCTIONS CIJ and RI for Problem 8 and find the deflection vector {Y} by use of either **FORTRAN** or **QuickBASIC** version of the program **OdeBvpFD**. Select appropriate values for the number of stations N so that the results obtained by this second-order approach can be compared to those by the fourth-order approach.
10. Use the central finite-difference method and an increment of $\Delta x = 1$ to find the y values at x = 1 and x = 2 when y is governed by the equation $d^2y/dx^2 + 3dy/dx - y = 2x - 3$ and satisfies the boundary conditions y = 0 at x = 0 and x = 3.
11. Use the central finite-difference method and an increment of $\Delta x = 1$ to find the y values at x = 1 and x = 2 when y is governed by the equation $d^2y/dx^2 + 3y = x-1$ and satisfies the boundary conditions y = 0 at x = 0 and x = 3.

12. It is known that $u = 0$ at $r = 2$ and $r = 5$ and that for $2<r<5$ u satisfies the equation $d^2u/dr^2 - r du/dr = -3$, use *central differences* to approximate both the first and second derivatives of u and an increment of r equal to 1 and then derive two equations relating the u values at $r = 3$ and $r = 4$ and solve them.
13. Apply **MATLAB** to solve Problem 3.
14. Apply **MATLAB** to solve the cable problem #4 listed under OdeBvpRK and using an increment of $\Delta x = 1$ m.
15. Apply **MATLAB** to solve Problem 10 by using an increment of $\Delta x = 0.05$.
16. Repeat Problem 13 except by application of **Mathematica**.
17. Repeat Problem 14 except by application of **Mathematica**.
18. Repeat Problem 15 except by application of **Mathematica**.

6.6 REFERENCES

1. C. R. Wylie, Jr., *Advanced Engineering Mathematics*, McGraw-Hill, New York, 1960, Chapter 6.
2. J. Water, "Methods of Numerical Integration Applied to System Having Trivial Function Evaluation," *ACM Communication*, Vol. 9, 1966, p. 293.
3. A. Higdon et al., *Mechanics of Materials*, John Wiley & Sons, New York, 1985, Chapter 7.
4. Y. C. Pao, *Elements of Computed-Aided Design and Manufacturing, CAD/CAM*, John Wiley & Sons, New York, 1984.
5. A. Higdon, E. H. Ohlsen, W. B. Stiles, J. A. Weese, and W. F. Riley, *Mechanics of Materials*, 4th Edition, John Wiley & Sons, New York, 1985.
6. W. Jaunzemis, *Continuum Mechanics*, MacMillan, New York, 1967, p. 365.
7. S. Timoshenko and D. H. Young, *Elements of Strength of Materials*, 5th Edition, Van Nostrand Reinhold Co., New York, 1968.
8. W. Jaunzemis, *Continuum Mechanics*, MacMillan, New York, 1967, p. 365.
9. J. L. Meriam and L. G. Kraige, *Engineering Mechanics, Volume One: Statics*, Third Edition, John Wiley & Sons, Inc., New York, 1992.

7 Eigenvalue and Eigenvector Problems

7.1 INTRODUCTION

There is a class of physical problems which lead to a governing ordinary differential equation containing an unknown parameter. As an example, consider the buckling of a slender rod subjected to an axial load P shown in Figure 1. The deflected shape y(x) is governed by the equation:[1]

$$\frac{d^2y}{dx^2} = \frac{M}{EI} = -\frac{Py}{EI} \tag{1}$$

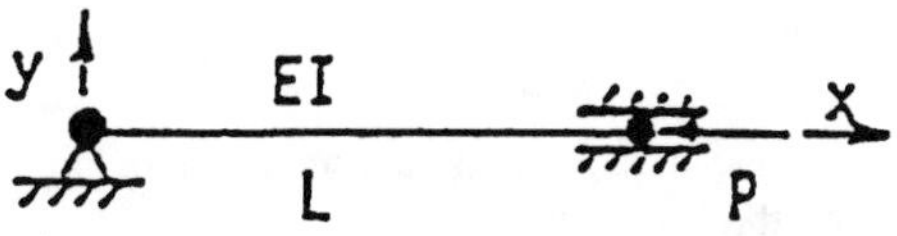

FIGURE 1. The buckling of a slender rod subjected to an axial load P.

where EI is the rigidity and M is the internal bending moment (in this case equal to -Py) of the rod at the section x. If the rod is supported at both ends such that the boundary conditions are:

$$y(x=0) = y(x=L) = 0 \tag{2}$$

The unknown parameter appearing in Equation 1 is P which is the load axially applied causing the rod to buckle. The problem is then to find P and the corresponding buckled shape y(x). If the value of EI is a constant for all x, this problem can be solved analytically. The buckling load can be shown to be $P = \pi^2 EI/L^2$. For the general case when EI is the function of x, numerical method has to be applied to obtain approximate solutions.

In this chapter, we will apply the finite-difference approximation to solve Equation 1. As will be presented in Section 7.2, the resulting matrix equation involving the buckled shape evaluated at N selected stations between the end supports of the rod will be of the standard form:

$$([A] - \lambda[I])\{Y\} = \{0\}, \quad \text{or,} \quad [A]\{Y\} = \lambda\{Y\} \tag{3}$$

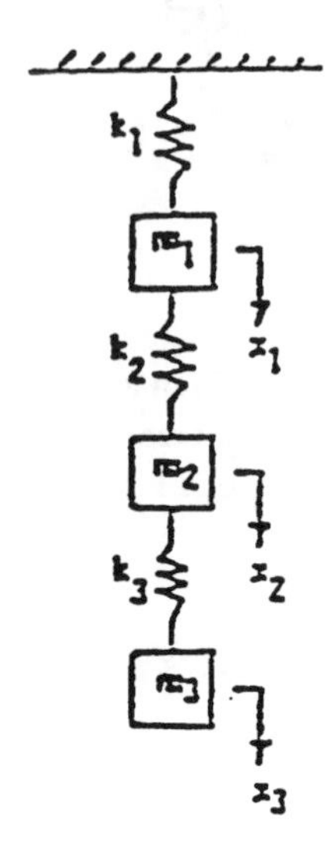

FIGURE 2. Another example of eigenvalue and eigenvector problem — the *vibration* of three masses connected by three springs.

where the matrix [A] will depend on the distances between the stations, {Y} contains the buckled amount of the rod at the stations, and λ is related to the unknown buckling load P. Equation 3 can be interpreted as knowing a matrix [A] and trying to find a proper vector {Y} when it is multiplied by [A], a scaled {Y} will result. This becomes the well-known eigenvector and eigenvalue problem because eigen means proper. λ and {Y} in Equation 3 are called the eigenvalue and eigenvector of [A], respectively. If N is the order of the matrix [A], there are N sets of eigenvalues and eigenvectors. In Section 7.3, how a polynomial from which all eigenvalues of a given matrix can be found as roots will be discussed.

As another example of eigenvalue and eigenvector problem, consider the *vibration* of three masses connected by three springs shown in Figure 2. If any one of these three masses is subjected to some disturbance such as the case when the mass m_3 is pulled down by a certain distance and then released, the whole system will then be vibrating! One will be interested in knowing at what frequency will they be oscillating up and down. To formulate the analysis, let us denote the displacements of the masses as $x_i(t)$ for i = 1 to 3 which are functions of time t. If the elastic constants of the three springs are denoted as k_i for i = 1,2,3, it can be shown[2] by application of the Newton's laws of motion that the governing differential equations for the displacements are:

$$m_1 \frac{d^2x_1}{dt^2} + (k_1 + k_2)x_1 - k_2x_2 = 0 \tag{4}$$

$$m_2 \frac{d^2x_2}{dt^2} - k_2x_1 + (k_2 + k_3)x_2 - k_3x_3 = 0 \tag{5}$$

$$m_3 \frac{d^2x_3}{dt^2} - k_3x_2 + k_3x_3 = 0 \tag{6}$$

If we assume that the masses are vibrating sinusoidally with a common frequency ω but with different amplitudes C_i, their displacements can then be expressed as:

$$x_i(t) = C_i \sin \omega t \quad \text{for} \quad i = 1,2,3 \tag{7}$$

Substituting Equation 7 into Equations 4 to 6, we obtain:

$$\left[(k_1 + k_2 - m_1\omega^2)C_1 - k_2c_2\right]\sin \omega t = 0$$

$$\left[-k_2A_1 + (k_2 + k_3 - m_2\omega^2)A_2 - k_3A_3\right]\sin \omega t = 0$$

and

$$\left[-k_2c_2 + (k_3 - m_3\omega^2)C_3\right]\sin \omega t = 0$$

In matrix form, the above equations can be written as:

$$\begin{bmatrix} k_1 + k_2 - m_1\omega^2 & -k_2 & 0 \\ -k_2 & k_2 + k_3 - m_2\omega^2 & -k_3 \\ 0 & -k_3 & k_3 - m_3\omega^2 \end{bmatrix} \begin{bmatrix} C_1 \\ C_2 \\ C_3 \end{bmatrix} \sin \omega t = \begin{bmatrix} 0 \\ 0 \\ 0 \end{bmatrix} \tag{8}$$

Since the amplitudes C_{1-3} and $\sin\omega t$ cannot be equal to zero which would have led to no motion at all, this leaves the only choice of requiring that the coefficient matrix be singular. In other words, its determinant must be equal to zero. The resulting equation is a cubic polynomial and enables us to solve for three roots which are the squared values of the frequencies (ω^2) of the vibrating system. For each frequency, we next need to know the associated amplitudes of the vibration. Equation 8 can be arranged into the standard form, Equation 3 by letting $\{Y\} = [C_1\ C_2\ C_3]^T$ and:

$$\lambda = \omega^2, \ [A] = \begin{bmatrix} (k_1 + k_2)/m_1 & -k_2/m_1 & 0 \\ -k_2/m_2 & (k_2 + k_3)/m_2 & -k_3/m_2 \\ 0 & -k_3/m_3 & -k_3/m_3 \end{bmatrix} \tag{9,10}$$

This example shows that the governing ordinary differential Equations 4 to 6 may not involve with an unknown parameter as in the buckling problem, but once the common frequency ω is introduced for the *free vibration* it becomes a standard eigenvalue and eigenvector problem described by the matrix [A].

Hence, the question becomes how to find the eigenvalues and their associated eigenvector of a prescribed matrix. The methods of solution are to be discussed in Sections 7.3 and 7.4 where the programs **CharacEq** and **EigenVec** are introduced. In Section 7.5, an iterative method for finding the eigenvector when an eigenvalue of a matrix is provided and program **EigenvIt** also will be presented. Prior to these discussions, in the next section we will first concern with how the matrices connected with the buckling and vibration problems are to be derived and demonstrate in advance how the programs **CharacEq**, **Bairstow**, **EigenVec**, and **EigenvIt** are to be employed for obtaining the eigenvalues and eigenvectors of these matrices.

7.2 PROGRAMS EigenODE.Stb AND EigenODE.Vib — FOR SOLVING STABILITY AND VIBRATION PROBLEMS

In order to obtain numerical solution of the buckling load and shape of the rod shown in Figure 1 in Section 7.1, the central-difference method introduced in Chapter 4 can be applied to approximate the second derivative term appearing in Equation 1 there. At a typical location along the rod, say $x = x_j$, Equation 1 can be approximated as:

$$\frac{d^2y}{dx^2} \doteq \frac{y_{j-1} - 2y_j + y_{j+1}}{h^2} = -\frac{Py_j}{(EI)_j} \tag{1}$$

where $(EI)_j$ is the rigidity of the rod at x_j and $y_j \equiv y(x = x_j)$ etc. This approach requires that the rod be investigated at N stations between the two supports which are labeled as x_0 and x_{N+1}. These stations are equally spaced so that the increment of x (stepsize h) is simply $h = \Delta x = L/(N + 1)$. As a result of such arrangement, the boundary conditions previously defined in Equation 2 now become:

$$y_0 = y_{N+1} = 0 \tag{2}$$

By writing out the equation for the first and last in-between stations, i.e., $j = 1$ and $j = N$, based on Equation 1 and the boundary conditions (2), the two simplified equations are, respectively:

$$\left[-2 + \frac{h^2P}{(EI)_1}\right]y_1 + y_2 = 0 \tag{3}$$

and

$$y_{N-1} + \left[-2 + \frac{h^2P}{(EI)_N}\right]y_N = 0 \tag{4}$$

Also Equation 1 can be rearranged into the form, for j = 2,3,…,N–1

$$y_{j-1} + \left[-2 + \frac{h^2 P}{(EI)_j}\right] y_j + y_{j+1} = 0 \quad (5)$$

where $(EI)_j$ is the rigidity of the beam at the jth station. By multiplying the jth equation by $-(EI)_j/h^2$ for j = 1,2,…,N, Equations 3 to 5 can be further simplified into the standard matrix form $[A-\lambda I]\{Y\} = \{0\}$ where $\{0\}$ is a null vector of order N, $\lambda = P$, $\{Y\} = [y_1\ y_2\ \ldots\ y_N]^T$, and $[A] = [a_{ij}]$ which for i,j = 1,2,…,N the elements are to be calculated with the formulas:

$$a_{ij} = \begin{cases} 2(EI)_i/h^2, & \text{for } i=j \\ -(EI)_i/h^2, & \text{for } i=j-1 \text{ or } i=j+1 \\ 0, & \text{elsewhere} \end{cases} \quad (6)$$

As a simple numerical example, consider the case of EI = 1, L = 1, and N = 2. We are seeking only the solution of displacements, y_1 and y_2 at two in-between points since the stepsize h = L/(N + 1) = 1/3. [A] is of order 2 by 2 and having elements $a_{11} = a_{22} = 18$ and $a_{12} = a_{21} = -9$ according to Equation 7. The eigenvalues of [AλI] can be easily obtained to be $\lambda_1 = 9$ and $\lambda_2 = 27$. The exact solution $P = \pi^2 EI/L^2$ in this case is $P = \lambda = \pi^2 = 9.87$, which indicates that λ_1 is off about 10% from the exact value. If N is increased to 3, h = 0.25 and the eigenvalues are $\lambda_1 = 9.4$, $\lambda_2 = 32$, and $\lambda_3 = 54.6$. The error in estimating the first buckling load is reduced to 100x(9.87–9.4)/9.87 = 4.76%.

Program EigenODE.Stb

Buckling problem belongs to a general class of *stability* problems, for which a program called EigenODE.Stb is developed to demonstrate how different increments or different number of stations can be adopted to continue improving the solution of eigenvalues and eigenvectors with the aid of programs EigenVec, Eigenvlt and Bairstow. The following shows the interactive application of this program.

FORTRAN Version

```
C     Program EigenODE.Stb - Buckling Problem governed by Ordinary
C                            Differential Equation
      DIMENSION C(10,10),EI(10)
      DATA EI/10*1./,RL/1./
      DO 25 N=2,5
      H=RL/(N+1)
```

```
      CALL CBUCKLE(N,10,H,EI,C)
      WRITE (*,50) N
      DO 15 I=1,N
   15 WRITE (*,55) (C(I,J),J=1,N)
   25 CONTINUE
   50 FORMAT(1X,'N = ',I5/1X,'C matrix is')
   55 FORMAT(8F10.5)
      END

      SUBROUTINE CBUCKLE(N,M,H,EI,C)
C     Generates [C] matrix for buckling problem.
      DIMENSION C(M,M),EI(N)
      DO 50 I=1,N
      DO 50 J=1,N
      C(I,J)=0.
      IF (I.EQ.J) C(I,I)=2*EI(I)/H**2
      IF ((I.EQ.(J-1)).OR.(I.EQ.(J+1))) C(I,J)= EI(I)/H**2
   50 CONTINUE
      RETURN
      END
```

Sample Applications

When program **EigenODE.Stb** is run for the buckling problem, the screen will show the coefficient [C] in the standard eigenvalue problem of the form $[A-\lambda I]\{Y\} = \{0\}$ where the values of the buckled shape y(x) computed in 2, 3, 4, and 5 stations between the supported ends of the rod are stored in the vector {Y} and λ is equal to the buckling load P. The resulting display is:

```
N =     2
C matrix is
 18.00000   9.00000
  9.00000  18.00000
N =     3
C matrix is
 32.00000  16.00000    .00000
 16.00000  32.00000  16.00000
   .00000  16.00000  32.00000
N =     4
C matrix is
 50.00000  25.00000    .00000    .00000
 25.00000  50.00000  25.00000    .00000
```

```
     .00000  25.00000  50.00000  25.00000
     .00000    .00000  25.00000  50.00000
 N =      5
 C matrix is
 71.99990  36.00000    .00000    .00000    .00000
 36.00000  71.99999  36.00000    .00000    .00000
   .00000  36.00000  71.99999  36.00000    .00000
   .00000    .00000  36.00000  71.99999  36.00000
   .00000    .00000    .00000  36.00000  71.99999
```

To find the eigenvalues of the above listed matrices, program **CharactEq** can be applied by interactively specifying the elements in these matrices to obtain the respective characteristic equations as

$$\lambda^2 - 36\lambda + 243 = 0,$$

$$\lambda^3 - 96\lambda^2 - 2560\lambda - 16384 = 0,$$

$$\lambda^4 - 200\lambda^3 + 13125\lambda^2 - 312500\lambda + 1953125 = 0, \quad \text{and}$$

$$\lambda^5 - 360\lambda^4 + 46656^3 - 2612736\lambda^2 + 5.878656\text{x}10^7\lambda - 3.62791\text{x}10^8 = 0$$

The eigenvalues for these equations can be found by application of the program **Bairstow**. The sets of eigenvalues for the first three equations are (9 and 27), (9.3726, 32, and 54.627), and (9.5492, 34.549, 65.451, and 90.451). The smallest eigenvalue in magnitude found for the fourth equation is 9.64569. It indicates that if the rod is partitioned into finer and finer increments, the numerical solution continue to improve in predicting the first buckling load from 9, 9.3726, 9.5492, to 9.64569 and converging to the exact value of $P = \lambda = \pi^2 = 9.8696$. For further improvement, the derivation of the characteristic equations of order 6 and higher is given as a homework problem for the reader to practice application of the programs EigenODE.Stb, CharacEq, and Bairstow.

PROGRAM EIGENODE.VIB

For a better understanding of the vibration problem also introduced in Section 7.1, let us assign values for the spring constants and masses to be $k_1 = k_2 = k_3 = 10$ lb/ft and $m_1 = m_2 = m_3 = 1$ lb-sec²/ft. The matrix [A] becomes:

$$[A] = \begin{bmatrix} 20 & -10 & 0 \\ -10 & 20 & -10 \\ 0 & -10 & 10 \end{bmatrix}$$

Indeed, [A] is singular. The determinant of $[A-\lambda I]$ gives the characteristic equation $\lambda^3-50\lambda^2 + 600\lambda-1000 = 0$. The roots can be obtained by application of the program Bairstow to be $\lambda = \omega^2 = 1.98$, 15.5, and 32.5. For $\lambda = 1.98$ the frequency ω is equal to 1.407 radians/second, the amplitude ratios are $C_2/C_1 = 1.80$ and $C_3/C_1 = 2.25$. The program EigenVib has been developed for generation of the matrix [A] when the values of the masses, m's, and the spring constants, k's, are provided.

FORTRAN VERSION

```
C     Program EigenODE.Vib - Vibration Problem governed by Ordinary
C                                Differential Equation
      DIMENSION A(3,3),RK(4),RM(3)
      DATA RK/3*10.,0./,RM/3*1./
      CALL CVIBRA(3,3,RK,RM,A)
      DO 15 I=1,3
   15 WRITE (*,55) (A(I,J),J=1,3)
   25 CONTINUE
   50 FORMAT(1X,'A matrix is')
   55 FORMAT(8F10.5)
      END

      SUBROUTINE CVIBRA(N,M,RK,RM,A)
C     Generates [A] matrix for vibration problem.
      DIMENSION A(M,M),RK(1),RM(1)
      DO 50 I=1,N
      DO 50 J=1,N
      A(I,J)=0.
      IF (I.EQ.J) A(I,I)=(RK(I)+RK(I+1))/RM(I)
      IF (I.EQ.(J-1)) A(I,J)= RK(J)/RM(I)
      IF (I.EQ.(J+1)) A(I,J)= RK(I)/RM(I)
   50 CONTINUE
      RETURN
      END
```

Sample Application

To demonstrate application of the program **EigenODE.Vib**, the numerical example for the vibration of three masses shown in Figure 1 in Section 7.1 is run to generate the matrix [A] and then the program **CharacEq** is used to obtain a characteristic equation $\lambda^3 - 50\lambda^2 - 600\lambda - 1000 = 0$. The program Bairstow enables its roots to be found as equal to 1.9806, 15.550, and 37.470.

It is also of interest to show an application of the programs **MatxInvD** and **EigenvIt** (to be introduced in Section 7.5) for inverting the matrix [C] and then iteratively finding the smallest eigenvalue in magnitude (which is related to the lowest natural frequency of the vibration). The resulting display from using these two programs is:

```
  * Program MatxInvD - Calculate inverse and determinant of a square matrix *

Enter the order of the matrix : 3

Enter the elements of the matrix row-by-row and press <Enter> key
  after entering each element :
? 20? 10? 0
? 10? 20? 10
? 0? 10? 10

Determinant =  999.9999

The inverse matrix is :

  1.000E-01 -1.000E-01  1.000E-01
 -1.000E-01  2.000E-01 -2.000E-01
  1.000E-01 -2.000E-01  3.000E-01

Program EigenvIt - Iterates the largest eigenvalue in magnitude
                   & associated eigenvector of an Nth-order matrix

Enter the order of the matrix, N ? 3

Enter elements of the matrix [M] row-by-row
     Press <Enter> key after entering each number.
? .1? -.1? .1
? -.1? .2? -.2
? .1? -.2? .3

Enter a trial eigenvector of order N
     Press <Enter> key after entering each number.
? 1? 0? 0

Iterations go as follows :

Trial #   Eigenvalue   Normalized Eigenvector    Error
 (Terminated when total error < 0.0001)
   1        0.17321  0.57735 -0/57735  0.57735  1.57735
   2        0.48305  0.35857 -0.59761  0.71714  0.37883
   3        0.50455  0.33165 -0.59222  0.73436  0.04953
   4        0.50489  0.32844 -0.59118  0.73663  0.00653
   5        0.50489  0.32804 -0.59103  0.73693  0.00084
   6        0.50489  0.32799 -0.59101  0.73697  0.00011
   7        0.50489  0.32799 -0.59101  0.73698  0.00001
```

The smallest eigenvalue in magnitude of the matrix [A] is therefore equal to 1/0.50489, or, 1.9806 same as obtained by application of the programs **CharacEq** and **Bairstow**.

MATLAB APPLICATIONS

A file **EigenvIt.m** for **MATLAB** has been developed and is listed and discussed in the program EigenvIt. This function is in the form of [EigenVec,Lambda] = EigenvIt(A,N,V0,NT,Tol). It accepts a matrix [A] of order N, an initial guessed eigenvector V0, and tries to find the eigenvector Eigenvec and eigenvalue Lambda iteratively until the sum of the absolute values of the differences of the components of two consecutive guessed eigenvectors is less than the specified tolerance Tol. The number of iterations is limited by the user to be no more than NT times. The reader should refer to the program **EigenvIt** for more details, here provide a simple example of using **EigenvIt.m**:

```
>> A=[2,0,3; 0,5,0; 3,02,]; V0=[1; 0; 0]; format compact
>> [EigenVec,Lambda]=EigenvIt(A,3,V0,10,0.0001)
 TryNumb =
      8
 V =
     0.7071
          0
     0.7071
 Lambda =
     5.0000
```

The display indicates that for the specified matrix [A] of order equal to 3, the largest eigenvalue in magnitude is equal to 5.0000 and its associated eigenvector is $[0.7071\ 0\ 0.7071]^T$ after 8 iterative steps. The iteration is terminated when the sum of the absolute values of the differences of the corresponding components of the guessed eigenvectors obtained during the seventh and eighth iterations is less than the specified tolerance 0.0001.

To find the smallest eigenvalue and its associated eigenvector by iteration, **EigenvIt.m** also can be applied effectively. Let us use the example in Sample Applications:

```
>> A=[20,10,0; 10,20,10; 0,10,10]; Ainv=inv(A); format compact
>> V0=[1; 0; 0]; [EigenVec,LambdaR]=EigenvIt(Ainv,3,V0,10,0.0001)
 TryNumb =
      7
 V =
     0.3280
    -0.5910
     0.7370
 LambdaR =
     0.5049
 Lambda =
     1.9806
```

Notice that **inv.m** of **MATLAB** has been applied to find the inverse of [A] and using it for **EigenvIt.m** to find the eigenvalue and eigenvector by iteration.

MATHEMATICA APPLICATIONS

For the buckling problem, **Mathematica** can be applied as follows:

```
In[1]: = Ns = 5; EI = 1.; L = 1.; H = L/(Ns + 1);

In[2]: = (Print["Number of Station = ", Ns, " EI = ", EI, " Length = ", L,
              "Delta L = ", H)

Out[2]: =

   Number of Station = 5 EI = 1. Length = 1. Delta L = 0.166667

In[3]: = (Do[Do[If[i == j, M[[i,j]] = 2.*EI/H^2,
                     If[i == (j + 1)||i = = (j–1), M[[i,j]] = EI/H^2, 0]],
                 {i,Ns}],{j,Ns}]); MatrixForm[M]

Out[3]//MatrixForm: =

   72.  36.  0.   0.   0.
   36.  72.  36.  0.   0.
   0.   36.  72.  36.  0.
   0.   0.   36.  72.  36.
   0.   0.   0.   36.  72.
```

In the next section, we will show how the characteristic equation for the above derived matrix [M] can be determined by application of **Mathematica** and subsequently how the eigenvalues and eigenvectors are to be obtained.

7.3 PROGRAM CHARACEQ — DERIVATION OF CHARACTERISTIC EQUATION OF A SPECIFIED SQUARE MATRIX

The program **CharacEq** is designed to generate the coefficients of the characteristic equation of an interactively specified square matrix by use of the *Feddeev-Leverrier* method. Such a characteristic equation is needed in the stability, vibration, and other so-called *eigenvalue* problems.[3] Readers interested in these problems should also refer to the discussions on the programs **EigenODE** and **EigenVec**. The former discusses how the square matrix is to be generated by finite-difference approximation of ordinary differential equation. The latter program delineates how the eigenvectors are to be found by a modified Gaussian elimination method for each eigenvalue and how the eigenvalues are to be solved from the characteristic equation by the program **Bairstow**. Here for derivation of the characteristic equation, let us denote the specified square matrix be [A] and its elements be $a_{i,j}$ for i,j = 1,2,…,n with n being the order of [A]. The Feddeev-Leverrier method first express the characteristic equation of [A] as:

$$(-1)^n(\lambda^n - p_1\lambda^{n-1} - p_2\lambda^{n-2} - \ldots - p_{n-1}\lambda - p_n) = 0 \quad (1)$$

where the coefficients p_1 through p_n are to be determined by the following recursive formulas:

$$p_k = \frac{1}{k} \text{ Trace of } [B]_k \qquad \text{for } k = 1, 2, \ldots, n \tag{2}$$

and

$$[B]_1 = [A] \quad \text{and} \quad [B]_k = [A]([B]_{k-1} - p_{k-1}[I]) \tag{3,4}$$

Equation 4 is to be applied for j = 2,3,…,n. **Trace**, appearing in Equation 2, of a square matrix is the sum of the diagonal elements. A specific, numerical example will help further explain the details involved in applying the formulas presented above. Consider a square matrix:

$$[A] = \begin{bmatrix} 0 & 2 & 3 \\ -10 & -1 & 2 \\ -2 & 4 & 7 \end{bmatrix} \tag{5}$$

Then, $[B]_1 = [A]$ and $p_1 = \text{Trace}([B]_1) = 0\text{–}1\text{+}7 = 6$. The other p's and [B]'s are to be calculated according to Equations 2 and 4, and finally the characteristic equation is to be expressed according to Equation 1 as:

$$[B]_2 = [A]([B]_1 - 6[I]) = \begin{bmatrix} 0 & 2 & 3 \\ -10 & -1 & 2 \\ -2 & 4 & 7 \end{bmatrix} \begin{bmatrix} -6 & 2 & 3 \\ -10 & -7 & 2 \\ -2 & 4 & 1 \end{bmatrix}$$

$$= \begin{bmatrix} -26 & -2 & 7 \\ 66 & -5 & -30 \\ -42 & -4 & 9 \end{bmatrix} \quad p_2 = \text{ Trace }([B]_2)/2 = (-26 - 5 + 9)/2 = -11$$

$$[B]_3 = [A]([B]_2 + 11[I]) = \begin{bmatrix} 0 & 2 & 3 \\ -10 & -1 & 2 \\ -2 & 4 & 7 \end{bmatrix} \begin{bmatrix} -15 & -2 & 7 \\ 66 & 6 & -30 \\ -42 & -4 & 20 \end{bmatrix}$$

$$= \begin{bmatrix} 6 & 0 & 0 \\ 0 & 6 & 0 \\ 0 & 0 & 6 \end{bmatrix} \quad p_3 = \text{ Trace }([B]_3)/3 = (6 + 6 + 6)/3 = 6$$

and finally, the characteristic equation is:

$$(-1)^3(\lambda^3 - p_1\lambda^2 - p_2\lambda - p_3) = -\lambda^3 + 6\lambda^2 - 11\lambda + 6 = 0$$

Both **QuickBASIC** and **FORTRAN** version of the program CharacEq have been made available for derivation of the characteristic equation based on the Feddeev-Leverrier method. The program listings are presented below along with some sample applications.

QuickBASIC Version

```
Program CharacEq - Generates Characteristic Equation of an nth-order matrix
                   [M] using Faddeev-Leverrier method.
                   SCREEN 2: CLEAR : KEY OFF: CLS
  PRINT "Program CharacEq - Generates Characteristic Equation of an";
  PRINT " nth-order matrix [M] using Faddeev-Leverrier method."
  PRINT : INPUT "Enter the order of the matrix, N "; N
  N1 = N - 1
  N2 = N - 2
  DIM A(N1), M(N, N), MK(N, N), MK1(N, N)
  PRINT : PRINT "Enter elements of the matrix [M] row-by-row"
  PRINT "     Press <Enter> key after entering each number."
  FOR I = 1 TO N
      FOR J = 1 TO N
          INPUT ; M(I, J)
          NEXT J: PRINT
      NEXT I
  A(N1) = 0!: FOR I = 1 TO N
                 A(N1) = A(N1) - M(I, I)
                 FOR J = 1 TO N
                     MK1(I, J) = M(I, J)
                     NEXT J
                 NEXT I
  FOR T = N2 TO 0 STEP -1
      K = N - T
      FOR I = 1 TO N
          FOR J = 1 TO N
              IF I = J THEN 175
              MK(I, J) = MK1(I, J)
              GOTO 180
175           MK(I, J) = MK1(I, I) + A(N - K + 1)
180           NEXT J
          NEXT I
      A(N - K) = 0
      FOR I = 1 TO N
          FOR J = 1 TO N
              MK1(I, J) = 0
              FOR Q = 1 TO N
                  MK1(I, J) = MK1(I, J) + M(I, Q) * MK(Q, J)
                  NEXT Q
              NEXT J
```

```
            A(N - K) = A(N - K) - MK1(I, I)
        NEXT I
        A(N - K) = A(N - K) / K
    NEXT T
PRINT : PRINT "The calculated coefficients of the ";
PRINT "characteristic equation are : "
PRINT "     (from highest- to lowest-order terms)": PRINT : PRINT 1;
FOR I = N1 TO 0 STEP -1
    PRINT A(I);
    NEXT I
PRINT : PRINT :          END
```

Sample Application

The display screen will show the following questions-and-answers and the computed results when the matrix [A] given in (5) is interactively entered as the matrix, for which its characteristic equation is to be obtained:

```
Program CharacEq - Generates Characteristic Equation of an nth-order
                   matrix [M] using Faddeev-Leverrier method.

Enter the order of the matrix, N ? 3

Enter elements of the matrix [M] row-by-row
     Press <Enter> key after entering each number.
? 0? 2? 3
? -10? -1? 2
? -2? 4? 7

The calculated coefficients of the characteristic equation are :
     (from the highest- to lowest-order terms)

 1 -6  11 -6
```

FORTRAN Version

```
C  Program CharacEq - Generates Characteristic Equation of an Nth-order
C                     matrix [M] using Faddeev-Leverrier method.
      DIMENSION A(51),M(50,50),MK(50,50),MK1(50,50)
      REAL M,MK,MK1
      WRITE (*,2)
    2 FORMAT(' Program CharacEq - Generates Characteristic Equation',
     *       ' of an Nth-order matrix [M]'/
     *       ' using Faddeev-Leverrier method.')
      WRITE (*,4)
    4 FORMAT(' Enter the order of the matrix, N :')
      READ (*,*) N
      N1=N-1
      N2=N-2
      WRITE (*,6)
    6 FORMAT(/' Enter elements of the matrix [M] row-by-row'/
     *        ' Press <Enter> key after entering each number.')
```

```
     *        ' Press <Enter> key after entering each number.')
      DO 8 I=1,N
             DO 8 J=1,N
  8                READ (*,*) M(I,J)
      A(1)=0.
      DO 12 I=1,N
             A(1)=A(1)-M(I,I)
             DO 12 J=1,N
 12                MK1(I,J)=M(I,J)
      DO 250 K=2,N
              DO 180 I=1,N
                      DO 180 J=1,N
                              IF (I.EQ.J) GO TO 175
                              MK(I,J)=MK1(I,J)
                              GO TO 180
175                           MK(I,J)=MK1(I,I)+A(K-1)
180           CONTINUE
              A(K)=0
              DO 200 I=1,N
                      DO 220 J=1,N
                              MK1(I,J)=0
                              DO 220 IQ=1,N
220                           MK1(I,J)=MK1(I,J)+M(I,IQ)*MK(IQ,J)
200           A(K)=A(K)-MK1(I,I)
250   A(K)=A(K)/K
      WRITE (*,300)
300   FORMAT(/' The calculated coefficients of the characteristic',
     *        ' equation are : '/' (from highest- to lowest-order',
     *        ' terms)'/)
      WRITE (*,*) 1
      DO 310 I=1,N
310   WRITE (*,*) A(I)
      END
```

Sample Application

```
Program CharacEq - Generates Characteristic Equation of an Nth-order matrix [M]
                   using Faddeev-Leverrier method.
Enter the order of the matrix, N :
3

Enter elements of the matrix [M] row-by-row
Press <Enter> key after entering each number
1
2
3
-10
0
2
-2
4
8

The calculated coefficients of the characteristic equation are :
(from highest- to lowest-order terms)

           1
   -9.0000000
   26.0000000
  -24.0000000
```

MATLAB Application

MATLAB has a file called **poly.m** which can be applied to obtain the characteristic equation of a specified square matrix. The following is an example of how to specify a square matrix of order 3, how **poly.m** is to be called, and the resulting display:

```
>> A=[0,2,3;-10,-1,2;-2,4,7]

   A =

        0     2     3

      -10    -1     2

       -2     4     7

>> p=poly(A)

   p =

        1.0000   -6.0000   11.0000   -6.0000
```

For the **FORTRAN** sample problem, we can have:

```
>> A=[1,2,3;-10,0,2;-2,4,8]

   A =

        1     2     3

      -10     0     2

       -2     4     8

>> p=poly(A)

   p =

        1.0000   -9.0000   26.0000  -24.0000
```

Here, we can apply plot.m and polyval of **MATLAB** to graphically explore the roots of this obtained polynomial $P(x) = x^3 - 9x^2 + 26x - 24 = 0$ by interactive entering:

```
>> x = [1:0.05:5]; y = polyval(p,x); plot(x,y), hold
>> XL = [1 5]; YL = [0 0]; plot(XL,YL)
```

The resulting curve is shown in Figure 3. Notice that the added horizontal line intercepts the polynomial curve, it helps indicate where the real roots are. To actually calculate the values of all roots, real or complex, the **roots.m** of **MATLAB** can be applied as follows:

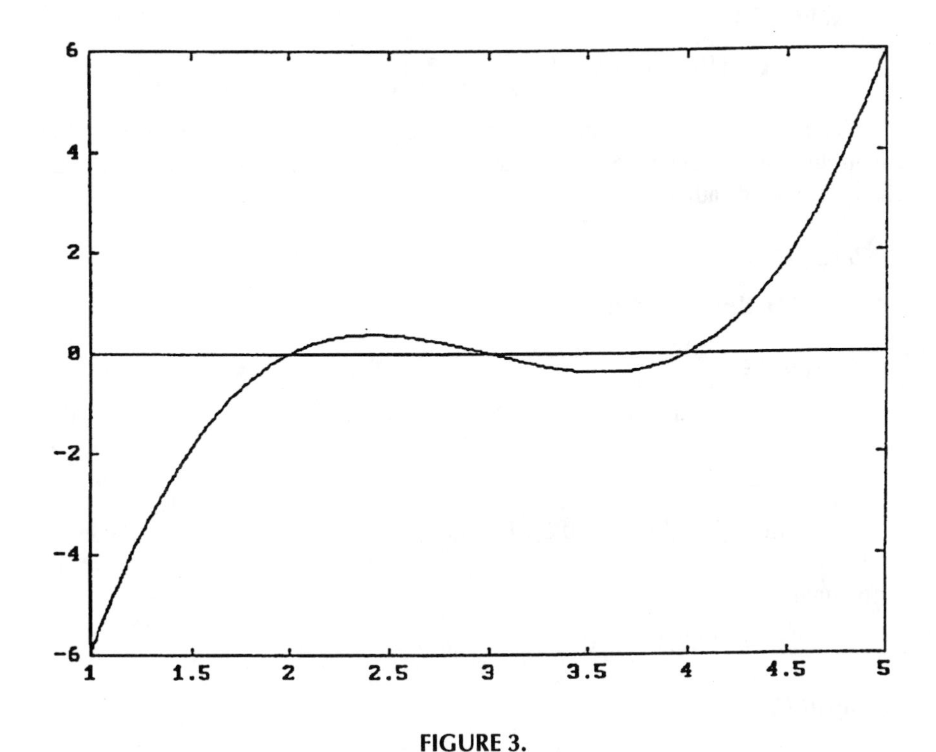

FIGURE 3.

```
>> Xroots=roots(p)

   Xroots =

            4.0000
            3.0000
            2.0000
```

These results complement well with those presented in Figure 3.

MATHEMATICA APPLICATIONS

For finding the characteristic equation of a given matrix, **Mathematica**'s function **Det** which derives the determinant of a specified matrix can be employed. To do so, the matrix should be entered first and then **Det** is to be called next.

Input[1]: =

m = {{0,2,3), {–10,–1,2}, {–2,4,7}}

Output[1] =

```
m = {{0,2,3), {–10,–1,2}, {–2,4,7}}
```

Notice that the elements in each row are separated by comma and enclosed by a pair of braces, and rows are separated also by comma. Next, we derive the characteristic equation of the matrix **m**.

Input[2]: =

```
Det[m — x IdentityMatrix[3]]
```

Output[2] =

$$6 - 11 X + 6 X^2 - X^3$$

Input[3]: =

```
m = {{1,2,3), {–10,0,2}, {–2,4,8}}
```

Output[3] =

```
m = {{1,2,3), {–10,0,2}, {–2,4,8}}
```

Input[4]: =

```
Det[m — x IdentityMatrix[3]]
```

Output[4] =

$$24 - 26 X + 9 X^2 - X^3$$

We may proceed to solve the characteristic roots as follows:

Input[5]: =

```
NSolve[24–26x + 9x^2x^3 = = 0,x]
```

Output[5] =

```
{{x -> 2.}, {x -> –3.}, {x -> 4.}}
```

Again, the polynomial can be plotted with:

Input[6]: =

```
Plot[x^3–9x^2 + 26x–24, {x,1,5},
    Frame->True}, AspectRatio->1]
```

Output[6] =

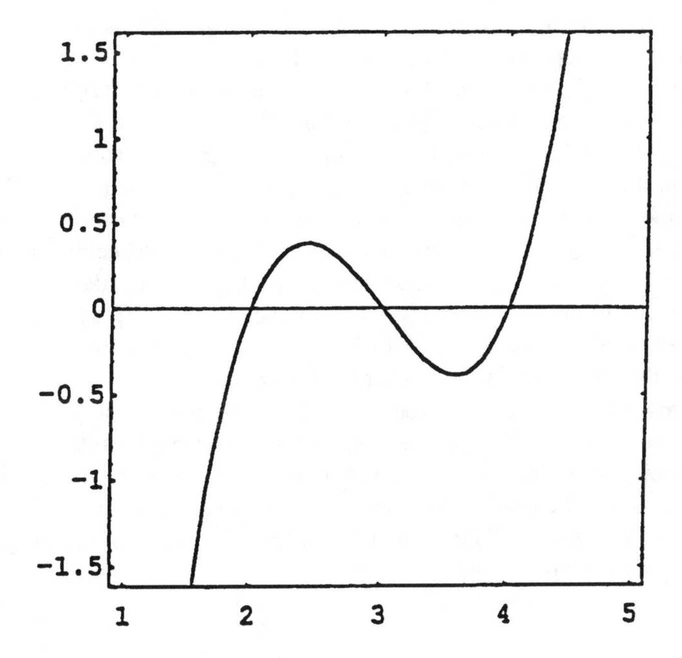

Notice that the graph intercepts the x axis at x = 2, x = 3, and x = 4.

7.4 PROGRAM EIGENVEC — SOLVING EIGENVECTOR BY GAUSSIAN ELIMINATION METHOD

The program **EigenVec** is designed to solve for the associated eigenvector {V} when an eigenvalue of a given square matrix [A] is specified. Eigenvalue and eigenvector problems are discussed in the programs **CharacEq** and **EigenODE**. Here, we describe how the Gaussian Elimination method can be modified for finding the eigenvector {V}. Since the eigenvector {V} satisfies the matrix equation:

$$([A]-\lambda[I])\{V\}=\{0\} \tag{1}$$

where [I] is the identity matrix of same order as [A]. Equation 1 is called *homogeneous* since the right-hand side is a null vector. This equation has nontrivial solution only if the determinant of the coefficient matrix [A]–λ[I] is equal to zero. In other words, the linear algebraic equations represented by Equation 1 are not all independent. The number of equations which are dependent on the other equations, is equal to the multiplicity of the specified λ. For example, if the matrix [A] is of order N and if the multiplicity of λ is M which means M characteristic roots are equal to λ, then there are M equations in Equation 1 are dependent on the other N-M equations.

When Gaussian Elimination method is applied for solving {V} from Equation 1, the normalization of the last equation cannot be carried out if λ has a multiplicity equal to 1 even with the pivoting provision in the program. This is because one of the N equation is dependent on the other N–1 equations. But, it suggests that we may assign the last component of {V} to be equal to an arbitrary constant c and express the other components of {V} in terms of c. This concept can be extended to the case when λ has a multiplicity of M. Since only N-M equations of (1) are independent, there are M independent solutions of {V}. To obtain the first solution, we assign the last component of {V} a value c_1 and the other last M–1 components of {V} equal to zero and then proceed to express the first N-M components of {V} in terms of c_1. To obtain the second solution, we assign the next to the last component of {V} a value c_2 and the other last M–1 components of {V} equal to zero and express the first N-M components of {V} in terms of c_2, and so on. The M solution of {V} can thus be expressed in terms of c_i for i = 1,2,…,M.

The program **EigenVec** is developed from modifying the program **Gauss** by following the above-explained procedure. This program requires the user to interactively specify the order N of [A], the elements of [A], a specified value of λ and its multiplicity M. The results produced by the program EigenVec are the M set of eigenvectors {V}. Both FORTRAN and QuickBASIC versions of this programs are listed below along with sample applications.

QuickBASIC Version

```
' * Program EigenVec - Solving for eigenvector of matrix M of order N when an eigenvalue is
'                      specified, using Gaussian Elimination method.
   CLEAR : CLS : KEY OFF
   PRINT "Program EigenVec - Solving for eigenvector of an Nth-order matrix [M]"
    PRINT " when an eigenvalue is specified, using Gaussian Elimination"; : PRINT " method."
   PRINT : INPUT "Enter the order (N) of the matrix [M] : ", N: DIM M(N,N),V(N)
   FOR I = 1 TO N: PRINT "Enter elements of [M], Row "; : PRINT USING "#### :"; I
       PRINT "  One at a time and press <Return> :";
       FOR J = 1 TO N: INPUT ; M(I, J): NEXT J: PRINT    : NEXT I
   INPUT "Enter the eigenvalue (Lambda) : ", LAMBDA
   INPUT "Is this a repeated eigenvalue?  Enter Y/N : ", A$
   IF A$ = "Y" THEN INPUT "How many times does it repeat? ", NR: GOTO 240
   NR = 1
240 FOR I = 1 TO N     : M(I,I) = M(I, I) - LAMBDA: NEXT I
   FOR K = 1 TO N - NR: JJ = K : K1 = K + 1 : BIG = ABS(M(K, K))      'Pivoting
       FOR I = K1 TO N:  T = ABS(M(I, K)): IF BIG < T THEN BIG=T: JJ=I: NEXT I
       IF JJ = K THEN 310
       FOR J =  1 TO N: T = M(JJ, J): M(JJ, J) = M(K, J):  M(K, J) = T: NEXT J
310    FOR J = K1 TO N: M(K, J) = M(K, J) / M(K, K): NEXT J     'Normalization
       M(K,K)= 1
       FOR I = 1 TO N     :    IF I = K THEN 360                  'Elimination
           FOR J = K1 TO N: M(I, J) = M(I,J)-M(K,J)*M(I,K): NEXT J: M(I,K)=0
360        NEXT I
       NEXT K
   PRINT : PRINT "Normalized Eigenvector(s) :"
   FOR L=1 TO NR: PRINT: FOR LR=N-NR+1 TO N: V(LR)=0: NEXT LR: V(N-L+1)=1: SUM=0
       FOR I=1 TO N-NR : V(I)=-M(I,N-L+1)  :   SUM=SUM+V(I)^2: NEXT I
       SUM=SUM+1: SQRSUM=SQR(SUM): FOR I=1 TO N: V(I)=V(I)/SQRSUM: PRINT V(I): NEXT I
       NEXT L
   END
```

Sample Application

```
Program EigenVec - Solving for eigenvector of an Nth-order matrix [M]
  when an eigenvalue is specified, using Gaussian Elimination method.

Enter the order (N) of the matrix [M] : 3
Enter elements of [M], Row    1 :
  One at a time and press <Enter> :? 3? 0? 2
Enter elements of [M], Row    2 :
  One at a time and press <Enter> :? 0? 5? 0
Enter elements of [M], Row    3 :
  One at a time and press <Enter> :? 2? 0? 3
Enter eigenvalue (Lambda) : 1
Is this a repeated eigenvalue?  Enter Y/N : N

Normalized Eigenvector(s) : {-.7071068   0   .7071968}
```

FORTRAN Version

```
C     Program EigenVec - Solving for eigenvector of matrix M of order N
C                        when an eigenvalue is specified,
C                        using Gaussian Elimination method.
      DIMENSION M(50,50),V(50)
      REAL LAMBDA,M
      WRITE (*,2)
    2 FORMAT(' Program EigenVec - Solving for eigenvector of an',
     *       ' Nth-order matrix [M]'/' when an eigenvalue is',
     *       ' specified, using Gaussian Elimination method.')
      WRITE (*,4)
    4 FORMAT(' Enter the order (N) of the matrix [M] : ',I5)
      READ (*,*) N
      DO 8 I=1,N
      WRITE (*,6) I
    6 FORMAT(' Enter elements of [M], Row ',I2)
    8 READ  (*,*) (M(I,J),J=1,N)
   10 FORMAT(' Press <Return> after entering each number : ')
      WRITE (*,12)
      READ  (*,* ) LAMBDA
   12 FORMAT(' Enter the eigenvalue (Lambda) : ')
      WRITE (*,14)
      READ  (*,* ) NR
   14 FORMAT(' Enter the multiplicity of this eigenvalue : ')
      DO 16 I=1,N
   16 M(I,I)=M(I,I)-LAMBDA
      LAST=N-NR
      DO 31 K=1,LAST
            JJ=K
            K1=K+1
C  Pivoting
            BIG=ABS(M(K,K))
            DO 18 I=K1,N
                  T=ABS(M(I,K))
                  IF (BIG.LT.T) GO TO 17
                  GO TO 18
   17             BIG=T
                  JJ=I
```

```
   18        CONTINUE
             IF (JJ.EQ.K) GO TO 20
             DO 19 J=1,N
                   T=M(JJ,J)
                   M(JJ,J)=M(K,J)
   19        M(K,J)=T
C  Normalization
   20        DO 21 J=K1,N
   21        M(K,J)=M(K,J)/M(K,K)
             M(K,K)=1
C  Elimination
             DO 29 I=1,N
                   IF (I.EQ.K) GO TO 29
                   DO 27 J=K1,N
   27              M(I,J)=M(I,J)-M(K,J)*M(I,K)
                   M(I,K)=0
   29        CONTINUE
   31 CONTINUE
      WRITE (*,35)
   35 FORMAT(' Normalized Eigenvector(s) :')
      DO 40 L=1,NR
            NMNR=N-NR
            WRITE (*,*)
            N1=N-NR+1
            DO 36 LR=N1,N
   36       V(LR)=0
            V(N-L+1)=1.
            SUM=0
            DO 38 I=1,NMNR
            V(I)=-M(I,N-L+1)
   38       SUM=SUM+V(I)**2
            SUM=SUM+1
            SQRSUM=SQRT(SUM)
            DO 39 I=1,N
            V(I)=V(I)/SQRSUM
   39       WRITE (*,41) V(I)
   40 CONTINUE
   41 FORMAT(E12.5)
      END
```

Sample Application

```
Program EigenVec - Solving for eigenvector of an Nth-order matrix [M]
  when an eigenvalue is specified, using Gaussian Elimination method.

Enter the order (N) of the matrix [M] :
3
Enter elements of [M], Row    1 :
3,0,2
Enter elements of [M], Row    2 :
0,5,0
Enter elements of [M], Row    3 :
2,0,3
Enter eigenvalue (Lambda) :
5
Enter the multiplicity of this eigenvalue :
2
Normalized Eigenvector(s) {
```

```
.70711E+00
.00000E+00
.70711E+00

.00000E+00
.10000E+01
.00000E+00
```

MATLAB Applications

MATLAB has a file called **eig.m** which can be applied for finding the eigenvalues and normalized vectors of a specified square matrix. To do so, we first interactively specify the elements of a matrix [A} and then ask for the eigenvalues Lambda and normalized eigenvectors EigenVec by entering (such as for the sample problem in the **FORTRAN** and **QuickBASIC** versions)

```
>> A = [3,0,2;0,5,0;2,03]; [EigenVec,Lambda] = eig(A)
```

It results in a display on the screen:

```
EigenVec =

         0    0.7071    0.7071
    1.0000         0         0
         0   -0.7071    0.7071

Lambda =

     5     0     0
     0     1     0
     0     0     5
```

Notice that the eigenvalues are listed in the diagonal of the matrix Lambda and the corresponding normalized eigenvectors are listed in the matrix EigenVec as columns. To list the eigenvalues in a vector Lambda, we could enter:

```
>> [Lambda] = eig(A)
```

The resulting display is:

```
Lambda =
     5
     1
     5
```

Mathematica Applications

Mathematica has functions **Eigenvalues** and **Eigenvectors** which can be applied to find the eigenvalues and eigenvectors, respectively, for a specified matrix as illustrated by the following example:

In[1]: = a = {{3,0,2},{0,5,0},{2,0,3}}; MatrixForm[a]

Out[1]//MatrixForm: =

$$\begin{matrix} 3 & 0 & 2 \\ 0 & 5 & 0 \\ 2 & 0 & 3 \end{matrix}$$

In[2]: = Eigenvalues[a]

Out[2]: =

{1, 5, 5}

In[3]: = Eigenvectors[a]

Out[3]: =

{–1, 0, 1}, {1, 0, 1}, {0, 1, 0}}

Notice that the computed eigenvectors are not normalized.

As another example, consider the matrix M generated in the program **EigenODE** for the buckling problem when the number of stations is equal to 5. To obtain the eigenvalues, the interactive application of **Mathematica** goes as:

In[4]: = Eigenvalues[M]

Out[4]: =

{134.354, 108., 72., 36., 9.64617}

Notice that the smallest eigenvalue is equal to 9.64617 which predicts the lowest buckling load. Since the exact solution is 9.8696, this further indicates that by continuously increasing the number of stations the smallest eigenvalue in magnitude will eventually converge to the expected value.

Principal Stresses and Planes

As another example of solving the eigenvalues and eigenvectors, consider the problem of determining the principal stresses at a point within a two-dimensional body which is subjected to in-plane loadings. If the normal stresses (σ_x and σ_y) and shear stresses ($\tau_{xy} = \tau_{yx}$), Figure 5, at that point are known, it is a common practice to graphically determine the principal stresses and principal planes, on which the principal stresses act by use of Mohr's circle.[4] But, here we demonstrate how the principal stresses and principal planes can be solved as the eigenvalues and eigenvectors, respectively, of a matrix [A] constructed using the values of σ_x, σ_y, and τ_{xy} as follows:

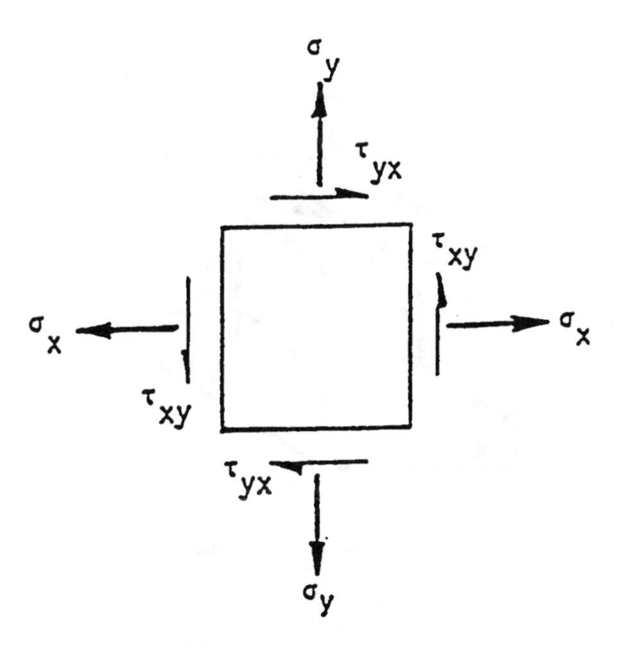

FIGURE 5. If the normal stresses (σ_x and σ_y) and shear stresses ($\tau_{xy} = \tau_{yx}$), are known, it is a common practice to graphically determine the principal stresses and principal planes, on which the principal stresses act by use of Mohr's circle.

$$[A] = \begin{bmatrix} \sigma_x & \tau_{xy} \\ \tau_{yx} & \sigma_y \end{bmatrix} \tag{2}$$

For the three-dimensional cases, the normal stresses σ_x, σ_y, and σ_z, and shear stresses τ_{xy}, τ_{yz}, and τ_{zx} ($\tau_{yx} = \tau_{xy}$, $\tau_{zy} = \tau_{yz}$, and $\tau_{xz} = \tau_{zx}$) are involved, Figure 6. Again, the Mohr's circle method can be applied to graphically solve for the principal stresses and the principal planes, on which they act.[6] But, as an extension of Equation 2, these principal stresses and principal planes can be determined as the eigenvalues and eigenvectors, respectively, of a matrix constructed using the values of the normal and shear stresses as follows:

$$[A] = \begin{bmatrix} \sigma_x & \tau_{xy} & \tau_{xz} \\ \tau_{yx} & \sigma_y & \tau_{yz} \\ \tau_{zx} & \tau_{zy} & \sigma_z \end{bmatrix} \tag{3}$$

Presented below are **MATLAB** solutions of two problems: (a) a two-dimensional case of $\sigma_x = 50$, $\sigma_y = -30$, and $\tau_{xy} = \tau_{yx} = -20$, and (b) a three-dimensional case of $\sigma_x = 25$, $\sigma_y = 36$, $\sigma_z = 49$, $\tau_{xy} = \tau_{yx} = -12$, $\tau_{yz} = \tau_{zy} = 8$, and $\tau_{zx} = \tau_{xz} = -9$, all in N/cm^2.

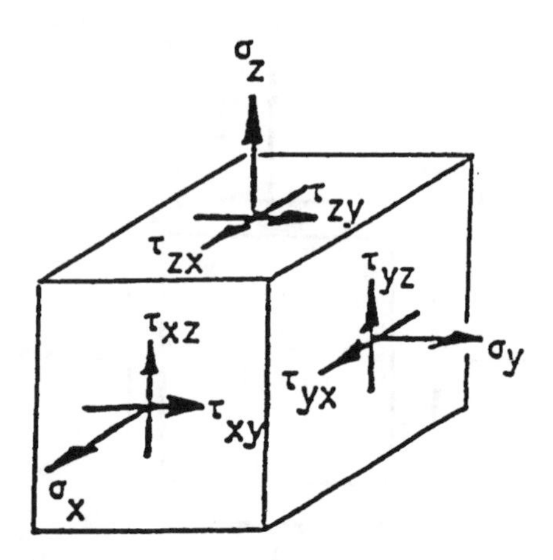

FIGURE 6. For the three-dimensional cases, the normal stresses σ_x, σ_y, and σ_z, and shear stresses τ_{xy}, τ_{yz}, and τ_{zx} ($\tau_{yx} = \tau_{xy}$, $\tau_{zy} = \tau_{yz}$, and $\tau_{xz} = \tau_{zx}$) are involved.

```
>> A=[50,-20;-20,-30]; format compact, [EigenV,Lambda]=eig(A)
   EigenV =
      -0.9732   -0.2298
       0.2298   -0.9732
   Lambda =
      54.7214         0
            0  -34.7214
>> A=[25,-12,-9;-12,36,8;-9,8,49]; [EigenV,Lambda]
   EigenV =
      0.8623    0.3277   -0.3860
      0.4921   -0.7218    0.4867
      0.1192    0.6096    0.7837
   Lambda =
     16.9085         0         0
           0   34.6906         0
           0         0   58.4009
```

Notice that for Problem (a), the result indicates that maximum principal stress equal to 54.7214 N/cm^2 is on a plane having an outward normal vector whose directional cosines are equal to –0.9732 and 0.2298. That is to say this principal plane has an outward normal vector making an angle of $\theta_{max} = 166.7°$ ($\cos\theta_{max} = -0.9732$ and $\cos[90°-\theta_{max}] = 0.2298$) measured counterclockwise from the x-axis. The minimum principal stress is found to be equal to –34.7217 N/cm^2 which is on a plane having an outward normal vector whose directional cosines are equal to –0.2298 and –0.9732, or at an angle equal to $\theta_{min} = -103.3°$ ($\cos\theta_{min} = 0.2298$ and

$\cos[90° - \theta_{min}] = -0.9732$). The two principal planes are perpendicular to each other. This can also be proven by taking the dot product of the two normalized eigenvectors: $(-0.9732i + 0.2298j) \bullet (-0.2298i - 0.9732j) = 0$.

Similar observation can be made from the results for Problem (b). The principal stresses are equal to 16.9085, 34.6906, and 58.4009 N/cm² and they on the planes having outward normal vectors $\mathbf{n}_1 = 0.8632\mathbf{i} + 0.4921\mathbf{j} + 0.1192\mathbf{k}$, $n_2 = 0.3277\mathbf{i} - 0.7218\mathbf{j} + 0.6096\mathbf{k}$, and $\mathbf{n}_3 = -0.3860\mathbf{i} + 0.4867\mathbf{j} + 0.7837\mathbf{k}$, respectively. It is easy to prove that these principal planes are indeed orthogonal by showing that $\mathbf{n}_1 \bullet \mathbf{n}_2 = \mathbf{n}_2 \bullet \mathbf{n}_3 = \mathbf{n}_1 \bullet \mathbf{n}_3 = 0$.

Quadratic Forms and Canonical Transformation

Another interesting application of the procedure involved in solving eigenvalues and eigenvectors of a square matrix is the canonical transformation of quadratic forms[6] for Consider a surface described by the equation:

$$25x^2 + 36y^2 + 49z^2 - 24xy - 18xz + 16yz = 100 \qquad (4)$$

The left-hand side is called a quadratic form in x, y, and z. The surface is an ellipsoid but it is difficult to make out what are the values of its major and minor axes, for which a transformation of the coordinate system is necessary for changing the quadratic form into a canonical one. By canonical form, it means that only the squared terms should remain. To find what transformation is needed, we first write Equation 4 in matrix form as:

$$\{V\}^T[A]\{V\} = 100 \qquad (5)$$

where:

$$\{V\} = \begin{bmatrix} x \\ y \\ z \end{bmatrix} \quad \text{and} \quad [A] = \begin{bmatrix} 25 & -12 & -9 \\ -12 & 36 & 8 \\ -9 & 8 & 49 \end{bmatrix} \qquad (6,7)$$

It can be shown[6] that if we find the eigenvalues and eigenvectors of [A] which in fact are already available in the answer to the previously discussed Problem (b), the coordinate system x-y-z can be transformed into another set of x′-y′-z′ system by using the so-called normalized modal matrix, [Q], formed with the normalized eigenvectors as its columns. That is:

$$\{V'\} = \begin{bmatrix} x' \\ y' \\ z' \end{bmatrix} = \begin{bmatrix} .8623 & .3277 & -.3860 \\ .4921 & -.7218 & .4867 \\ .1192 & .6096 & .7837 \end{bmatrix} \begin{bmatrix} x \\ y \\ z \end{bmatrix} = [Q]\{V\} \qquad (8)$$

Since [Q] has the property of $[Q]^{-1} = [Q]^T$, the quadratic form can then be written as $\{V\}^T[A]\{V\} = ([Q]^{-1}\{V'\})^T[A]([Q]^{-1}\{V'\}) = \{V'\}^T[Q][A][Q]^T\{V'\} = \{V'\}[D]\{V'\}$ where [D] is a diagonal matrix having the eigenvalues of [A] (16.9085, 34.6906, and 58.4009) along its diagonal. Hence, the resulting canonical form for the surface defined by Equation 4 can now be expressed as:

$$16.9085x'^2 + 34.6906y'^2 + 58.4009z'^2 = 10^2 \tag{9}$$

z'-axis and minor axes equal to 10/(16.9085)- and 10/(34.6906)- along x- and y-axes, respectively. According to Equation 8, the orientations of the x'-, y'-, and z'-axes can be determined by using the rows of [Q]. For example, a unit vector I' along x'-axis is equal to 0.9623i + 0.3277j–0.3860k where i, j, and k are unit vectors along x-, y-, and z-axes, respectively. That means, the angles between x'-axis and x-, y-, and z-axes by be obtained easily as:

$$\theta_{x'x} = \cos^{-1}(.8623) = 30.42^\circ, \quad \theta_{x'y} = \cos^{-1}(.3277) = 70.87^\circ$$

and

$$\theta_{x'z} = \cos^{-1}(-.3860) = 112.7^\circ$$

Similarly, we can find $\theta_{y'x} = \cos^{-1}(.4921) = 60.52°$, $\theta_{y'y} = \cos^{-1}(-.7218) = 136.2°$, and $\theta_{y'z} = \cos^{-1}(.4867) = 60.88°$, and $\theta_{z'x} = \cos^{-1}(.1192) = 83.15°$, $\theta_{z'y} = \cos^{-1}(.6096) = 52.44°$, and $\theta_{z'z} = \cos^{-1}(.7837) = 38.40°$.

To verify the above assertions, **MATLAB** is again applied to obtain:

```
>> format compact, [A]=[25,-12,-9;-12,36,8;-9,8,49]
A =
    25   -12    -9
   -12    36     8
    -9     8    49

>>[EigenV,Lambda]=eig(A)
EigenV =
    0.8623    0.3277   -0.3860
    0.4921   -0.7218    0.4867
    0.1192    0.6096    0.7837
Lambda =
   16.9085         0         0
         0   34.6906         0
         0         0   58.4009
>>EigenV'*A*EigenV
ans =
   16.9085    0.0000   -0.0000
    0.0000   34.6906   -0.0000
   -0.0000   -0.0000   58.4009
```

```
>>[ThetaDeg]=180/pi*acos(EigenV)
ThetaDeg =
   30.4189   78.8738  112.7064
   60.5217  136.2040   60.8788
   83.1557   52.4379   38.4009
```

The arc-cosine function acos of **MATLAB** is employed above and pi (= 3.14159) also has been used for converting the results in radians into degrees.

7.5 PROGRAM EIGENVIT — ITERATIVE SOLUTION OF THE EIGENVALUE AND EIGENVECTOR

The program **EigenvIt** is designed to iteratively solve for the largest eigenvalue in magnitude λ_{max} and its associated eigenvector {V} of a given square matrix [A]. Eigenvalue and eigenvector problems are discussed in the programs CharacEq, EigenODE, and EigenVec. Since the eigenvector {V} satisfies the matrix equation:

$$[A]\{V\} = \lambda\{V\} \tag{1}$$

which indicates that if we make a successful guess of {V} then when it is multiplied by the matrix [A] the product should be a scaled version of {V} and the scaling vector is the eigenvalue. Of course, it is not easy to guess correctly what this vector {V} is. But, we may devise a successive guessing scheme and hope for eventual convergence toward the needed solution. In order to make the procedure better organized, let us use *normalized* vectors, that is, to require all guessed eigenvectors to have a length equal to unity.

The iterative scheme may be written as, for k = 0,1,2,...

$$[A]\left\{V^{(k)}\right\} = \lambda^{(k)}\left\{V^{(k+1)}\right\} \tag{2}$$

where k is the iteration counter and $\{V^{(0)}\}$ is the initial guess of the eigenvector for [A]. The iteration is to be terminated when the differences in every components (denoted in lower case of V) of $\{V^{(k+1)}\}$ and $\{V^{(k)}\}$ are sufficiently small. Or, mathematically

$$\left|v_i^{(k+1} - v_i^{(k)}\right| < \varepsilon \tag{3}$$

for i = 1,2,...,N and N being the order of [A]. ϵ in Equation 3 is a predetermined tolerance of accuracy. As can be mathematically proven.[7] this iterative process leads to the largest eigenvalue in magnitude, λ_{max}, and its associated eigenvectors. If it is the smallest eigenvalue in magnitude, λ_{min}, and its associated eigenvector {V} that are necessary to be found, the iterative procedure can also be applied but instead of

[A] its inverse $[A]^{-1}$ should be utilized. The iteration should use the equation, for k = 0,1,2,...

$$[A]^{-1}\{V^{(k)}\} = \alpha^{(k)}\{V^{(k+1)}\} \quad (4)$$

When α_{max} is found, the required smallest eigenvalue in magnitude is to be computed as:

$$\lambda_{min} = 1/\alpha_{max} \quad (5)$$

The program **EigenvIt** has been developed following the concept explained above. Both **QuickBASIC** and **FORTRAN** versions are made available and listed below along with sample applications.

QuickBASIC Version

```
' Program EigenvIt - Iterative determination of the largest Eigenvalue in
'                    magnitude and associated eigenvector of a given matrix (M)
    SCREEN 2: CLEAR : KEY OFF: CLS
    PRINT : PRINT "Program EigenvIt - Iterates the largest eigenvalue in magnitude"
    PRINT "                  & associated eigenvector of an Nth-order matrix"
    PRINT : INPUT "Enter the order of the matrix, N "; N: DIM M(N, N), V(N), VN(N)
    PRINT : PRINT "Enter elements of the matrix [M] row-by-row"
    PRINT "     Press <Enter> key after entering each number."
    FOR I = 1 TO N
        FOR J = 1 TO N
            INPUT ; M(I, J)
            NEXT J
        PRINT
        NEXT I
    PRINT : PRINT "Enter a trial eigenvector of order N"
    PRINT "     Press <Enter> key after entering each number."
    FOR I = 1 TO N
        INPUT ; V(I)
        NEXT I
                      * Iteration *
    CLS : PRINT "Iterations go as follows :" : PRINT
      PRINT "Trial #   Eigenvalue   Normalized Eigenvector   Error"
      PRINT " (Terminated when total error < 0.0001)"
    FOR T = 1 TO 100 : L = 0
        FOR I = 1 TO N
            VN(I) = 0
            FOR J=1 TO N
                VN(I)=VN(I)+M(I,J)*V(J)
                NEXT J: L=L+VN(I)^2
            NEXT I
        L = SQR(L) * SGN(V(1) / VN(1)): E = 0
        FOR I=1 TO N
            E=E+ABS(ABS(V(I))-ABS(VN(I)/L)): V(I)=VN(I)/L
            NEXT I
        PRINT USING "####"; T; : PRINT USING "       ##.#####"; L;
        FOR P = 1 TO N
            PRINT USING " ##.#####"; V(P);
            NEXT P
        PRINT USING " ##.#####"; E
        IF E < .0001 THEN 380
        NEXT T
    PRINT : PRINT "Iteration fails after 100 trials": END
380 PRINT : END
```

Sample Application

```
Program EigenvIt - Iterates the largest eigenvalue in magnitude
                   & associated eigenvector of an Nth-order matrix

Enter the order of the matrix, N ? 3

Enter elements of the matrix [M] row-by-row
     Press <Enter > key after entering each number.
? 2? 0? 3
? 0? 5? 0
? 3? 0? 2

Enter a trial eigenvector of order N
     Press <Enter> key after entering each number.
? 1? 0? 0

Iterations go as follows :

Trial #   Eigenvalue   Normalized Eigenvector    Error
  (Terminated when total error < 0.0001)
   1        3.60555  0.55470  0.00000  0.83205  1.27735
   2        4.90682  0.73480  0.00000  0.67828  0.33387
   3        4.99616  0.70143  0.00000  0.71274  0.06784
   4        4.99985  0.70824  0.00000  0.70597  0.01358
   5        4.99995  0.70688  0.00000  0.70733  0.00272
   6        5.00000  0.70715  0.00000  0.70706  0.00054
   7        5.00000  0.70710  0.00000  0.70712  0.00011
   8        5.00000  0.70711  0.00000  0.70710  0.00002
```

FORTRAN Version

```
C  Program EigenvIt - Iterative determination of the largest Eigenvalue in
C           magnitude and associated eigenvector of a given matrix [M].
      DIMENSION M(50,50),V(50),VN(50)
      REAL L,M
      WRITE (*,2)
    2 FORMAT(' Program EigenvIt - Iterates the largest eigenvalue in ',
     *       'magnitude'/'                    & associated eigenvector ',
     *       'of an Nth-order matrix')
      WRITE (*,5)
      READ  (*,*) N
    5 FORMAT(' Enter the order of the matrix, N : ')
      WRITE (*,8)
    8 FORMAT(' Enter elements of the matrix [M] row-by-row')
      DO 11 I=1,N
   11 READ (*,*) (M(I,J),J=1,N)
      WRITE (*,15)
   15 FORMAT(' Enter a trial eigenvector of order N')
C     DO 20 I=1,N
   20 READ (*,*) (V(I),I=1,N)
C
C  * Iteration *
C
```

```
      WRITE (*,25)
   25 FORMAT('1Iterations go as follows :'/,' Trial #  Eigenvalue ',
     *       '  Normalized Eigenvector   Error'/' (Terminated when ',
     *       'total error < 0.0001)')
      DO 35 IT=1,100
              L=0
              DO 27 I=1,N
                     VN(I)=0
                     DO 26 J=1,N
   26                VN(I)=VN(I)+M(I,J)*V(J)
   27         L=L+VN(I)**2
              L=SQRT(L)
              IF ((V(1)/VN(1)).LT.0.) GO TO 28
              GO TO 285
   28         L=-L
  285         E=0
              DO 29 I=1,N
                     E=E+ABS(ABS(V(I))-ABS(VN(I)/L))
   29         V(I)=VN(I)/L
              WRITE (*,31) IT,L,(V(IP),IP=1,N),E
   31         FORMAT(I5,F14.5,10(F9.5))
              IF (E.LT.0.0001) GO TO 38
   35 CONTINUE
      WRITE (*,36)
   36 FORMAT(/' Iteration fails after 100 trials')
      GO TO 40
   38 WRITE (*,*)
   40 END
```

Sample Application

```
Program EigenvIt - Iterates the largest eigenvalue in magnitude
                   & associated eigenvector of an Nth-order matrix
Enter the order of the matrix, N :
3
Enter elements of the matrix [M] row-by-row
3, 0, 2
0, 5, 0
2, 0, 3
Enter a trial eigenvector of order N
1, 0, 0
Iterations go as follows :
Trial #  Eigenvalue   Normalized Eigenvector    Error
  (Terminated when total error < 0.0001)
   1        3.60555  0.83205  0.00000  0.55470  .72265
   2        4.90682  0.73480  0.00000  0.67828  .22083
   3        4.99616  0.71274  0.00000  0.70143  .04521
   4        4.99985  0.70824  0.00000  0.70597  .00905
   5        4.99995  0.70733  0.00000  0.70688  .00181
   6        5.00000  0.70715  0.00000  0.70706  .00036
   7        5.00000  0.70712  0.00000  0.70710  .00007
```

MATLAB Application

A **EigenvIt.m** file can be created and added to **MATLAB** m files for iterative solution of the eigenvalue largest in magnitude and its associated normalized eigenvector of a given square matrix. It may be written as follows:

```
function [EigenVec,Lambda]=EigenvIt(A,N,V0,NT,Tol)
        % Iterates eigenvalue (Lambda) and eigenvector (EigenVec) of [A]
        %   of order N using an initial guess of EigenVec=V0.
        % Terminates when the sum of the absolute values of differences
        %   of the respective components of two consecutive guessed
        %   eigenvectors < Tol.
        % Will be tried NT times.
    for TryNumb=1:NT
        V=A*V0; Lambda=sqrt(V'*V); VN=V./Lambda;
        D=0.; for I=1:N; D=D+abs(VN(I)-V0(I)); end
        if D<Tol; EigenVec=VN; TryNumb, break
           else V0=VN;
           end
        end
        if TryNumb >= NT
           'Iteration fails after the specified number of trials.'
           end
```

The arguments of **EigenvIt** are explained in the comment statements which in **MATLAB** start with a character %. As illustrations of how this function can be applied, two examples are given below.

```
>> A=[2,0,3; 0,5,0; 3,0,2]; V0=[1; 0; 0];
>> [V,Lambda]=EigenvIt(A,3,V0,10,0.0001)

 TryNumb =
      8
 V =
     0.7071
          0
     0.7071
 Lambda =
     5.0000
>> A=[2,0,3; 0,5,0; 3,0,2]; V0=[1; 0; 0];
>> [V,Lambda]=EigenvIt(A,3,V0,2,0.0001)
 ans =
 Iteration fails after the specified number of trials.
 V =
      []
 Lambda =
     4.9068
```

Notice that the first attempt allows 10 iterations but the answer has been obtained after 8 trials whereas the second attempt allowing only two trials fails to converge and V is printed as a blank vector.

In fact, the iteration can be carried out without a M file. To resolve the above problem, **MATLAB** commands can be repeatedly entered by interactive operations as follows:

```
>> format compact, Ntry=0, A=[2,0,3;0,5,0;3,0,2], V=[1;0;0]
Ntry =
     0
A =
     2     0     3
     0     5     0
     3     0     2
V =
     1
     0
     0
>> Ntry=Ntry+1, V=A*V; Lambda=sqrt(V'*V), V=V/Lambda
Ntry =
     1
Lambda =
    3.6056
V =
    0.5547
         0
    0.8321
>> Ntry=Ntry+1, V=A*V; Lambda=sqrt(V'*V), V=V/Lambda
Ntry =
     2
Lambda =
    4.9068
V =
    0.7348
         0
    0.6783
>> Ntry=Ntry+1, V=A*V; Lambda=sqrt(V'*V), V=V/Lambda
Ntry =
     3
Lambda =
    4.9962
V =
    0.7014
         0
    0.8321
>> Ntry=Ntry+1, V=A*V; Lambda=sqrt(V'*V), V=V/Lambda
Ntry =
     4
Lambda =
    4.9998
V =
    0.7082
         0
    0.7060
>> Ntry=Ntry+1, V=A*V; Lambda=sqrt(V'*V), V=V/Lambda
Ntry =
     5
Lambda =
    5.0000
```

```
V =
    0.7069
         0
    0.7073
>> Ntry=Ntry+1, V=A*V; Lambda=sqrt(V'*V), V=V/Lambda
Ntry =
     6
Lambda =
    5.0000
V =
    0.7072
         0
    0.7071
>> Ntry=Ntry+1, V=A*V; Lambda=sqrt(V'*V), V=V/Lambda
Ntry =
     7
Lambda =
    5.0000
V =
    0.7071
         0
    0.7071
```

Notice that the command **format compact** makes the printout lines to be closely spaced. The iteration converges after 7 trials. This interactive method of continuous iteration for finding the largest eigenvalue and its associated eigenvector is easy to follow, but the repeated entering of the statement ">> Ntry = Ntry + 1...V = V/Lambda" in the interactive execution is a cumbersome task. To circumvent this situation, one may enter:

```
>> A= [2,0,3;0,5,0;3,0,2]; V= [1;0;0]; format compact
>> for Ntry= 1:100, Ntry, V= A*V; Lambda= sqrt(V'*V), V= V/Lambda, pause, end
```

The **pause** command enables each iterated result to be viewed. To continue the trials, simply press any key. The total number of trials is arbitrarily limited at 100; the actual need is 7 trials as indicated by the above printed results. Viewer can terminate the iteration by pressing the <Ctrl> and <Break> keys simultaneously after satisfactorily seeing the 7th, converged results of Lambda = 5 and V = [0.7071; 0; 0.7071] being displayed on screen.

To iteratively find the smallest eigenvector and its associated, normalized eigenvector of [A] according to Equations 4 and 5, we apply the iterative method to $[A]^{-1}$ by entering

```
>> Ainv= inv(A); V= [1;0;0];
>> for Ntry= 1:100, Ntry, V= Ainv*V; Lambda= 1/sqrt(V'*V), V= V*Lambda, pause, end
```

The resulting display is:

```
Ntry =
     1
Lambda =
     1.3868
V =
   -0.7069
         0
    0.7073
Ntry =
     2
Lambda =
     1.0190
V =
    0.7348
         0
   -0.6783
Ntry =
     3
Lambda =
     1.0008
V =
   -0.7014
         0
    0.7127
Ntry =
     4
Lambda =
     1.0000
V =
    0.7082
         0
   -0.7060
```

```
Ntry =
     5
Lambda =
     1.0000
V =
   -0.5547
         0
    0.8321
Ntry =
     6
Lambda =
     1.0000
V =
    0.7072
         0
   -0.7071
Ntry =
     7
Lambda =
     1.0000
V =
   -0.7071
         0
    0.7071
Ntry =
     8
Lambda =
     1.0000
V =
    0.7071
         0
   -0.7071
```

For saving space, the last four iterations are listed in the column on the right. The components of the last two iterated normalized eigenvectors are numerically equal but differ in sign, it suggests that the eigenvalue is actually equal to –1.0000 instead of 1.0000.

Mathematica Applications

The **While** command of **Mathematica** can be effectively applied here for iteration of eigenvalue and eigenvector. It has two arguments, the first is a testing expression when the condition is true the statement(s) specified in the second argument should then be implemented. When the condition is false, the **While** statement should be terminated. To obtain the largest eigenvalue in magnitude, we proceed directly with a given matrix as follows:

Input[1]: = A = {{2,0,3},{0,5,0},{3,0,2}}; V = {1,0,0}; I = 0;

Input[2]: = While[i<8, I = I + 1; VN = A.V; Lambda = Sqrt[VN.VN];
Print["Iteration # ",I," ","Lambda = ",N[Lambda],
" {V} =",V = N[VN/Lambda]]]

Output[2]: =

Iteration # 1 Lambda = 3.60555 {V} = {0.5547, 0, 0.83205}
Iteration # 2 Lambda = 4.90682 {V} = {0.734803, 0, 0.67828}
Iteration # 3 Lambda = 4.99616 {V} = {0.701427, 0, 0.712741}
Iteration # 4 Lambda = 4.99985 {V} = {0.708237, 0, 0.709575}
Iteration # 5 Lambda = 4.99999 {V} = {0.70688, 0, 0.707333}
Iteration # 6 Lambda = 5. {V} = {0.707152, 0, 0.707062}
Iteration # 7 Lambda = 5. {V} = {0.707098, 0, 0.707016}
Iteration # 8 Lambda = 5. {V} = {0.707109, 0, 0.707105}

Notice that the testing condition is whether the running iteration counter I is less than 8 or not. This setup enables up to eight iterations to be conducted. The function N in *Input[2]* requests the value of the variable inside the brackets to be given in numerical form. For example, when the value of Lambda is displayed as sqrt[4], it will be displayed as 2.00000 if the input is N[Lambda]. VA·VB is the dot product of VA and VB. In *Input[2]* some sample printouts of the character strings specified inside a pair of parentheses are also demonstrated.

For iterating the smallest eigenvalue in magnitude, we work on the inverse of this given matrix as follows:

Input[3]: = Ainv = Inverse[A]; V = {1,0,0}; I = 0;

Input[4]: = While[i<8, I = I + 1; VN = Ainv.V; Lambda = Sqrt[VN.VN];
Print["Iteration # ",I," ","Lambda = ",N[Lambda],
" {V} = ",V = N[VN/Lambda]]]

Output[4]: =

Iteration # 1 Lambda = 0.72111 {V} = {–0.5547, 0, 0.83205}
Iteration # 2 Lambda = 0.981365 {V} = {0.734803, 0, –0.67828}
Iteration # 3 Lambda = 0.999233 {V} = {–0.701427, 0, 0.712741}
Iteration # 4 Lambda = 0.999969 {V} = {0.708237, –0, 0.709575}
Iteration # 5 Lambda = 0.999999 {V} = {–0.70688, 0, 0.707333}
Iteration # 6 Lambda = 1. {V} = {0.707152, 0, –0.707062}
Iteration # 7 Lambda = 1. {V} = {–0.707098, 0, 0.707016}
Iteration # 8 Lambda = 1. {V} = {0.707109, 0, –0.707105}

Inverse is a **Mathematica** function which inverts a specified matrix. The above printout of eight iterations shows that {V} continues to change its sign. This is an indication that the eigenvalue carries a minus sign.

7.6 PROBLEMS

PROGRAMS EIGENODE.STB AND EIGENODE.VIB

1. Apply the program **EigenODE.Stb** for the cases N = 6, 7, and 8 to obtain the coefficient matrix [C] in the standard eigenvalue problem of ([C]-λ[I]){Y} = {0}.
2. Apply the program **CharacEq** to obtain the characteristic equations of order 6, 7, and 8 for the matrices [C] derived in Problem 1.
3. Apply the program **Bairstow** to find the roots for the characteristic equationsderived in Problem 2. If necessary, change this program to allow interactive input of the u and v values for the guessing quadratic factor $\lambda^2 + u\lambda + v$. This enhancement will be helpful if the iteration fails to converge.
4. The program **EigenODE.Vib** has been arranged for solving the general problem of having N masses, m_1–m_N, in series connected by N + 1 springs with stiffnesses k_1–k_{N+1}. Apply it for the case when the three masses shown in Figure 1 are connected by four springs with the fourth spring attached to the ground. Use $m_1 = m_2 = m_3 = 1$ and $k_1 = k_2 = k_3 = k_4 = 10$.
5. Apply the program **CharacEq** to find the characteristic equation for the vibration problem described in Problem 4.
6. Apply the program **MatxInvD** to invert the matrix obtained in Problem 4 and then apply the program **EigenvIt** to iteratively determine its smallest eigenvalue in magnitude and associated eigenvector.
7. Extend the vibrating system described in Problem 4 to four masses and five springs and then implement the application of the programs **MatxInvD** and **EigenvIt** as described in Problems 5 and 6, respectively.
8. Apply the programs **CharacEq**, **Bairstow**, and **EigenVec** to find the characteristic equation, eigenvalues, and associated eigenvectors for the matrix derived in Problem 4, respectively.
9. Same as Problem 8 except for a four masses and five springs system.
10. An approximate analysis of a three-story building is described in Problem 7 in the program EigenvIt. Derive the governing differential equations for the swaymotions x_i for i = 1,2,3 and then show that the stiffness matrix [K] and mass matrix [M] are indeed as those given there.

CHARACEQ

1. Apply Feddeev-Leverrier method to find the characteristic equation of the matrix:

$$\begin{bmatrix} 1 & 2 & 3 \\ 4 & 5 & 6 \\ 7 & 8 & 9 \end{bmatrix}$$

2. Apply Feddeev-Leverrier method to find the characteristic equation of the matrix:

$$\begin{bmatrix} 5 & 0 & 1 \\ -10 & 6 & 0 \\ -2 & 0 & 7 \end{bmatrix}$$

3. Apply Feddeev-Leverrier method to find the characteristic equation of the matrix:

$$\begin{bmatrix} 2 & 2 & 3 \\ -10 & 1 & 2 \\ -2 & 4 & 9 \end{bmatrix}$$

4. Apply the program **CharacEq** for solving Problems 1 to 3.
5. Apply Feddeev-Leverrier method to find the characteristic equation of the matrix:

$$\begin{bmatrix} 5 & 0 & 1 \\ 2 & 6 & 0 \\ 0 & 3 & 7 \end{bmatrix}$$

6. Apply poly.m of **MATLAB** to Problems 1 to 3 and 5.
7. Find the roots of the polynomials found in Problem 6 by application of **roots.m** of **MATLAB**.
8. Apply **plot.m** of **MATLAB** for the polynomials obtained in Problem 6.
9. Apply the function **det** of **Mathematica** to derive the characteristic equation for the matrix given in Problem 1.
10. Apply the function **det** of **Mathematica** to derive the characteristic equation for the matrix given in Problem 2.
11. Apply the function **det** of **Mathematica** to derive the characteristic equation for the matrix given in Problem 3.
12. Apply the function det of **Mathematica** to derive the characteristic equation for the matrix given in Problem 5.

EIGENVEC

1. Run the **QuickBASIC** version of the program **EigenVec** for the sample case used in the **FORTRAN** version.
2. Apply the program **EigenVec** to find the eigenvector corresponding to the eigenvalue equal to 4.41421 for the matrix:

$$\begin{bmatrix} 2 & 0 & 3 \\ 0 & 3 & 0 \\ 1 & 0 & 4 \end{bmatrix}$$

3. Apply the program **CharacEq** to find the characteristic equation for matrix:

$$\begin{bmatrix} 1 & 2 & 3 \\ 4 & 5 & 6 \\ 7 & 8 & 10 \end{bmatrix}$$

   and then apply the program **Bairstow** to find the eigenvalues. Finally, apply the program **EigenVec** to find the eigenvectors.
4. Apply the program **CharacEq** to find the characteristic equation for the matrix:

$$\begin{bmatrix} 5 & 0 & 1 \\ -10 & 6 & 0 \\ -2 & 0 & 7 \end{bmatrix}$$

   and then apply the program **Bairstow** to find the eigenvalues. Finally, apply the program **EigenVec** to find the eigenvectors.
5. Apply the program **CharacEq** to find the characteristic equation for the matrix:

$$\begin{bmatrix} 2 & 2 & 3 \\ -10 & 1 & 2 \\ -2 & 4 & 9 \end{bmatrix}$$

   and then apply the program Bairstow to find the eigenvalues. Finally, apply the program **EigenVec** to find the eigenvectors.
6. The eigenvalues for the following matrix have been found to be equal to 9.3726, 32 and 54.627:

$$\begin{bmatrix} 32 & 16 & 0 \\ 16 & 32 & 16 \\ 0 & 16 & 32 \end{bmatrix}$$

Find the associated eigenvector by applying the program **EigenVec**.

7. The eigenvalues of the following matrix have been found to be equal to 9.5492, 34.549, 64.451, and 90.451:

$$\begin{bmatrix} 50 & 25 & 0 & 0 \\ 25 & 50 & 25 & 0 \\ 0 & 25 & 50 & 25 \\ 0 & 0 & 25 & 50 \end{bmatrix}$$

Find the associated eigenvector by applying the program **EigenVec**.

8. Find the eigenvalue and associated eigenvector of the matrix:

$$\begin{bmatrix} 76 & 36 & 0 & 0 & 0 \\ 36 & 72 & 36 & 0 & 0 \\ 0 & 36 & 72 & 36 & 0 \\ 0 & 0 & 36 & 72 & 36 \\ 0 & 0 & 0 & 36 & 72 \end{bmatrix}$$

9. Swaying motion of a three-story building is described in Problem 7 in the program **EigenvIt**. Use the data there to form the matrix [A] which is equal to $[K]^{-1}$ [M].
   Apply the programs **CharacEq** and **Bairstow** to find all three eigenvalues and then apply the program **EigenVec** to find the associated eigenvectors.
10. Apply the function **eig.m** of **MATLAB** to find all eigenvalues of the matrices given in Problems 2 to 8.
11. Apply the functions **eigenvalues** and **eigenvectors** of **Mathematica** to find all eigenvalues of the matrices given in Problems 2 to 8.

EIGENVIT

1. Using an initial, guessed eigenvector $\{V\} = [1\ 0\ 0]^T$, perform four iterative steps to find the largest eigenvalue in magnitude and its associated normalized vector of the matrix:

$$[A] = \begin{bmatrix} 2 & 2 & 3 \\ -10 & 1 & 2 \\ -2 & 4 & 9 \end{bmatrix}$$

2. Using an initial, guessed eigenvector {V} = $[1\ 0\ 0]^T$, perform four iterative steps to find the largest eigenvalue in magnitude and its associated normalized vector of the matrix:

$$[A]=\begin{bmatrix}2 & 0 & 1\\0 & 3 & 0\\1 & 0 & 4\end{bmatrix}$$

3. Apply the program **EigenvIt** to find the largest eigenvalue in magnitude and its associated normalized eigenvector of the matrix:

$$[A]=\begin{bmatrix}5 & 3\\1 & 3\end{bmatrix}$$

   Next, apply the program **MatxInvD** to find the inverse of [A] which is to be entered as input for program **EigenvIt** to iterate the smallest eigenvalue in magnitude and its associated normalized eigenvector for [A]. Compare the results with the analytical solution of $\lambda_{smallest} = 2$ and $\lambda_{largest} = 6$.
4. Apply the program **MatxInvD** to find the inverse of the matrix [A] given in Problem 1 and then apply the program **EigenvIt** to find the smallest eigenvalue in magnitude and its associated normalized eigenvector of [A]. For checking the values of $\lambda_{smallest}$ obtained here and $\lambda_{largest}$ obtained in Problem 1, derive the characteristic equation of [A] by use of the program CharacEq and solve it by application of the program Bairstow.
5. Same as Problem 4 but for the matrix [A] given in Problem 2.
6. Apply **poly.m**, **roots.m**, **polyval.m**, **plot.m**, and **xlabel** and **ylabel** to obtain a plot of the characteristic equation of the matrix [A] given in Problem 1, shown in Figure 7, to know the approximate locations of the characteristic roots.
7. For a 3-floor building as sketched in the left side of Figure 8, an approximate calculation of its natural frequencies can be attempted by using a **lumped** approach which represents each floor with a mass and the stiffnesses of the supporting columns by a spring as shown in the right side of Figure 8. If the swaying motion of the floors are expressed as $x_i = X_i \sin\omega t$ for i = 1,2,3 where ω is the natural frequency and X_i are the amplitudes, it can be shown that ω and $\{X\} = [X_1\ X_2\ X_3]^T$ satisfy the matrix equation $[K]\{X\} = \omega^2[M]\{X\}$, in which the mass matrix [M] and stiffness matrix [K] are formed by the masses and spring constants as follows:

$$[M]=\begin{bmatrix}m_1 & 0 & 0\\0 & m_2 & 0\\0 & 0 & m_3\end{bmatrix} \quad \text{and} \quad [K]=\begin{bmatrix}k_1 & -k_2 & 0\\-k_2 & k_1+k_2 & -k_3\\0 & -k_3 & k_2+k_3\end{bmatrix}$$

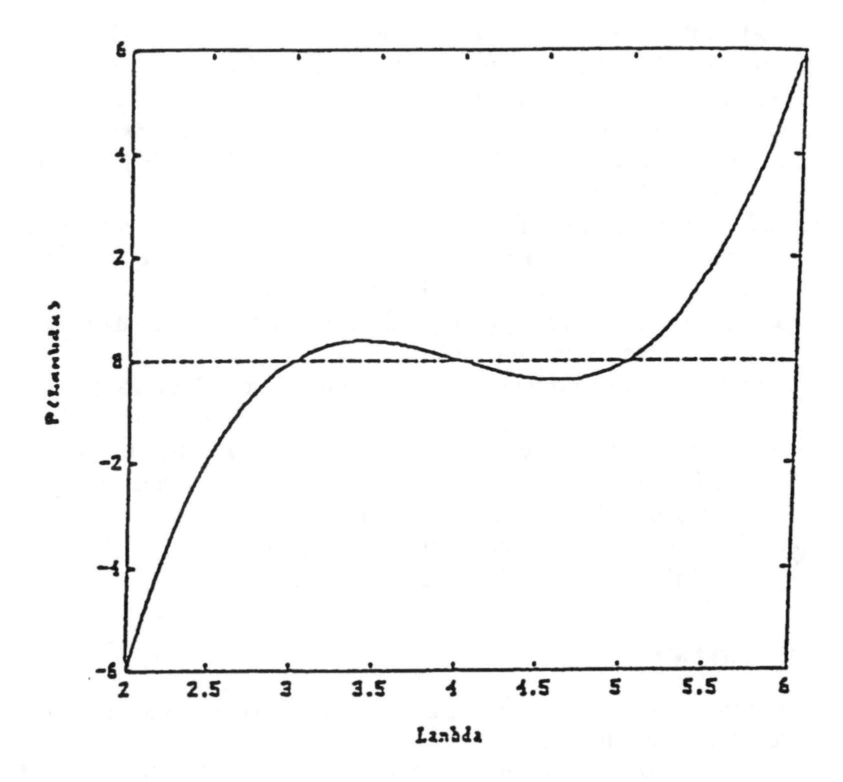

FIGURE 7. Problem 6.

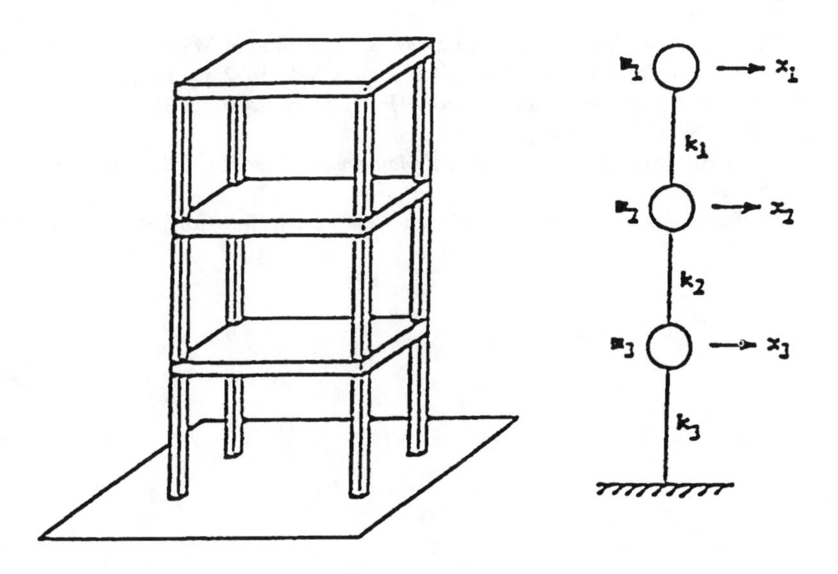

FIGURE 8. Problem 7.

To find the lowest natural frequency ω_{min}, the program EigenvIt can be applied to obtain the λ_{max} from the matrix equation $[A]\{X\} = \lambda\{X\}$ where the matrix [A] is equal to $[K]^{-1}[M]$ and $= \omega^2$. ω_{min} is equal to $1/\lambda_{max}$. Determine the numeric value of ω_{min} for the case when $m_1 = 8x10^5$, $m_2 = 9x10^5$, and $m_3 = 1x10^6$ all in N-sec²/m, and $k_1 = 3x10^8$, $k_2 = 4x10^8$, and $k_3 = 5x10^8$ all in N/m.

8. Referring to Figure 2 in the program **EigenVec**, iteratively determine the maximum and minimum principal stresses and their associated principal planes at a point where the two-dimensional normal and shear stresses are $\sigma_x = 50$, $\sigma_y = -30$, and $\tau_{xy} = \tau_{yx} = -20$ all in N/cm². Compare the results with those obtained in the program EigenVec.
9. Same as Problem 8, except for a three-dimensional case of $\sigma_x = 25$, $\sigma_y = 36$, $\sigma_z = 49$, $\tau_{xy} = \tau_{yx} = -12$, $\tau_{yz} = \tau_{zy} = 8$, and $\tau_{zx} = \tau_{xz} = -9$, all in N/cm².
10. Apply **MATLAB** to invert the matrix [A] given in Problem 1 and then apply **EigenvIt.m** to iterate the eigenvalue which is the smallest in magnitude and also the associated eigenvector.
11. Same as Problem 10 but for the matrix [A] given in Problem 2.
12. Apply **Mathematica** to solve Problems 10 and 11.

7.7 REFERENCES

1. W. F. Riley and L. Zachary, *Introduction to Mechanics of Materials*, Wiley & Sons, Inc., New York, 1989.
2. K. N. Tong, *Theory of Mechanical Vibration*, Wiley & Sons, Inc., New York, 1960.
3. Y. C. Pao, "A General Program for Computer Plotting of Mohr's Circle," *Computers and Structures*, V. 2, 1972, pp. 625–635. This paper discusses various sources of how eigenvalue problems are formed and also methods of analytical, computational, and graphical solutions.
4. Y. C. Pao, "A General Program for Computer Plotting of Mohr's Circle," (for two-dimensional cases), *Computers and Structures*, V. 2, 1972, pp. 625–635.
5. F. B. Seely and J. O. Smith, *Advanced Mechanics of Materials*, Second Edition, John Wiley, New York, 1957, pp. 59–64.
6. F. B. Hilebrand, *Methods of Applied Mathematics*, Prentice-Hall, Englewood Cliffs, NJ, 1960.
7. S. Perlis, *Theory of Matrices*, Addison-Wesley Publishing Company, Reading, MA, 1952.

8 Partial Differential Equations

8.1 INTRODUCTION

Different engineering disciplines solve different types of problems in their respective fields. For mechanical engineers, they may need to solve the temperature change within a solid when it is heated by the interior heat sources or due to a rise or decrease of its *boundary* temperatures. For electrical engineers, they may need to find the voltages at all circuit joints of a computer chip board. Temperature and voltage are the variables in their respective fields. Hence, they are called *field variables*. It is easy to understand that the value of the field variable is *space-dependent* and *time-dependent*. That is to say, that we are interested to know the *spatial* and *temporal* changes of the field variable. Let us denote the field variable as ϕ. and let the independent variables be x_i which could be the time t, or, the space coordinates as x, y, and z. In order not to overly complicate the discussion, we introduce the general two-dimensional partial differential equation which governs the field variable in the form of:

$$\begin{aligned} A(x_1,x_2)\frac{\partial^2\phi}{\partial x_1^2}+B(x_1,x_2)\frac{\partial^2\phi}{\partial x_1\partial x_2}+C(x_1,x_2)\frac{\partial^2\phi}{\partial x_2^2} \\ =F\left(x_1,x_2,\phi,\frac{\partial\phi}{\partial x_1},\frac{\partial\phi}{\partial x_2}\right) \end{aligned} \tag{1}$$

where the coefficient functions A, B, and C in the general cases are dependent on the variables x_1 and x_2, and the right-hand-side function F, called *forcing function* may depend not only on the independent variables x_1 and x_2 but may also depend on the first derivatives of ϕ. There are innumerable of feasible solutions for Equation 1. However, when the initial and/or boundary conditions are specified, only particular solution(s) would then be found appropriate.

In this chapter, we will discuss three simple cases when A, B, and C are all constants. The first case is a two-dimensional, *steady-state heat conduction* problem involving temperature as the field variable and only the spatial distribution of the temperature needs to be determined, Equation 1 is reduced to a *parabolic* partial differential equation named after *Poisson* and *Laplace* when the forcing function F is not equal to, or, equal to zero, respectively. This is a case when the coefficient functions in Equation 1 are related by the condition $B^2-4AC<0$.

The second case is a one-dimensional, *transient heat conduction* problem. Again, the field variable is the temperature which is changing along the longitudinal x-axis of a straight rod and also in time. That is, x_1 becomes x and x_2 become the time t. Equation 1 is reduced to an *elliptical* partial differential equation. This is a case when $B^2–4AC = 0$.

The third case is the study of the vibration of a tightened string. The field variable is the lateral deflection of this string whose shape is changing in time. Equation 1 is reduced to a *hyperbolic* partial differential equation. If x is the longitudinal axis of the string, then same as in the second case, the two independent variables are x and t. This is a case when $B^2–4AC>0$.

The reason that these problems are called parabolic, elliptical, and hyperbolic is because their characteristic curves have such geometric features. Readers interested in exploring these features should refer to a textbook on partial differential equations.

Details will be presented regarding how the forward, backward, and central differences discussed in Chapter 4 are to be applied for approximating the first and second derivative terms appearing in Equation 1. Repetitive algorithms can be devised to facilitate programming for straight-forward computation of the spatial and temporal changes of the field variable. Numerical examples are provided to illustrate how these changes can be determined by use of either **QuickBASIC**, **FORTRAN**, **MATLAB**, or, **Mathematica** programs.

Although explanation of the procedure for numerical solution of these three types of problems is given only for the simple one- and two-dimensional cases, but its extension to the higher dimension case is straight forward. For example, one may attempt to solve the transient heat conduction problem of a thin plate by having two space variables instead of one space variable for a long rod. The steady-state heat conduction problem of a thin plate can be extended for the case of a three-dimensional solid, and the string vibration problem can be extended to a two-dimensional membrane problem.

8.2 PROGRAM PARABPDE — NUMERICAL SOLUTION OF PARABOLIC PARTIAL DIFFERENTIAL EQUATIONS

The program **ParabPDE** is designed for numerically solving engineering problems governed by parabolic partial differential equation in the form of:

$$\frac{\partial \phi}{\partial t} = a\frac{\partial^2 \phi}{\partial x^2} \tag{1}$$

and ϕ is a function of t and x and satisfies a certain set of supplementary conditions. Equation 1 is called a parabolic partial differential equation. For example, ϕ could be the temperature, T, of a longitudinal rod shown in Figure 1 and the parameter a in Equation 1 could be equal to $k/c\rho$ where k, c, and ρ are the thermal conductivity, specific heat, and specific weight of the rod, respectively. To make the problem more specific, the rod may have an initial temperature of 0°F throughout and it is completely insulated around its lateral surface and also at its right end. If its left end is to be maintained at 100°F beginning at the time t = 0, then it is of interest to know

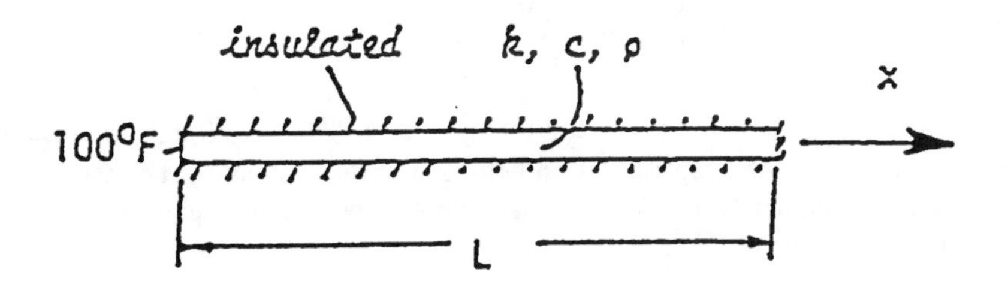

FIGURE 1. ϕ could be the temperature, T, of a longitudinal rod.

how the temperatures along the entire length of the rod will be changing as the time progresses. This is therefore a transient heat conduction problem. One would like to know how long would it take to have the entire rod reaching a uniform temperature of 100°F.

If the rod is made of a single material, k/cρ would then be equal to a constant. Analytical solution can be found for this simple case.[1] For the general case that the rod may be composed of a number of different materials and the physical properties k, c, and ρ would not only depend on the spatial variable x but may also depend on the temporal variable t. The more complicated the variation of these properties in x and t, the more likely no analytical solution is possible and the problem can only be solved numerically. The finite-difference approximation of Equation 1 can be achieved by applying the forward difference for the first derivative with respective to t and central difference for the second derivative with respect to x as follows (for t at t_i and x at x_j):

$$\frac{\partial T}{\partial t} \doteq \frac{T_{i+1,j} - T_{i,j}}{\Delta t} \quad \text{and} \quad \frac{\partial^2 T}{\partial x^2} \doteq \frac{T_{i,j-1} - 2T_{i,j} + T_{i,j+1}}{(\Delta x)^2}$$

If k/c ρ is changing in time and also changing from one location to another, we could designate it as $a_{i,j}$. As a consequence, Equation 1 can then be written as:

$$\frac{T_{i+1,j} - T_{i,j}}{\Delta t} = a_{i,j} \frac{T_{i,j-1} - 2T_{i,j} + T_{i,j+1}}{(\Delta x)^2} \tag{2}$$

Since the initial temperature distribution T is known, the above expression suggests that for a numerical solution we may select an appropriate increment in t, Δt, and the temperature be determined at a finite number of stations, N. It is advisable to have these stations be equally spaced so that the increment Δx is equal to L/(N–1) where L is the length of the rod, and the instants are to be designated as $t_1 = 0$, $t_2 = t, \ldots, t_i = (i-1)\Delta t$, and the stations as $x_1 = 0$, $x_2 = \Delta x, \ldots, x_j = (j-1)\Delta x, \ldots$, and $x_N = (N-1)\Delta x = L$. The task at hand is then to find $T(t_i,x_j)$ for i = 1,2,... and j = 1,2,...,N. It can be noticed from Equation 2 that the there is only one temperature at t_{i+1} and can be expressed in terms of those at the preceding instant t_i as:

$$T_{i+1,j} = T_{i,j} + \frac{a_{i,j}\Delta t}{(\Delta x)^2}\left(T_{i,j-1} - 2T_{i,j} + T_{i,j+1}\right) \quad (3)$$

Equation 3 is to be used for j = 2 through j = N–1. For the last station, j = N, which is insulated,the temperatures on both side of this station can be assumed to be equal (the station N + 1 is a fictitious one!). The modified equation for this particular station is:

$$T_{i+1,N} = T_{i,N} + \frac{2a_{i,N}\Delta t}{(\Delta x)^2}\left(T_{i,N-1} - T_{i,N}\right) \quad (4)$$

For generating the temperatures of the rod at N stations for any specified time increment Δt until the temperatures are almost all equal to 100°F throughout, the program ParabPDE has been applied. It is listed below along with a typical printout of the results.

FORTRAN Version

```
C  Program ParabPDE - Parabolic Partial Differential Equation
C                     solved numerically using a transient
C                     heat-conduction problem.
C  For a simple illustrat'n, let kdt/c(rho)(dx)**2=1 AND N=11.
      DIMENSION T(11),TN(11)
      DATA T/100.,10*0./,TN(1)/100./
      DATA RK,C,RHO,DT,DX,N/0.037,0.212,168.,1.,.1,11/
      C1=RK/C/RHO*DT/(DX)**2
      NM1=N-1
      WRITE (*,1)
    1 FORMAT(' At what temperature differential (for all',
     *        ' stations in degree F should'
     *       /'   the computation be terminated?')
      READ (*,*) TDF
      WRITE (*,2)
    2 FORMAT(3X,'X1',5X,'X2',5X,'X3',5X,'X4',5X,'X5',5X,'X6',
     *       5X,'X7',5X,'X8',5X,'X9',4X,'X10',4X,'X11')
      WRITE (*,5) (T(IP),IP=1,N)
    5 FORMAT(11F7.1)
      TM=0.
    8 DO 15 J=2,N
   15 TN(J)=T(J)+C1*(T(J-1)-2*T(J)+T(J+1))
      TN(N)=T(N)+2*C1*(T(N-1)-T(N))
      WRITE (*,5) (TN(IP),IP=1,N)
      TM=TM+DT
      DO 25 J=2,N
      IF (ABS(TN(J)-T(J)).GT.TDF) GOTO 30
   25 CONTINUE
      WRITE (*,27) TM
   27 FORMAT(' It takes ',E12.5,' seconds.')
      GOTO 99
   30 DO 35 J=2,N
   35 T(J)=TN(J)
      GOTO 8
   99 END
```

Sample Output

```
 At what temperature differential (for all stations in degree F should
   the computation be terminated?
1
   X1       X2       X3       X4       X5       X6       X7       X8       X9      X10      X11
 100.0    65.7     37.5     18.5      7.8      2.8       .9       .2       .1       .0       .0
 It takes  .24000E+02 seconds.

 At what temperature differential (for all stations in degree F should
   the computation be terminated?
0.5
   X1       X2       X3       X4       X5       X6       X7       X8       X9      X10      X11
 100.0    75.5     53.2     34.9     21.3     12.0      6.2      3.0      1.4       .6       .4
 It takes  .50000E+02 seconds.

 At what temperature differential (for all stations in degree F should
   the computation be terminated?
0.1
   X1       X2       X3       X4       X5       X6       X7       X8       X9      X10      X11
 100.0    93.9     88.0     82.3     77.1     72.5     68.5     65.3     63.0     61.6     61.1
 It takes  .46400E+03 seconds.

 At what temperature differential (for all stations in degree F should
   the computation be terminated?
0.05
   X1       X2       X3       X4       X5       X6       X7       X8       X9      X10      X11
 100.0    97.0     94.0     91.2     88.5     86.2     84.2     82.6     81.5     80.8     80.5
 It takes  .73500E+03 seconds.

 At what temperature differential (for all stations in degree F should
   the computation be terminated?
0.01
   X1       X2       X3       X4       X5       X6       X7       X8       X9      X10      X11
 100.0    99.4     98.8     98.2     97.7     97.2     96.9     96.5     96.3     96.2     96.1
 It takes  .13650E+04 seconds.

 At what temperature differential (for all stations in degree F should
   the computation be terminated?
0.005
   X1       X2       X3       X4       X5       X6       X7       X8       X9      X10      X11
 100.0    99.7     99.4     99.1     98.9     98.6     98.4     98.3     98.1     98.1     98.1
 It takes  .16360E+04 seconds.

 At what temperature differential (for all stations in degree F should
   the computation be terminated?
0.001
   X1       X2       X3       X4       X5       X6       X7       X8       X9      X10      X11
 100.0    99.9     99.9     99.8     99.8     99.7     99.7     99.7     99.6     99.6     99.6
 It takes  .22640E+04 seconds.

 At what temperature differential (for all stations in degree F should
   the computation be terminated?
0.0005
   X1       X2       X3       X4       X5       X6       X7       X8       X9      X10      X11
 100.0    99.9     99.9     99.9     99.9     99.8     99.8     99.8     99.8     99.8     99.8
 It takes  .25370E+04 seconds.

 At what temperature differential (for all stations in degree F should
   the computation be terminated?
0.0001
   X1       X2       X3       X4       X5       X6       X7       X8       X9      X10      X11
 100.0   100.0    100.0    100.0    100.0    100.0    100.0    100.0    100.0    100.0    100.0
 It takes  .31570E+04 seconds.
```

QuickBASIC Version

```
'   Program PDEParab - Parabolic Partial Differential Equation solved numerically
'                      using a transient heat-conduction problem.
'   For a simple illustrat'n, let kdt/c(rho)(dx)**2=1 AND N=11.
        DIM T(11), TN(11)
        T(1)=100!
        FOR I=2 TO 11
            T(I)=0!
            NEXT I
        TN(1)=100!
        READ RK,C,RHO,DT,DX,N
        DATA 0.037,0.212,168.,1.,.1,11
        C1 = RK / C / RHO * DT / (DX) ^ 2
        NM1 = N - 1
        PRINT " At what temperature differential (for all";
        PRINT " stations)', in degree F should"
        INPUT "   the computation be terminated? ", TDF
        PRINT SPC(3);"X1";SPC(5);"X2";SPC(5);"X3";SPC(5);"X4";
        PRINT SPC(5);"X5";SPC(5);"X6";SPC(5);"X7";SPC(5);"X8";
        PRINT SPC(5);"X9";SPC(4);"X10";SPC(4);"X11"
        FOR ip=1 TO N
            PRINT USING "#####.#";T(ip);
            NEXT ip
        PRINT
        TM = 0!
 8  FOR J = 2 TO N - 1
         TN(J) = T(J) + C1*(T(J-1) - 2*T(J)+T(J+1))
         NEXT J
    TN(N) = T(N) + 2 * C1 * (T(N - 1) - T(N))
    FOR ip=1 TO N
         PRINT USING "#####.#"; TN(ip);
         NEXT ip
    PRINT
    TM    = TM + DT
    FOR J = 2 TO N
         IF (ABS(TN(J) - T(J)) > TDF) THEN 30
25       NEXT J
         PRINT USING " It takes ##.#####^^^^ seconds."; TM
         GOTO 99
30  FOR J    = 2 TO N
         T(J) = TN(J)
         NEXT J
         GOTO 8
99  END
```

MATLAB Applications

A **MATLAB** version of **ParabPDE** can be created easily by converting the **QuickBASIC** program. The m file may be arranged as follows:

```
          function T=ParabPDE(N)
% Parabolic Partial Differential Eq. solved numerically using a transient heat-conduction
%   problem.  For a simple illustration, let kdt/c(rho)(dx)**2=1 AND N=11.
  T(1) =100; for I=2:N, T(I)=0; end
  TN(1)=100; Rk=0.037; C=0.212; Rho=168; Dt=1; Dx=0.1;
  C1    = Rk/C/Rho*Dt/Dx^2; NM1=N-1;
  fprintf('At what temperature differential (for all stations), in degree F')
  TDF=input('  should the computation be terminated? ');
  TM = 0; ExitFlag=0;
  while ExitFlag==0; Jump=0;
        for J=2:N-1
            TN(J)=T(J)+C1*(T(J-1)-2*T(J)+T(J+1));
        end
        TN(N)=T(N)+2*C1*(T(N-1)-T(N));
        TM=TM+Dt;
```

```
        for J=2:N
            if abs(TN(J)-T(J))>TDF, Jump=1; break
            end
        end
        if Jump==0, fprintf('It takes %12.5e seconds.\n',TM), ExitFlag=1; break
           else T=TN;
        end
end
```

For solving the sample transient temperature problem, this m file can be called and interactive **MATLAB** instructions can be entered through keyboard to obtain the temperature distribution of the rod at various times:

```
>> format compact, T=feval('a:ParabPDE',11)
 At what temperature differential (for all stations), in degree F
     should the computation be terminated? 1
 It takes 2.40000e+001 seconds.
 T =
   Columns 1 through 7
   100.0000    65.0005    36.5231    17.5928     7.2408     2.5459     0.7659
   Columns 8 through 11
     0.1976     0.0438     0.0085     0.0027
>> Tsave(:,1)=T';
>> T=feval('a:ParabPDE',11)
 At what temperature differential (for all stations), in degree F
     should the computation be terminated? 0.1
 It takes 4.62000e+002 seconds.
 T =
   Columns 1 through 7
   100.0000    93.8964    87.9431    82.2870    77.0671    72.4122    68.4367

   Columns 8 through 11
    65.2385    62.8962    61.4673    60.9871
>> Tsave(:,2)=T';
>> T=feval('a:ParabPDE',11)
 At what temperature differential (for all stations), in degree F
     should the computation be terminated? 0.01
 It takes 1.36100e+003 seconds.
 T =
   Columns 1 through 7
   100.0000    99.3897    98.7944    98.2288    97.7069    97.2413    96.8438
   Columns 8 through 11
    96.5239    96.2896    96.1467    96.0987
>> Tsave(:,3)=T';
>> T=feval('a:ParabPDE',11)
 At what temperature differential (for all stations), in degree F
     should the computation be terminated? 0.001
 It takes 2.26000e+003 seconds.
 T =
   Columns 1 through 7
   100.0000    99.9390    99.8794    99.8229    99.7707    99.7241    99.6844
   Columns 8 through 11
    99.6524    99.6290    99.6147    99.6099
>> Tsave(:,4)=T';
>> T=feval('a:ParabPDE',11)
 At what temperature differential (for all stations), in degree F
     should the computation be terminated? 0.0001
 It takes 3.15900e+003 seconds.
 T =
   Columns 1 through 7
   100.0000    99.9939    99.9879    99.9823    99.9771    99.9724    99.9684
   Columns 8 through 11
    99.9652    99.9629    99.9615    99.9610
>> plot(Tsave), xlabel('Station Number'), ylabel('Temperature, in Fahrenheits')
>> text(3.5,30,'t=24 sec  '), text(7.5,60,'t=462 sec ')
>> text(4  ,95,'t=1361 sec'), text(9  ,96,'t>2260 sec')
```

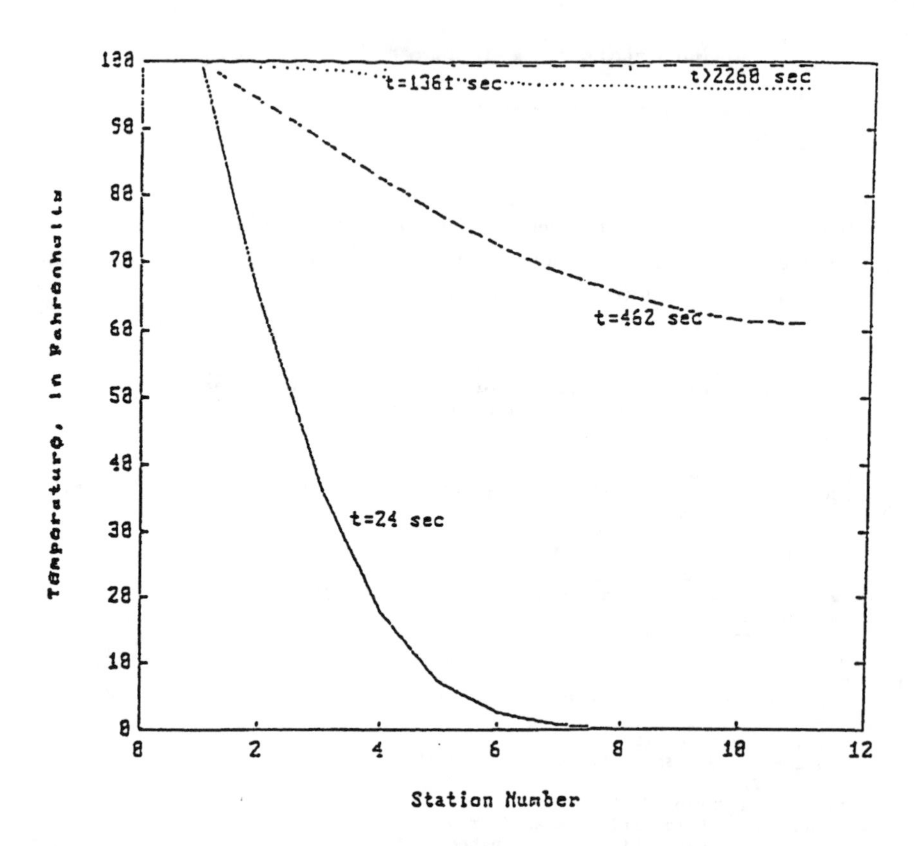

FIGURE 2. A composite graph with axes labels and markings of the curves by making use of the **MATLAB** commands **xlabel, ylabel,** and **text.**

Notice that results of temperature distributions which have been terminated using five differentials of 1, 0.1, 0.01, 0.001, and 0.0001 Fahrenheit are saved in Tsave and then later plotted. Since T is a row matrix, the transpose of T, T′, is stored in an appropriate column of Tsave. If the temperatures were rounded, the rod reaches a uniform temperature distribution of 100°F when the required temperature differential is selected to be 0.001°F. If the fifth curve for the temperature differential equal to 0.0001 is plotted, it will be too close to the forth curve and also the plot function provides only four line types (solid, broken, dot, and center lines), the fifth set of results is therefore not saved in Tsave. Figure 2 shows a composite graph with axes labels and markings of the curves by making use of the MATLAB commands xlabel, ylabel, and text.

MATHEMATICA APPLICATIONS

The heat conduction problem of an insulated rod previously discussed in the versions for **FORTRAN, QuickBASIC,** and **MATLAB** can be solved by application of **Mathematica** as follows:

```
In[1]: = (n = 11; rk = 0.037; c = 0.212; rho = 168; dt = 1; dx = 0.1;
                     c1 = rk/c/rho*dt/dx^2; nm1 = n–1;)

In[2]: = (t = Table[0,{n}]; tn = Table[0,{n}]; t[[1]] = 100; tn[[1]] = 100; tdf = 1;
        tm = 0; flag1 = 0;

In[3]: = (While[flag1 = = 0, Do[tn[[j]] = t[[j]] + c1*(t[[j–1]]–2*t[[j]] + t[[j + 1]]),
                     {j,2,nm1}];
        tn[[n]] = t[[n]] + 2*c1*(t[[nm1]]-t[[n]]);tm = tm + dt;flag1 = 1;
            Do[If[Abs[tn[[I]]-t[[I]]]>tdf, flag1 = 0; Break,
                     Continue],{i,n}];
            Do[t[[I]] = tn[[I]],{i,n}]]; Print["t = ",tm]; Print[N[tn,3]])

Out[3] = t = 24
   {100., 65.7, 37.5, 18.5, 7.83, 2.85, 0.892, 0.241, 0.0561, 0.0116, 0.00394}
```

The temperature distribution of the heated rod after 24 seconds is same as obtained by the **FORTRAN**, **QuickBASIC**, and **MATLAB** versions when every component of two consecutive temperature distribution, kept as {t} and {tn}, differ no more than the allowed set value of tdf = 1 degree in *In[2]*. Any component has a difference exceeding the tdf value will cause **flag1** to change from a value of 1 to 0 and the **Break** command in the second Do loop in *In[2]* to exit and to continue the iteration. **flag1** is created to control the **While** command which determines when the iteration should be terminated. The N[tn,3] instructs the components of {tn} be printed with 3 significant figures.

When tdf is changed to a value of 0.5, **Mathematica** can again be applied to yield

```
In[4]: = t = (Table[0,{n}]; tn = Table[0,{n}]; t[[1]] = 100; tn[[1]] = 100;
        tdf = 0.5; tm = 0; flag1 = 0;)

In[5]: = (While[flag1 = = 0,Do[tn[[j]] = t[[j]] + c1*(t[[j–1]]–2*t[[j]] + t[[j + 1]]),
                     {j,2,nm1}];
        tn[[n]] = t[[n]] + 2*c1*(t[[nm1]]-t[[n]]);tm = tm + dt;flag1 = 1;
        Do[If[Abs[tn[[I]]-t[[I]]]>tdf, flag1 = 0; Break,
            Continue],{i,n}];
        Do[t[[I]] = tn[[I]],{i,n}]]; Print["t = ",tm]; Print[N[tn,3]])

Out[5]: = t = 49
   {100., 75.5, 53.2, 34.9, 21.3, 12., 6.24, 3., 1.35, 0.617, 0.413}
```

Notice that **Mathematica** takes only 49 seconds, one second less than that required for the **FORTRAN**, **QuickBASIC**, and **MATLAB** versions. The reason is that **Mathematica** keeps more significant digits in carrying out all computations. To show more on the effect of changing the tdf value, the following **Mathematica** runs are provided:

```
In[6]: = t = (Table[0,{n}]; tn = Table[0,{n}]; t[[1]] = 100; tn[[1]] = 100;
        tdf = 0.1; tm = 0; flag1 = 0;)

In[7]: = (While[flag1 == 0, Do[tn[[j]] = t[[j]] + c1*(t[[j–1]]–2*t[[j]] + t[[j + 1]]),
                    {j,2,nm1}];
                  tn[[n]] = t[[n]] + 2*c1*(t[[nm1]]-t[[n]]);
                  tm = tm + dt; flag1 = 1;
                  Do[If[Abs[tn[[I]]-t[[I]]]>tdf, flag1 = 0; Break,
                                  Continue],{i,n}];
                  Do[t[[I]] = tn[[I]],{i,n}]]; Print["t = ",tm];
                    Print[N[tn,3]])

Out[7]: = t = 462
   {100., 93.9, 88., 82.3, 77.1, 72.5, 68.5, 65.3, 63., 61.6, 61.1}

In[8]: = t = (Table[0,{n}]; tn = Table[0,{n}]; t[[1]] = 100; tn[[1]] = 100;
        tdf = 0.05; tm = 0; flag1 = 0;)

In[9]: = (While[flag1 == 0, Do[tn[[j]] = t[[j]] + c1*(t[[j–1]]–2*t[[j]] + t[[j + 1]]),
                  {j,2,nm1}];
                 tn[[n]] = t[[n]] + 2*c1*(t[[nm1]]-t[[n]]);
                 tm = tm + dt; flag1 = 1;
                 Do[If[Abs[tn[[I]]-t[[I]]]>tdf, flag1 = 0; Break,
                                 Continue],{i,n}];
                 Do[t[[I]] = tn[[I]],{i,n}]]; Print["t = ",tm];
                   Print[N[tn,3]])

Out[9]: = t = 732
   {100., 97., 94., 91.2, 88.5, 86.2, 84.2, 82.6, 81.5, 80.8, 80.5}

In[10]: = t = (Table[0,{n}]; tn = Table[0,{n}]; t[[1]] = 100; tn[[1]] = 100;
        tdf = 0.0001; tm = 0; flag1 = 0;

In[11]: = (While[flag1 == 0, Do[tn[[j]] = t[[j]] + c1*(t[[j–1]]–2*t[[j]] + t[[j + 1]]),
                    {j,2,nm1}];
                  tn[[n]] = t[[n]] + 2*c1*(t[[nm1]]-t[[n]]);
                  tm = tm + dt; flag1 = 1;
                  Do[If[Abs[tn[[I]]-t[[I]]]>tdf, flag1 = 0; Break,
                                  Continue],{i,n}];
          Do[t[[I]] = tn[[I]],{i,n}]]; Print["t = ",tm]; Print[N[tn,3]])

Out[11]: = t = 3159
   {100., 100., 100., 100., 100., 100., 100., 100., 100., 100., 100.}
```

For tdf = 0.1 and tdf = 0.05, **Mathematica** continues to take lesser time than the other version; but when tdf = 0.0001, **Mathematica** needs two additional seconds. The reason is that seven significant figures are required in the last case, rounding may have resulted in earlier termination of the iteration when the **FORTRAN**, Quick-**BASIC**, and **MATLAB** versions are employed.

8.3 PROGRAM RELAXATN — SOLVING ELLIPTICAL PARTIAL DIFFERENTIAL EQUATIONS BY RELAXATION METHOD

The program **Relaxatn** is designed for solving engineering problems which are governed by elliptical partial differential equation of the form:

$$\frac{\partial^2 \phi}{\partial x^2} + \frac{\partial^2 \phi}{\partial y^2} = F(x, y) \tag{1}$$

where ϕ is called the field function and F(x,y) is called forcing function. When the steady-state heat conduction of a two-dimensional domain is considered,[2] then the field function becomes the temperature distribution, T(x,y), and the forcing function becomes the heat-source function, Q(x,y). If the distribution of T is influenced only by the temperatures at the boundary of the domain, then Q(x,y) = 0 and Equation 1 which is often called a Poisson equation is reduced to a Laplace equation:

$$\frac{\partial^2 T}{\partial x^2} + \frac{\partial^2 T}{\partial y^2} = 0 \tag{2}$$

The second-order, central-difference formulas (read the program **DiffTabl**) can be applied to approximate the second derivatives in the above equation at an arbitrary point $x = x_i$ and $y = y_j$ in the domain as:

$$\left.\frac{\partial^2 T}{\partial x^2}\right|_{\text{at } x_i, y_j} \doteq \frac{T_{i-1,j} - 2T_{i,j} + T_{i+1,j}}{(\Delta x)^2}$$

and

$$\left.\frac{\partial^2 T}{\partial y^2}\right|_{\text{at } x_i, y_j} \doteq \frac{T_{i,j-1} - 2T_{i,j} + T_{i,j+1}}{(\Delta y)^2}$$

By substituting both of the above equations into Equation 2 and taking equal increments in both x- and y-directions, the reduced equation is, for $\Delta x = \Delta y$,

$$T_{i,j} = \frac{1}{4}\left(T_{i-1,j} + T_{i+1,j} + T_{i,j-1} + T_{i,j+1}\right) \tag{3}$$

The result is expected when the temperature distribution reaches a steady state because it states that the temperature at any point should be equal to the average of its surrounding temperatures.

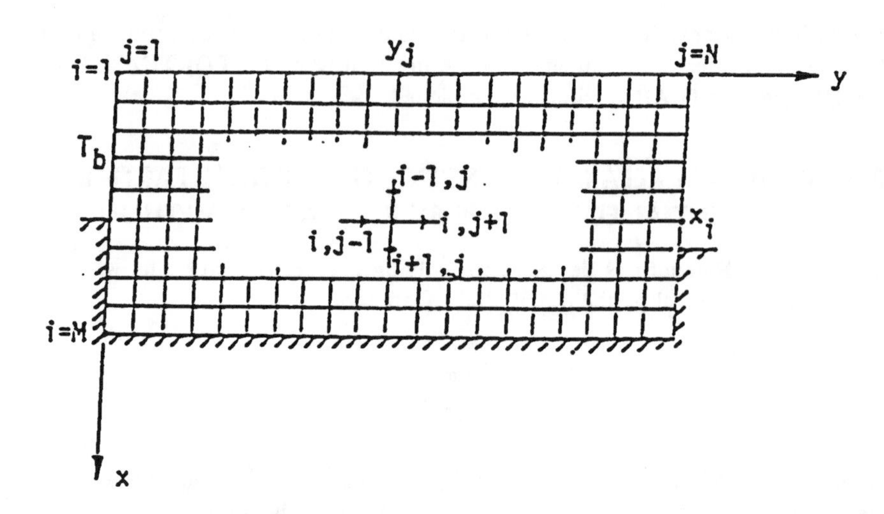

FIGURE 3. A plate, which initially (at time t = 0) had a temperature equal to 0°F throughout, insulated along a portion of its boundary and suddenly heated at its upper left boundary to maintain a linearly varying temperature T_b.

Before we proceed further, it is appropriate at this time to introduce a numerical case. Suppose that a plate which initially (at time t = 0) has a temperature equal to 0°F throughout and is insulated along a portion of its boundary, is suddenly heated at its upper left boundary to maintain a linearly varying temperature T_b as shown in Figure 3. If this heating process is to be maintained, we are then interested in knowing the temperature distribution, changed from its initial state of uniformly equal to 0°F if given sufficient time to allow it to reach an equilibrium (steady) state. Numerically, we intend to calculate the temperatures, denoted as a matrix [T], at a selected number of locations. Therefore, the plate is first divided into a gridwork of M rows and N columns along the x- and y-directions, respectively, as indicated in Figure 1. The directions of x- and y-axes are so selected for the convenience of associating them with the row and column of the temperature matrix [T] which is of order M by N. The values of M and N should be so decided such that the increments Δx and Δy are equal in order to apply Equation 3. To be more specific, let M = 10 and N = 20 and the linear temperature variation along the upper left boundary T_b be:

$$T_{i,1} = 10(i-1) \quad \text{for} \quad i = 1,2,\ldots,6 \tag{4}$$

Here, $T_{i,j}$ is to be understood as the temperature at the location (x_i,y_j). Equation 4 describes the temperature along the left boundary $y = y_1$ but only for x_i in the range of i = 1 to i = 6.

Since the temperatures at the stations which are on the insulated boundaries of the plate are also involved, these unknown temperatures need to be treated differently. By an insulated boundary, it means that there is no heat transfer normal to the boundary. Since the heat flow is depended on the temperature difference across that insulated boundary, mathematically it requires that $\partial T/\partial n = 0$ there, n being the normal direction. At the vertical boundaries $x = x_M$, we have $\partial T/\partial n = \partial T/\partial x = 0$ since x is the normal direction. Based on the central difference and considering two increments in the x direction, at a y_jth station we can have:

$$\left.\frac{\partial T}{\partial x}\right| \text{at } \left(x_M, y_j\right) \doteq \frac{T_{M+1,j} - T_{M-1,j}}{2(\Delta x)} = 0$$

or

$$T_{M+1,j} = T_{M-1,j} \tag{5}$$

Since x_{M+1} is below the bottom boundary of the plate shown in Figure 3, there is no need to calculate the temperatures there, however Equation 5 enables the temperatures at the stations along the bottom boundary of the plate $T_{M,j}$ for j = 1 to N to be averaged. Returning to Equation 4, we notice that if Equation 5 is substituted into it, the resulting equation which relates only to *three* neighboring temperatures is:

$$T_{M,j} = \frac{1}{4}\left(2T_{M-1,j} + T_{M,j-1} + T_{M,j+1}\right) \quad \text{for} \quad j = 2,3,\ldots,N-1 \tag{6}$$

Notice that j = 1 and j = N are not covered in Equation 6. These two cases concerning the insulated stations at the left and right bottom corners of the plate will be discussed after we address the two vertical, insulated boundaries $y = y_1$ and $y = y_N$.

For the boundaries $y = y_1$, $\partial T/\partial n$ becomes $\partial T/\partial y$. Again, we can apply the central difference for double y increments to obtain:

$$\left.\frac{\partial T}{\partial y}\right| \text{at } \left(x_1, y_1\right) \doteq \frac{T_{i,0} - T_{i,2}}{2(\Delta y)} = 0$$

or

$$T_{i,0} = T_{i,2} \tag{7}$$

Thus, the modified Equation 3 for the left insulated boundary is:

$$T_{i,1} = \frac{1}{4}\left(T_{i-1,1} + T_{i+1,1} + 2T_{i,2}\right) \quad \text{for} \quad i = 7,8,9 \tag{8}$$

In a similar manner, we can derive for the right insulated boundary $y = y_N$

$$\left.\frac{\partial T}{\partial y}\right| \text{at } \left(x_i, y_N\right) \doteq \frac{T_{i,N+1} - T_{i,N-1}}{2(\Delta y)} = 0$$

or

$$T_{i,N+1} = T_{i,N-1} \tag{9}$$

and

$$T_{i,N} = \frac{1}{4}\left(T_{i-1,N} + T_{i+1,N} + 2T_{i,N-1}\right) \quad \text{for } i = 8,9 \tag{10}$$

Having derived Equations 6, 8, and 10, it is easy to deduce the two special equation for the corner insulated stations to be:

$$T_{M,1} = \frac{1}{2}\left(T_{M,2} + T_{M-1,1}\right) \tag{11}$$

and

$$T_{M,N} = \frac{1}{2}\left(T_{M,N-1} + T_{M-1,N}\right) \tag{12}$$

We have derived all equations needed for averaging the temperature at any station of interest including those at the insulated boundaries by utilizing those at its neighboring stations. It suggests that a continuous upgrading process can be developed which assumes that the neighboring temperatures are known. This so-called *relaxation* method starts with an initial assumed distribution of temperature $[T^{(0)}]$ and continues to use Equations 3 and 6 to 12 until the differences at all locations are small enough. **Mathematically**, the process terminates when:

$$\sum_{i=1}^{M}\sum_{j=1}^{N}\left|T_{i,j}^{(k+1)} - T_{i,j}^{(k)}\right| < \varepsilon \tag{13}$$

where ε is a prescribed tolerance of accuracy and $k = 0,1,2,\ldots$ is the number of sweeps in upgrading the temperature distribution. Superscripts $(k + 1)$ and (k) refer to the improved and previous distributions, respectively. The order of sweep will affect how the temperatures should be upgraded. For example, if the temperatures are to be re-averaged from top to bottom and left to right, referring to Figure 1, then Equation 3 is to be modified as:

$$T_{i,j}^{(k+1)} = \frac{1}{4}\left(T_{i-1,j}^{(k+1)} + T_{i+1,j}^{(k)} + T_{i,j-1}^{(k+1)} + T_{i,j+1}^{(k)}\right) \tag{14}$$

Notice that the neighboring temperatures in the row above, $i-1$, and in the column to the left, $j-1$, have already been upgraded while those in the row below, $i + 1$, and in the column to the right, $j + 1$, are yet to be upgraded. Similar modifications are to be made to Equations 6 to 12 during relaxation.

The program **Relaxatn** is developed according to the relaxation method described above. For solving the problem shown in Figure 3, both **FORTRAN** and **QuickBASIC** versions are made available for interactively specifying the tolerance. Sample results are presented below.

FORTRAN VERSION

```
C  Program Relaxatn - Steady-state temperature distribution determined
C                     by application of relaxation method.
      DIMENSION T(10,20)
      DATA T/200*0./
C  Initially all temperatures are equal to zero degree F.
C  The left, upper boundary is heated.
      WRITE (*,101)
  101 FORMAT(1X,'Program Relaxatn - Relaxation method applied for ',
     *          'solving heat conduction problem')
      WRITE (*,103)
  103 FORMAT(1X,'Enter the tolerance, epsilon, for termination : ')
      READ (*,*) Epsilon
      DO 1 I=2,6
    1 T(I,1)=(I-1)*10.
C  Relaxation proceeds.
      NR=1
    2 WRITE (*,3) NR
    3 FORMAT(1X,'Sweep #',I5)
      DO 25 I=1,10
   25 WRITE (*,30) (T(I,J),J=1,20)
   30 FORMAT(20F4.0)
      D=0
      DO 4 I=2,6
      DO 4 J=2,19
      TS=T(I,J)
      T(I,J)=.25*(T(I-1,J)+T(I+1,J)+T(I,J-1)+T(I,J+1))
    4 D=D+ABS(TS-T(I,J))
      DO 8 I=7,9
      TS=T(I,1)
      T(I,1)=.25*(T(I-1,1)+T(I+1,1)+2*T(I,2))
      D=D+ABS(TS-T(I,1))
      DO 6 J=2,19
      TS=T(I,J)
      T(I,J)=.25*(T(I-1,J)+T(I+1,J)+T(I,J-1)+T(I,J+1))
    6 D=D+ABS(TS-T(I,J))
      IF (I.EQ.7) GOTO 8
      TS=T(I,20)
      T(I,20)=.25*(T(I-1,20)+T(I+1,20)+2*T(I,19))
      D=D+ABS(TS-T(I,20))
    8 CONTINUE
      TS=T(10,1)
      T(10,1)=.5*(T(9,1)+T(10,2))
      D=D+ABS(TS-T(10,1))
      DO 10 J=2,19
      TS=T(I,J)
      T(10,J)=.25*(2*T(9,J)+T(10,J-1)+T(10,J+1))
   10 D=D+ABS(TS-T(10,J))
      TS=T(10,20)
      T(10,20)=.5*(T(9,20)+T(10,19))
      D=D+ABS(TS-T(10,20))
      IF (D.LT.EPSILON) GOTO 20
      NR=NR+1
      GO TO 2
```

```
C  Program Relaxatn - Steady-state temperature distribution determined
C                     by application of relaxation method.
      DIMENSION T(10,20)
      DATA T/200*0./
C  Initially all temperatures are equal to zero degree F.
C  The left, upper boundary is heated.
      WRITE (*,101)
  101 FORMAT(1X,'Program Relaxatn - Relaxation method applied for ',
     *          'solving heat conduction problem')
      WRITE (*,103)
  103 FORMAT(1X,'Enter the tolerance, epsilon, for termination : ')
      READ (*,*) Epsilon
      DO 1 I=2,6
    1 T(I,1)=(I-1)*10.
C  Relaxation proceeds.
      NR=1
    2 WRITE (*,3) NR
    3 FORMAT(1X,'Sweep #',I5)
      DO 25 I=1,10
   25 WRITE (*,30) (T(I,J),J=1,20)
   30 FORMAT(20F4.0)
      D=0
      DO 4 I=2,6
      DO 4 J=2,19
      TS=T(I,J)
      T(I,J)=.25*(T(I-1,J)+T(I+1,J)+T(I,J-1)+T(I,J+1))
    4 D=D+ABS(TS-T(I,J))
      DO 8 I=7,9
      TS=T(I,1)
      T(I,1)=.25*(T(I-1,1)+T(I+1,1)+2*T(I,2))
      D=D+ABS(TS-T(I,1))
      DO 6 J=2,19
      TS=T(I,J)
      T(I,J)=.25*(T(I-1,J)+T(I+1,J)+T(I,J-1)+T(I,J+1))
    6 D=D+ABS(TS-T(I,J))
      IF (I.EQ.7) GOTO 8
      TS=T(I,20)
      T(I,20)=.25*(T(I-1,20)+T(I+1,20)+2*T(I,19))
      D=D+ABS(TS-T(I,20))
    8 CONTINUE
      TS=T(10,1)
      T(10,1)=.5*(T(9,1)+T(10,2))
      D=D+ABS(TS-T(10,1))
      DO 10 J=2,19
      TS=T(I,J)
      T(10,J)=.25*(2*T(9,J)+T(10,J-1)+T(10,J+1))
   10 D=D+ABS(TS-T(10,J))
      TS=T(10,20)
      T(10,20)=.5*(T(9,20)+T(10,19))
      D=D+ABS(TS-T(10,20))
      IF (D.LT.EPSILON) GOTO 20
      NR=NR+1
      GO TO 2
```

Sample Results

The program **Relaxatn** is first applied for an interactively entered value of equal to 100. Only one relaxation needs to be implemented as shown below. The temperature distribution for Sweep #1 is actually the initial assumed distribution. One cannot assess how accurate this distribution is. The second run specifies that be equal to 1. The results show that 136 relaxation steps are required. For giving more insight on how the relaxation has proceeded, Sweeps #10, #30, #50, #100, and #137 are presented for interested readers. It clearly indicates that a tolerance of equal to 100 is definitely inadequate.

```
Program Relaxatn - Relaxation method applied for solving heat conduction problem

Enter the tolerance, epsilon, for termination :
100

Sweep #    1
 0.  0.  0.  0.  0.  0.  0.  0.  0.  0.  0.  0.  0.  0.  0.  0.  0.  0.  0.  0.
10.  0.  0.  0.  0.  0.  0.  0.  0.  0.  0.  0.  0.  0.  0.  0.  0.  0.  0.  0.
20.  0.  0.  0.  0.  0.  0.  0.  0.  0.  0.  0.  0.  0.  0.  0.  0.  0.  0.  0.
30.  0.  0.  0.  0.  0.  0.  0.  0.  0.  0.  0.  0.  0.  0.  0.  0.  0.  0.  0.
40.  0.  0.  0.  0.  0.  0.  0.  0.  0.  0.  0.  0.  0.  0.  0.  0.  0.  0.  0.
50.  0.  0.  0.  0.  0.  0.  0.  0.  0.  0.  0.  0.  0.  0.  0.  0.  0.  0.  0.
 0.  0.  0.  0.  0.  0.  0.  0.  0.  0.  0.  0.  0.  0.  0.  0.  0.  0.  0.  0.
 0.  0.  0.  0.  0.  0.  0.  0.  0.  0.  0.  0.  0.  0.  0.  0.  0.  0.  0.  0.
 0.  0.  0.  0.  0.  0.  0.  0.  0.  0.  0.  0.  0.  0.  0.  0.  0.  0.  0.  0.
 0.  0.  0.  0.  0.  0.  0.  0.  0.  0.  0.  0.  0.  0.  0.  0.  0.  0.  0.  0.
Sweep #    2
 0.  0.  0.  0.  0.  0.  0.  0.  0.  0.  0.  0.  0.  0.  0.  0.  0.  0.  0.  0.
10.  3.  1.  0.  0.  0.  0.  0.  0.  0.  0.  0.  0.  0.  0.  0.  0.  0.  0.  0.
20.  6.  2.  0.  0.  0.  0.  0.  0.  0.  0.  0.  0.  0.  0.  0.  0.  0.  0.  0.
30.  9.  3.  1.  0.  0.  0.  0.  0.  0.  0.  0.  0.  0.  0.  0.  0.  0.  0.  0.
40. 12.  4.  1.  0.  0.  0.  0.  0.  0.  0.  0.  0.  0.  0.  0.  0.  0.  0.  0.
50. 16.  5.  1.  0.  0.  0.  0.  0.  0.  0.  0.  0.  0.  0.  0.  0.  0.  0.  0.
13.  7.  3.  1.  0.  0.  0.  0.  0.  0.  0.  0.  0.  0.  0.  0.  0.  0.  0.  0.
 3.  3.  1.  1.  0.  0.  0.  0.  0.  0.  0.  0.  0.  0.  0.  0.  0.  0.  0.  0.
 1.  1.  1.  0.  0.  0.  0.  0.  0.  0.  0.  0.  0.  0.  0.  0.  0.  0.  0.  0.
 0.  1.  0.  0.  0.  0.  0.  0.  0.  0.  0.  0.  0.  0.  0.  0.  0.  0.  0.  0.
Stop - Program terminated.

Program Relaxatn - Relaxation method applied for solving heat conduction problem

Enter the tolerance, epsilon, for termination :
1

Sweep #   10

 0.  0.  0.  0.  0.  0.  0.  0.  0.  0.  0.  0.  0.  0.  0.  0.  0.  0.  0.  0.
10.  7.  5.  3.  2.  1.  0.  0.  0.  0.  0.  0.  0.  0.  0.  0.  0.  0.  0.  0.
20. 14.  9.  6.  3.  2.  1.  0.  0.  0.  0.  0.  0.  0.  0.  0.  0.  0.  0.  0.
30. 20. 13.  8.  5.  3.  1.  1.  0.  0.  0.  0.  0.  0.  0.  0.  0.  0.  0.  0.
40. 26. 16.  9.  5.  3.  2.  1.  0.  0.  0.  0.  0.  0.  0.  0.  0.  0.  0.  0.
50. 27. 16. 10.  6.  3.  2.  1.  1.  0.  0.  0.  0.  0.  0.  0.  0.  0.  0.  0.
25. 19. 13.  8.  5.  3.  2.  1.  1.  0.  0.  0.  0.  0.  0.  0.  0.  0.  0.  0.
14. 13.  9.  7.  4.  3.  2.  1.  1.  0.  0.  0.  0.  0.  0.  0.  0.  0.  0.  0,
10.  9.  7.  5.  4.  2.  2.  1.  1.  0.  0.  0.  0.  0.  0.  0.  0.  0.  0.  0.
 8.  8.  7.  5.  4.  2.  2.  1.  1.  0.  0.  0.  0.  0.  0.  0.  0.  0.  0.  0.

Sweep #   30

 0.  0.  0.  0.  0.  0.  0.  0.  0.  0.  0.  0.  0.  0.  0.  0.  0.  0.  0.  0.
10.  8.  6.  4.  3.  2.  2.  1.  1.  0.  0.  0.  0.  0.  0.  0.  0.  0.  0.  0.
20. 15. 12.  8.  6.  4.  3.  2.  1.  1.  1.  0.  0.  0.  0.  0.  0.  0.  0.  0.
30. 23. 17. 12.  9.  6.  4.  3.  2.  1.  1.  1.  0.  0.  0.  0.  0.  0.  0.  0.
40. 29. 20. 14. 10.  7.  5.  4.  3.  2.  1.  1.  1.  0.  0.  0.  0.  0.  0.  0.
50. 31. 22. 16. 11.  8.  6.  4.  3.  2.  1.  1.  1.  0.  0.  0.  0.  0.  0.  0.
31. 26. 20. 15. 12.  9.  6.  5.  3.  2.  2.  1.  1.  0.  0.  0.  0.  0.  0.  0.
23. 21. 18. 14. 11.  9.  6.  5.  3.  2.  2.  1.  1.  1.  0.  0.  0.  0.  0.  0.
20. 19. 16. 14. 11.  9.  7.  5.  4.  3.  2.  1.  1.  1.  0.  0.  0.  0.  0.  0.
19. 18. 16. 14. 11.  9.  7.  5.  4.  3.  2.  1.  1.  1.  0.  0.  0.  0.  0.  0.

Sweep #   50

 0.  0.  0.  0.  0.  0.  0.  0.  0.  0.  0.  0.  0.  0.  0.  0.  0.  0.  0.  0.
10.  8.  6.  5.  4.  3.  2.  2.  1.  1.  1.  0.  0.  0.  0.  0.  0.  0.  0.  0.
```

```
20. 16. 12.  9.  7.  5.  4.  3.  2.  2.  1.  1.  1.  0.  0.  0.  0.  0.  0.  0.
30. 23. 18. 13. 10.  8.  6.  4.  3.  2.  2.  1.  1.  1.  1.  0.  0.  0.  0.  0.
40. 29. 22. 16. 12.  9.  7.  5.  4.  3.  2.  2.  1.  1.  1.  0.  0.  0.  0.  0.
50. 33. 24. 18. 14. 11.  8.  6.  5.  4.  3.  2.  2.  1.  1.  1.  0.  0.  0.  0.
33. 28. 23. 18. 14. 11.  9.  7.  5.  4.  3.  2.  2.  1.  1.  1.  0.  0.  0.  0.
26. 24. 21. 18. 15. 12.  9.  7.  6.  4.  3.  3.  2.  1.  1.  1.  1.  0.  0.  0.
23. 22. 20. 17. 14. 12. 10.  8.  6.  5.  4.  3.  2.  1.  1.  1.  1.  0.  0.  0.
22. 22. 20. 17. 14. 12. 10.  8.  6.  5.  4.  3.  2.  2.  1.  1.  1.  0.  0.  0.

Sweep #  100

 0.  0.  0.  0.  0.  0.  0.  0.  0.  0.  0.  0.  0.  0.  0.  0.  0.  0.  0.  0.
10.  8.  7.  5.  4.  3.  3.  2.  2.  1.  1.  1.  1.  1.  0.  0.  0.  0.  0.  0.
20. 16. 13. 10.  8.  6.  5.  4.  3.  3.  2.  2.  1.  1.  1.  1.  0.  0.  0.  0.
30. 24. 18. 14. 11.  9.  7.  6.  5.  4.  3.  2.  2.  2.  1.  1.  1.  0.  0.  0.
40. 30. 23. 18. 14. 11.  9.  7.  6.  5.  4.  3.  3.  2.  2.  1.  1.  1.  0.  0.
50. 34. 25. 20. 16. 13. 11.  9.  7.  6.  5.  4.  3.  2.  2.  1.  1.  1.  0.  0.
35. 30. 25. 21. 17. 14. 12.  9.  8.  6.  5.  4.  3.  3.  2.  2.  1.  1.  1.  0.
29. 27. 24. 20. 17. 15. 12. 10.  8.  7.  6.  5.  4.  3.  2.  2.  1.  1.  1.  1.
26. 25. 23. 20. 17. 15. 12. 10.  9.  7.  6.  5.  4.  3.  2.  2.  2.  1.  1.  1.
25. 24. 23. 20. 17. 15. 13. 11.  9.  7.  6.  5.  4.  3.  2.  2.  2.  1.  1.  1.

Sweep #  137

 0.  0.  0.  0.  0.  0.  0.  0.  0.  0.  0.  0.  0.  0.  0.  0.  0.  0.  0.  0.
10.  8.  7.  5.  4.  3.  3.  2.  2.  1.  1.  1.  1.  1.  0.  0.  0.  0.  0.  0.
20. 16. 13. 10.  8.  7.  5.  4.  3.  3.  2.  2.  1.  1.  1.  1.  0.  0.  0.  0.
30. 24. 18. 14. 12.  9.  8.  6.  5.  4.  3.  2.  2.  2.  1.  1.  1.  0.  0.  0.
40. 30. 23. 18. 14. 12. 10.  8.  6.  5.  4.  4.  3.  2.  2.  1.  1.  1.  0.  0.
50. 34. 26. 20. 16. 13. 11.  9.  7.  6.  5.  4.  3.  2.  2.  1.  1.  1.  0.  0.
35. 30. 25. 21. 17. 15. 12. 10.  8.  7.  6.  5.  4.  3.  3.  2.  2.  1.  1.  0.
29. 27. 24. 21. 18. 15. 13. 11.  9.  7.  6.  5.  4.  3.  3.  2.  2.  1.  1.  1.
26. 26. 23. 21. 18. 15. 13. 11.  9.  8.  6.  5.  4.  4.  3.  2.  2.  1.  1.  1.
26. 25. 23. 21. 18. 16. 13. 11.  9.  8.  7.  5.  4.  4.  3.  2.  2.  2.  1.  1.

Stop - Program terminated.
```

QuickBASIC Version

```
' Program Relaxatn - Steady-state temperature distribution determined
'                     by application of relaxation method.
     CLS : CLEAR : KEY OFF: DIM T(10,20)
     FOR I = 1 TO 10: FOR J = 1 TO 20: T(I,J) = 0!: NEXT J: NEXT I
' Initially all temperatures are equal to zero degree F.
' The left, upper boundary is heated.
  PRINT "Program Relaxatn - Relaxation method applied for solving heat conduction problem "
  INPUT "Enter the tolerance, epsilon, for termination : ", EPSILON
     FOR I = 2 TO 6: T(I,1) = (I-1)*10!: NEXT I
' Relaxation proceeds.
     NR = 1
2    PRINT : PRINT "Sweep #", NR
     FOR I = 1 TO 10: FOR J = 1 TO 20: PRINT USING " ##."; T(I,J); : NEXT J: NEXT I
     D = 0
     FOR I=2 TO 6
         FOR J=2 TO 19: TS=T(I,J)
             T(I,J)=.25*(T(I-1,J)+T(I+1,J)+T(I,J-1)+T(I,J+1))
             D = D + ABS(TS - T(I,J)): NEXT J:                NEXT I
     FOR I = 7 TO 9: TS = T(I, 1)
         T(I,1) = .25 * (T(I-1,1) + T(I+1,1) + 2 * T(I,2)): D = D + ABS(TS-T(I,1))
         FOR J  = 2 TO 19: TS=T(I,J): T(I,J)=.25*(T(I-1,J)+T(I+1,J)+T(I,J-1)+T(I,J+1))
             D  = D + ABS(TS - T(I,J)): NEXT J: IF I = 7 THEN 8
```

```
            TS = T(I,20): T(I,20) = .25 * (T(I-1,20)+T(I+1,20)+2*T(I,19))
            D  = D + ABS(TS - T(I,20))
8           NEXT I: TS = T(10,1): T(10,1)=.5*(T(9,1)+T(10,2)): D=D+ABS(TS-T(10,1))
         FOR J = 2 TO 19: TS=T(I,J): T(10,J)=.25*(2*T(9,J)+T(10,J-1)+T(10,J+1))
            D = D + ABS(TS - T(10,J)): NEXT J
         TS = T(10,20): T(10,20)=.5*(T(9,20)+T(10,19)): D=D+ABS(TS-T(10,20))
         IF D < EPSILON THEN 20 ELSE NR=NR+1: GOTO 2
20 END
```

Sample Results

```
Program Relaxatn - Relaxation method applied for solving heat conduction problem

Enter the tolerance, epsilon, for termination : 10

Sweep #         43
  0.  0.  0.  0.  0.  0.  0.  0.  0.  0.  0.  0.  0.  0.  0.  0.  0.  0.  0.  0.
 10.  8.  6.  5.  3.  3.  2.  1.  1.  1.  1.  0.  0.  0.  0.  0.  0.  0.  0.  0.
 20. 16. 12.  9.  7.  5.  4.  3.  2.  1.  1.  1.  1.  0.  0.  0.  0.  0.  0.  0.
 30. 23. 17. 13. 10.  7.  5.  4.  3.  2.  2.  1.  1.  1.  0.  0.  0.  0.  0.  0.
 40. 29. 21. 16. 12.  9.  7.  5.  4.  3.  2.  1.  1.  1.  1.  0.  0.  0.  0.  0.
 50. 32. 23. 17. 13. 10.  8.  6.  4.  3.  2.  2.  1.  1.  1.  0.  0.  0.  0.  0.
 33. 28. 22. 17. 14. 11.  8.  6.  5.  4.  3.  2.  1.  1.  1.  1.  0.  0.  0.  0.
 25. 24. 20. 17. 14. 11.  9.  7.  5.  4.  3.  2.  2.  1.  1.  1.  0.  0.  0.  0.
 22. 21. 19. 16. 14. 11.  9.  7.  5.  4.  3.  2.  2.  1.  1.  1.  0.  0.  0.  0.
 21. 21. 19. 16. 14. 11.  9.  7.  5.  4.  3.  2.  2.  1.  1.  0.  0.  0.  0.  0.
Press any key to continue
```

Irregular Boundaries

Practically, there are cases where the domain of heat conduction have boundaries which are quite irregular geometrically as illustrated in Figure 4. For such cases, the equation derived based on the relaxation method, Equation 3, which states that the temperature at any point has the average value of those at its four neighboring points if they are equally apart, has to be modified. The modified equation can be derived using a simple argument applied in both x and y directions. For example, consider the temperature at the point G, T_G, in Figure 4. First, let us investigate the horizontal, y direction (for convenience of associating x and y with the row and column indices i and j, respectively as in Figure 2). We observe that T_G is affected more by the temperature at the point C, T_C, than by that at the point I, T_I because the point C is closer to the point G than the point I. Since the closer the point, the greater the influence, based on linear variation of the temperature we can then write:

$$T_G = \frac{\Delta y}{\Delta y + \beta'\Delta y} T_C + \frac{\beta'\Delta y}{\Delta y + \beta\Delta y} T_I = \frac{1}{1+\beta'} T_C + \frac{\beta'}{1+\beta'} T_I \tag{15}$$

where the increment from point I to point G is the regular increment Δy while that between G and C is less and equal to $\beta'\Delta y$ with β' having a value between 0 and 1. Similarly, along the vertical, x direction and considering the points B, G, and H and a regular increment Δx, we can have:

$$T_G = \frac{\Delta x}{\Delta x + \alpha'\Delta x} T_B + \frac{\alpha'\Delta x}{\Delta x + \alpha'\Delta x} T_H = \frac{1}{1+\alpha'} T_B + \frac{\alpha'}{1+\alpha'} T_H \tag{16}$$

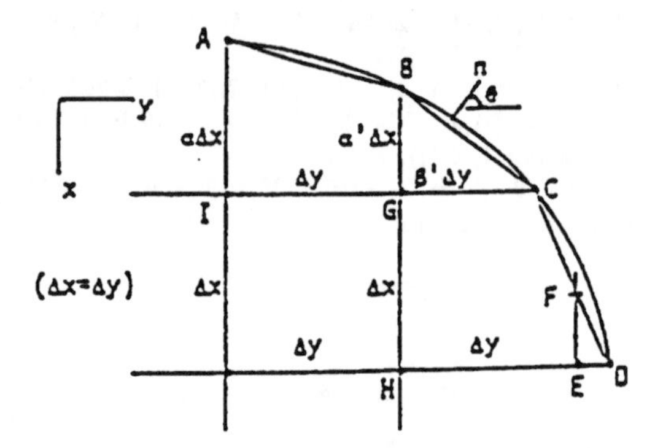

FIGURE 4. There are cases where the domain of heat conduction have boundaries which are quite irregular geometrically.

where like β', α' has a value between 0 and 1. As often is the case, the regular increments Δx and Δy are taken to be equal to each other for the simplicity of computation. Equations 15 and 16 can then be combined and by taking both x and y directions into consideration, an averaging approach leads to:

$$T_G = \frac{1}{2}\left(\frac{1}{1+\alpha'}T_B + \frac{\alpha'}{1+\alpha'}T_H + \frac{1}{1+\beta'}T_C + \frac{\beta'}{1+\beta'}T_I\right) \tag{17}$$

For every group of five points such as B, C, G, H, and I in Figure 4 situated at any irregular boundary, the values of α' and β' have to be measured and Equation 17 is to be used during the relaxation process if the boundary temperatures are known.

If some points along an irregular boundary are insulated such as the points B and C in Figure 4, we need to derive new formula to replace Equation 6 or Equation 10. The insulated condition along BC requires $\partial T \partial n = 0$ where n is the direction normal to the cord BC when the arc BC is approximated linearly. If the values of ‘ and ‘ are known, we can replace the condition $\partial T/\partial n = 0$ with

$$\frac{\partial T}{\partial n} = \frac{\partial T dX}{\partial X dn} + \frac{\partial T dY}{\partial Y dn} = \frac{\partial T}{\partial X}\sin\theta + \frac{\partial T}{\partial Y}\cos\theta = \alpha'\frac{\partial T}{\partial X} + \beta'\frac{\partial T}{\partial Y} = 0 \tag{18}$$

The remainder of derivation is left as a homework problem.

MATLAB Application

A **Relaxatn.m** file can be created to perform interactive **MATLAB** operations and generate plots of the temperature distributions during the course of relaxation. This file may be prepared as follows:

```
function [SumOfDs,T]=Relaxatn(T)
        SumOfDs=0;
        for i=2:6
            for j=2:19
                Tsave=T(i,j);
                T(i,j)=.25*(T(i-1,j)+T(i+1,j)+T(i,j-1)+T(i,j+1));
                SumOfDs=abs(Tsave-T(i,j));
            end
        end
        for i=7:9
            Tsave=T(i,1);
            T(i,1)=.25*(T(i-1,1)+T(i+1,1)+2*T(i,2));
            SumOfDs=abs(Tsave-T(i,1));
            for j=2:19
                Tsave=T(i,j);
                T(i,j)=.25*(T(i-1,j)+T(i+1,j)+T(i,j-1)+T(i,j+1));
                SumOfDs=abs(Tsave-T(i,j));
            end
            if i>7 Tsave=T(i,20);
                   T(i,20)=.25*(T(i-1,20)+T(i+1,20)+2*T(i,19));
                   SumOfDs=abs(Tsave-T(i,20));
            end
        end
        Tsave=T(10,1);
        T(10,1)=.5*(T(9,1)+T(10,2));
        SumOfDs=abs(Tsave-T(10,1));
        for j=2:19
            Tsave=T(10,j);
            T(10,j)=.25*(2*T(9,j)+T(10,j-1)+T(10,j+1));
            SumOfDs=abs(Tsave-T(10,j));
        end
        Tsave=T(10,20);
        T(10,20)=.5*(T(9,20)+T(10,19));
        SumOfDs=abs(Tsave-T(10,20));
```

This file can be applied to solve the sample problem run by first specifying the boundary temperatures described in Equation 4 to obtain an initial distribution by entering the **MATLAB** instructions:

```
>> T=zeros(10,20); for I=2:6, T(i,1)=(I-1)*10; end, format compact, NR=1;
>> fprintf('Sweep # %3.0f \n',NR), T
   Sweep #    1
   T =
   Columns 1 through 12

      0     0     0     0     0     0     0     0     0     0     0     0
     10     0     0     0     0     0     0     0     0     0     0     0
     20     0     0     0     0     0     0     0     0     0     0     0
     30     0     0     0     0     0     0     0     0     0     0     0
     40     0     0     0     0     0     0     0     0     0     0     0
     50     0     0     0     0     0     0     0     0     0     0     0
      0     0     0     0     0     0     0     0     0     0     0     0
      0     0     0     0     0     0     0     0     0     0     0     0
      0     0     0     0     0     0     0     0     0     0     0     0
      0     0     0     0     0     0     0     0     0     0     0     0
```

```
Columns 13 through 20
     0     0     0     0     0     0     0     0
     0     0     0     0     0     0     0     0
     0     0     0     0     0     0     0     0
     0     0     0     0     0     0     0     0
     0     0     0     0     0     0     0     0
     0     0     0     0     0     0     0     0
     0     0     0     0     0     0     0     0
     0     0     0     0     0     0     0     0
     0     0     0     0     0     0     0     0
     0     0     0     0     0     0     0     0
```

The fprintf command enables a label be added, in which the format %3.0f requests 3 columns be provided without the decimal point for printing the value of NR, and \n requests that next printout should be started on a new line. The Relaxatn.m can now be utilized to perform the relaxations. Let first perform one relaxation by entering

```
>> NR = NR + 1; fprintf('Sweep # %3.0f \n',NR), [D,T] = feval(A:Relaxatn',T]
```

The resulting display of the error defined in Equation 13 and the second temperature distribution is:

```
Sweep #   2
D =
   7.0137e-008
T =
Columns 1 through 7
         0         0         0         0         0         0         0
   10.0000    2.5000    0.6250    0.1563    0.0391    0.0098    0.0024
   20.0000    5.6250    1.5625    0.4297    0.1172    0.0317    0.0085
   30.0000    8.9063    2.6172    0.7617    0.2197    0.0629    0.0179
   40.0000   12.2266    3.7109    1.1182    0.3345    0.0993    0.0293
   50.0000   15.5566    4.8169    1.4838    0.4546    0.1385    0.0419
   12.5000    7.0142    2.9678    1.1104    0.3912    0.1324    0.0436
    3.1250    2.5348    1.3731    0.6209    0.2530    0.0964    0.0350
    0.7813    0.8290    0.5505    0.2929    0.1365    0.0582    0.0233
    0.3906    0.5512    0.4033    0.2473    0.1300    0.0616    0.0271
Columns 8 through 14
         0         0         0         0         0         0         0
    0.0006    0.0002    0.0000    0.0000    0.0000    0.0000    0.0000
    0.0023    0.0002    0.0002    0.0000    0.0000    0.0000    0.0080
    0.0050    0.0014    0 0004    0.0001    0.0000    0.0000    0.0000
    0.0086    0.0025    0.0007    0.0002    0.0001    0.0000    0.0000
    0.0126    0.0038    0.0011    0.0003    0.0001    0.0000    0.0000
    0.0141    0.0045    0.0014    0.0004    0.0001    0.0000    0.0000
    0.0123    0.0042    0.0014    0.0005    0.0001    0.0000    0.0000
    0.0089    0.0033    0.0012    0.0004    0.0001    0.0000    0.0000
    0.0112    0.0044    0.0017    0.0006    0.0002    0.0001    0.0000
Columns 15 through 20
         0         0         0         0         0         0
    0.0000    0.0000    0.0000    0.0000    0.0000         0
    0.0000    0.0000    0.0000    0.0000    0.0000         0
    0.0000    0.0000    0.0000    0.0000    0.0000         0
    0.0000    0.0000    0.0000    0.0000    0.0000         0
    0.0000    0.0000    0.0000    0.0000    0.0000         0
    0.0000    0.0000    0.0000    0.0000    0.0000         0
    0.0000    0.0000    0.0000    0.0000    0.0000    0.0000
    0.0000    0.0000    0.0000    0.0000    0.0000    0.0000
    0.0000    0.0000    0.0000    0.0000    0.0000    0.0000
```

In case that we need to have the 30th temperature distribution by performing 29 consecutive relaxations, we enter:

```
>> for NR = 3:30;  [D,T] = feval(A:Relaxatn',T];  end
>> fprintf('Sweep # %3.0f \n',NR),D,T
```

The resulting display of the error defined in Equation 13 and the 30th temperature distribution is:

```
    Sweep #   30
D =
    0.0082
T =
Columns 1 through 7
        0         0         0         0         0         0         0
  10.0000    7.8240    5.8971    4.3264    3.1130    2.2070    1.5452
  20.0000   15.4564   11.5265    8.4017    6.0310    4.2789    3.0045
  30.0000   22.5767   16.5012   11.9001    8.5147    6.0506    4.2671
  40.0000   28.5013   20.2189   14.4310   10.3309    7.3837    5.2490
  50.0000   31.4309   21.7354   15.5950   11.3147    8.2054    5.9135
  31.0236   25.8271   20.0781   15.2923   11.5089    8.6376    6.2913
  23.0526   21.2258   17.9193   14.4421   11.2972    0.6354    6.5749
  19.5060   18.6290   16.4481   13.7563   11.0644    8.6354    6.5749
  18.5571   17.8914   16.0265   13.5837   11.0448    8.6947    6.6665
Columns 8 through 14
        0         0         0         0         0         0         0
   1.0695    0.7318    0.4949    0.3307    0.2182    0.1422    0.0914
   2.0881    1.4358    0.9762    0.6559    0.4353    0.2853    0.1845
   2.9830    2.0645    1.4132    0.9560    0.6387    0.4213    0.2742
   3.7013    2.5840    1.7836    1.2163    0.8189    0.5441    0.3567
   4.2188    2.9802    2.0778    1.4298    0.9707    0.6500    0.4293
   4.5613    3.2612    2.2985    1.5968    1.0934    0.7379    0.4909
   4.7682    3.4517    2.4581    1.7229    1.1891    0.8083    0.5414
   4.8998    3.5816    2.5715    1.8153    1.2609    0.8623    0.5809
   4.9975    3.6720    2.6490    1.8784    1.3105    0.9001    0.6091
Columns 15 through 20
        0         0         0         0         0         0
   0.0580    0.0361    0.0220    0.0126    0.0058         0
   0.1177    0.0738    0.0451    0.0259    0.0120         0
   0.1760    0.1111    0.0683    0.0395    0.0185         0
   0.2305    0.1465    0.0908    0.0529    0.0250         0
   0.2794    0.1790    0.1119    0.0662    0.0320         0
   0.3219    0.2078    0.1315    0.0799    0.0414         0
   0.3574    0.2326    0.1494    0.0948    0.0606    0.0447
   0.3858    0.2529    0.1643    0.1074    0.0749    0.0650
   0.4063    0.2676    0.1751    0.1159    0.0828    0.0739
```

To obtain a plot of this temperature distribution after the initial temperature distribution has been relaxed 29 times, with gridwork and title as shown in Figure 5a, the interactive **MATLAB** instructions entered are:

```
>> V= 0:1:50; contour(T,V'), grid, title('* After 30 relaxations *')
```

Notice that 51 contours having values 0 through 50 with an increment of 1 defined in the row matrix V. In Figure 5a, the contour having a value equal to 0 is

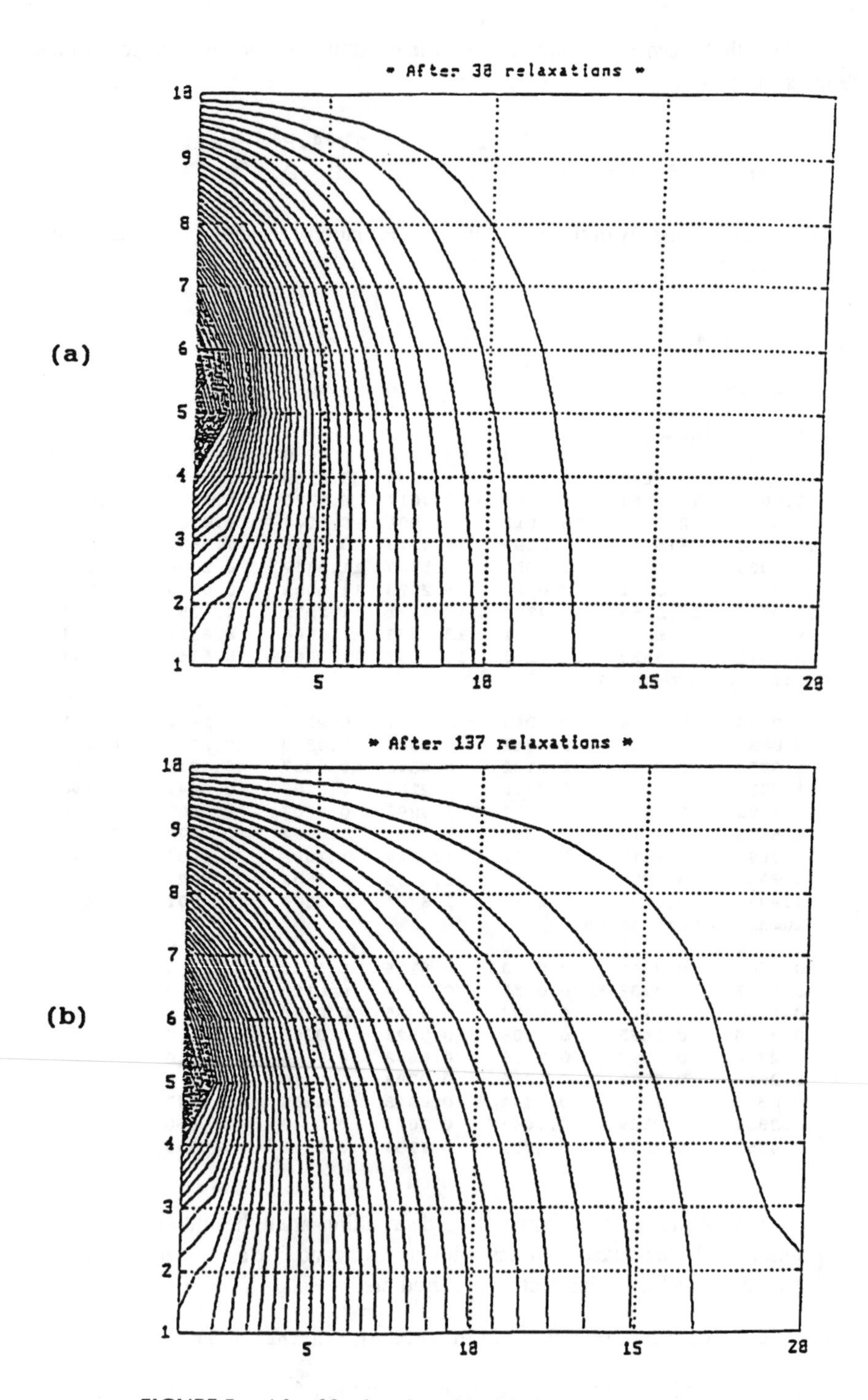

FIGURE 5. After 38 relaxations (a) and after 137 relaxations (b).

along the upper edge (Y = 10) and right edge (X = 20), the first *curved* contour of the right has a value equal to 1, and the values of the contours are increased from right to left until the point marked “5” which has a temperature equal to 50 is reached. It should be noted that along the left edge, the uppermost point marked “10” has a temperature equal to zero and the temperatures are increased linearly (as for the initial conditions) to 50 at the point marked with “5”, and from that point down to the point marked “1” the entire lower portion of the left edge is insulated.

For obtaining the 137th sweep, we can continue to call the service **Relaxatn.m** by similarly applying the **MATLAB** instructions as follows:

```
>> for NR = 31:137; [D,T] = feval(A:Relaxatn',T]; end
>> fprintf('Sweep # %3.0f \n',NR),D,T
```

The resulting display of the error defined in Equation 13 and the 137th temperature distribution is:

```
Sweep # 137
D =
    0.0039
T =
Columns 1 through 7
       0          0          0          0          0          0          0
 10.0000     8.2092     6.6034     5.2663     4.1979     3.3592     2.7027
 20.0000    16.2354    12.9413    10.2680     8.1709     6.5415     5.2714
 30.0000    23.7946    18.6638    14.7005    11.6842     9.3736     7.5788
 40.0000    30.2843    23.2266    18.1958    14.5030    11.7020     9.5196
 50.0000    34.1232    25.7729    20.3657    16.4440    13.4267    11.0289
 34.9702    30.4465    25.3896    21.0650    17.9476    14.5495    12.1007
 29.0063    27.3172    24.2898    21.0253    17.9476    15.1931    12.7873
 26.4439    25.5428    23.4447    20.8176    18.0951    15.5084    13.1623
 25.7091    24.9832    23.1469    20.7428    18.1272    15.6043    13.2838
Columns 8 through 14
       0          0          0          0          0          0          0
  2.1860     1.7755     1.4457     1.1780     0.9585     0.7764     0.6235
  4.2718     2.4760     2.8351     2.3135     1.8846     1.5384     1.2289
  6.1633     5.0311     4.1147     3.3652     2.7466     2.2313     1.7974
  7.7842     6.3834     5.2398     4.2979     3.5164     2.8631     2.3120
  9.0865     7.4942     6.1784     5.0849     4.1717     3.4053     2.7578
 10.0569     8.3464     6.9123     5.7084     4.6961     3.8432     3.1218
 10.7141     8.9421     7.4354     6.1586     5.0785     4.1464     3.3929
 11.0913     9.2934     7.7490     6.4315     5.3122     4.3639     3.5615
 11.2170     9.4124     7.8565     6.5259     5.3937     4.4336     3.6212
Columns 15 through 20
       0          0          0          0          0          0
  0.4931     0.3792     0.2770     0.1819     0.0904          0
  0.9731     0.7498     0.5490     0.3616     0.1802          0
  1.4266     1.1028     0.8111     0.5375     0.2696          0
  1.8410     1.4302     1.0601     0.7107     0.3619          0
  2.2050     1.7248     1.2945     0.8874     0.4688          0
  2.5078     1.9790     1.5129     1.0807     0.6278          0
  2.7380     2.1811     1.7060     1.3007     0.9651     0.7498
  2.8836     2.3131     1.8388     1.4589     1.1884     1.0729
  2.9358     2.3610     1.8876     1.5164     1.2645     1.1687
```

Figure 5b shows the 137th temperature distribution when the interactive **MATLAB** instructions entered are:

```
>> V= 0:1:50; contour(T,V'), grid, title('* After 137 relaxations *')
```

Notice that area near the insulated boundaries at the right-lower corner has finally reached a steady-state temperature distribution, i.e., changes of the entire temperature distribution will be insignificant if more relaxations were pursued.

Mathematica Applications

To apply the relaxation method for finding the steady-state temperature distribution of the heated plate already solved by the **FORTRAN**, **QuickBASIC**, and **MATLAB** versions, here we make use of the **Do**, **If**, and **While** commands of **Mathematica** to generate similar results through the following interactive operations:

```
In[1]: = t= Table[0,{10},{20}]; eps = 100; nr = 0; d = eps + 1;

In[2]: = Do[t[[i,1]] = (I–1)*10,{i,2,6}];

In[3]: = (While[d>eps, d = 0;nr = nr + 1;
     Do[Do[ts = t[[i,j]]; t[[i,j]] = .25*(t[[I–1,j]] + t[[I + 1,j]]
          + t[[i,j–1]] + t[[i,j + 1]]);
          d = Abs[ts-t[[i,j]]] + d;,{j,2,19}],{i,2,6}];
       Do[ts = t[[i,1]]; t[[i,1]] = .25*(t[[I–1,1]] + t[[I + 1,1]]
          + 2*t[[i,2]]); d = Abs[ts-t[[i,1]]] + d;
         Do[ts = t[[i,j]]; t[[i,j]] = .25*(t[[I–1,j]] + t[[I + 1,j]]
         + t[[i,j–1]] + t[[i,j + 1]]);
         d = Abs[ts-t[[i,j]]] + d;,{j,2,19}];
         If[i = = 7,Continue, ts = t[[i,20]];
         t[[i,20]] = .25*(t[[I–1,20]] + t[[I + 1,20]] + 2*t[[i,19]]);
         d = Abs[ts-t[[i,20]]] + d;],{i,7,9}];
         ts = t[[10,1]]; t[[10,1]] = .5*(t[[9,1]] + t[[10,2]]);
         d = Abs[ts-t[[10,1]]] + d;
         Do[ts = t[[10,j]]; t[[10,j]] = .25*(2*t[[9,j]] + t[[10,j–1]]
             + t[[10,j + 1]]); d = Abs[ts-t[[10,j]]] + d;,{j,2,19}];
           ts = t[[10,20]]; t[[10,20]] = .5*(t[[9,20]] + t[[10,19]]);
           d = Abs[ts-t[[10,20]]] + d;])

In[4]: = Print["Sweep #",nr]; Round[N[t,2]]

Out[4] = Sweep #2

     {{ 0, 0, 0, 0, 0, 0, 0, 0, 0, 0, 0, 0, 0, 0, 0, 0, 0, 0, 0, 0},
     {10, 4, 1, 0, 0, 0, 0, 0, 0, 0, 0, 0, 0, 0, 0, 0, 0, 0, 0, 0},
     {20, 9, 3, 1, 0, 0, 0, 0, 0, 0, 0, 0, 0, 0, 0, 0, 0, 0, 0, 0},
```

{30, 13, 5, 2, 1, 0, 0, 0, 0, 0, 0, 0, 0, 0, 0, 0, 0, 0, 0, 0},
{40, 18, 7, 3, 1, 0, 0, 0, 0, 0, 0, 0, 0, 0, 0, 0, 0, 0, 0, 0},
{50, 20, 8, 3, 1, 0, 0, 0, 0, 0, 0, 0, 0, 0, 0, 0, 0, 0, 0, 0},
{17, 11, 5, 2, 1, 0, 0, 0, 0, 0, 0, 0, 0, 0, 0, 0, 0, 0, 0, 0},
{ 6, 5, 3, 1, 1, 0, 0, 0, 0, 0, 0, 0, 0, 0, 0, 0, 0, 0, 0, 0},
{ 2, 2, 1, 1, 0, 0, 0, 0, 0, 0, 0, 0, 0, 0, 0, 0, 0, 0, 0, 0},
{ 1, 1, 1, 1, 0, 0, 0, 0, 0, 0, 0, 0, 0, 0, 0, 0, 0, 0, 0, 0}}

Notice that *In[2]* initializes the boundary temperatures, nr keeps the count of how many sweeps have been performed, and the function Round is employed in *In[4]* to round the temperature value to a two-digit integer. When the total temperature differences, d, is limited to eps = 100 degrees, the t values obtained after two sweeps are slightly different from those obtained by the other versions, this is again because **Mathematica** keeps more significant digits in all computation steps than those in the **FORTRAN**, **QuickBASIC**, and **MATLAB** programs. By changing the eps value from 100 degrees to 1 degree, **Mathematica** also takes 137 sweeps to converge as in the **FORTRAN**, **QuickBASIC**, and **MATLAB** versions:

In[5]: = t = Table[0,{10},{20}]; eps = 1; nr = 0; d = eps + 1;

In[6]: = Do[t[[i,1]] = (I–1)*10,{i,2,6}];

In[7]: = Print["Sweep #",nr]; Round[N[t,2]]

Out[7] = Sweep #137

{{ 0, 0, 0, 0, 0, 0, 0, 0, 0, 0, 0, 0, 0, 0, 0, 0, 0, 0, 0, 0},
{10, 8, 7, 5, 4, 3, 3, 2, 2, 1, 1, 1, 1, 1, 0, 0, 0, 0, 0, 0},
{20, 16, 13, 10, 8, 7, 5, 4, 3, 3, 2, 2, 2, 1, 1, 1, 1, 0, 0, 0},
{30, 24, 19, 15, 12, 9, 8, 6, 5, 4, 3, 3, 2, 2, 1, 1, 1, 1, 0, 0},
{40, 30, 23, 18, 15, 12, 10, 8, 6, 5, 4, 4, 3, 2, 2, 1, 1, 1, 0, 0},
{50, 34, 26, 20, 16, 13, 11, 9, 7, 6, 5, 4, 3, 3, 2, 2, 1, 1, 0, 0},
{35, 30, 25, 21, 17, 15, 12, 10, 8, 7, 6, 5, 4, 3, 3, 2, 2, 1, 1, 0},
{29, 27, 24, 21, 18, 15, 13, 11, 9, 7, 6, 5, 4, 3, 3, 2, 2, 1, 1, 1},
{26, 26, 23, 21, 18, 16, 13, 11, 9, 8, 6, 5, 4, 4, 3, 2, 2, 1, 1, 1},
{26, 25, 23, 21, 18, 16, 13, 11, 9, 8, 7, 5, 4, 4, 3, 2, 2, 2, 1, 1}}

The above results all agree with those obtained earlier.

Warping Analysis of a Twisted Bar with Noncircular Cross Section

As another example of applying the relaxation method for engineering analysis, consider the case of a long bar of uniform rectangular cross section twisted by the two equal torques (T) at its ends. The cross section of the twisted bar becomes warped as shown in Figure 6. If z-axis is assigned to the longitudinal direction of the bar, to find the amount of warping at any point (x,y) of the cross sectioned surface, denoted as W(x,y), the relaxation method can again be employed because

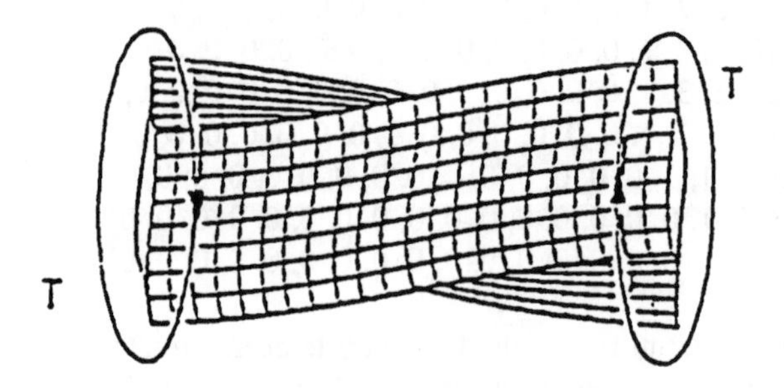

FIGURE 6. In the case of a bar of uniform rectangular cross section twisted by the two equal torques (T) at its ends, the cross section of the twisted bar becomes warped as shown.

W(x,y) is governed by the Laplace equation.[3] Due to anti-symmetry of W(x,y), that is W(–x,y) = W(x,–y) = –W(x,y), only one-fourth of the cross section needs to be analyzed. Let us consider a rectangular region $0 \le x \le a$ and $0 \le y \le b$. It is obvious that the anti-symmetry leads to the conditions W(x,0) = 0 and W(0,y) = 0 along two of the four linear boundaries. To derive the boundary conditions along the right side x = a and the upper side y = b, we have to utilize the relationships between the warping function W(x,y) and shear stresses τ_{xz} and τ_{yz} which are:

$$\tau_{xz} = G\theta\left(\frac{\partial W}{\partial x} - y\right) \quad \text{and} \quad \tau_{yz} = G\theta\left(\frac{\partial W}{\partial y} + x\right) \tag{19,20}$$

where G and θ are the shear rigidity and twisting angle of the bar, respectively. Along the boundaries x = a and y = b, the shear stress should be tangential to the lateral surface of the twisted bar requires:

$$\left(\frac{\partial W}{\partial x} - y\right)\frac{dy}{ds} - \left(\frac{\partial W}{\partial y} + x\right)\frac{dx}{ds} = 0 \tag{21}$$

where s is the variable changed along the boundary. Along the upper boundary y = b, dy/ds = 0 and dx/ds = 1 and along the right boundary x = a, dx/ds = 0 and dy/ds = 1. Consequently, Equation 19 reduces to:

$$\frac{\partial W}{\partial y} = -x \quad \text{for } y = b \quad \text{and} \quad \frac{\partial W}{x} = y \quad \text{for } x = a \tag{22,23}$$

To apply the relaxation method for solving W(x,y) which satisfies the Laplace equation for $0<x<a$ and $0<y<b$ and the boundary conditions W(0,y) = W(x,0) = 0

and Equations 22 and 23, let us partition the rectangular region $0 \le x \le a$ and $0 \le y \le b$ into a network of MxN using the increments $\Delta x = a/(M-1)$ and $\Delta y = b/(N-1)$. In finite-difference forms, Equations 22 and 23 yield, respectively:

$$W(x_i, y = y_N \equiv b) = W(x_i, b - \Delta y) - x_i \Delta y \quad \text{for } i = 2,3,\ldots,M-1 \tag{24}$$

and

$$W(x = x_M \equiv a, y_j) = W(a - \Delta x, y_j) + y_j \Delta x \quad \text{for } j = 2,3,\ldots,N-1 \tag{25}$$

where $x_i = (i-1)\Delta x$ and $y_j = (j-1)\Delta y$. Equations 24 and 25 can be combined to yield a finite-difference formula for relaxing the corner point (x_M, y_N). That is:

$$W(x_M \equiv a, y_N \equiv b) = .5\left[W(a, y_{N-1}) + W(x_{M-1}, b) + b\Delta x - a\Delta y\right] \tag{26}$$

Program **Relaxatn.m** has been modified to develop a new m file **Warping.m** for the warping analysis which uses input values for a, b, M, N, and which is needed in Equation 13 for termination of the relaxation. **MATLAB** statements for sample application of **Warping.m** are listed below along with the m-file itself. The **mesh** plot of the warping surface W(x,y) obtained after 131 sweeps of relaxation by initially assuming W(x,y) = 0 throughout the entire region and an ϵ value equal to 100 is shown in Figure 7.

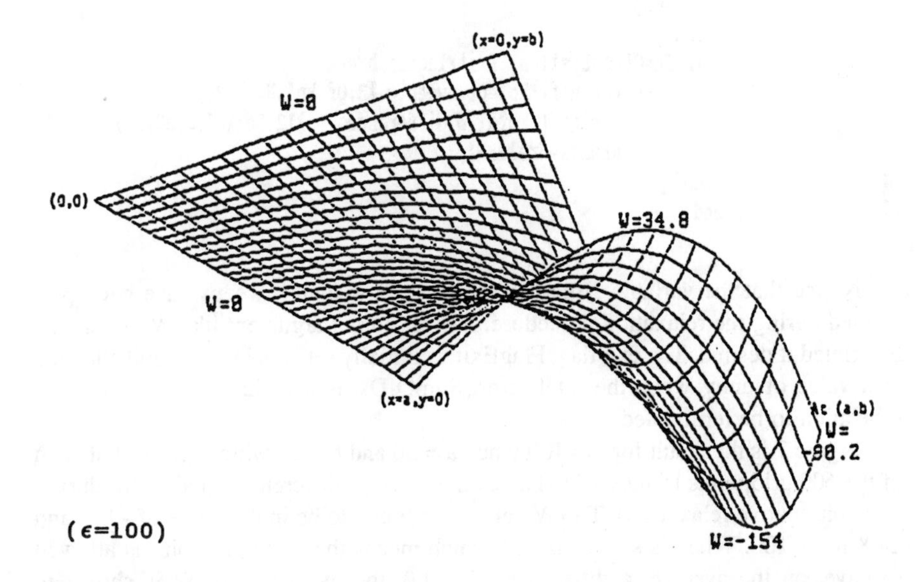

FIGURE 7. The result for ϵ = 100 when a = 30 and b = 20 means that each of the 600 gridpoints is allowed to have, on average, a difference equal to 1/6 during two consecutive relaxations.

```
>>W=zeros(31,21); [W,n]=warping(30,20,31,21,100,W), mesh(W)

function [W,Nrelax]=Warping(a,b,M,N,Epsilon,W)
        dx=a/(M-1); dy=b/(N-1); Nrelax=1; ExitFlag=0;
        while ExitFlag==0;
          SumOfDs=0; Wsave=W(M,N);
          %  Eq. (26)
             W(M,N)=.5*(W(M,N-1)+W(M-1,N)+b*dx-a*dy);
          SumOfDs=SumOfDs+abs(Wsave-W(M,N));
          for jr=2:N-1; j=N+1-jr;
             Wsave=W(M,j);
          %  Eq. (25)
             W(M,j)=W(M-1,j)+(j-1)*dy*dx;
             SumOfDs=SumOfDs+abs(Wsave-W(M,j));
          end
          for ir=2:M-1; i=M+1-ir;
             Wsave=W(i,N);
          %  Eq. (24)
             W(i,N)=W(i,N-1)-(i-1)*dx*dy;
             SumOfDs=SumOfDs+abs(Wsave-W(i,N));
             for jr=2:N-1; j=N+1-jr;
                Wsave=W(i,j);
             %  Eq. (8)
                W(i,j)=.25*(W(i-1,j)+W(i+1,j)+W(i,j-1)+W(i,j+1));
                SumOfDs=SumOfDs+abs(Wsave-W(i,j));
             end
          end
          if SumOfDs<Epsilon, ExitFlag=1; break
             else fprintf('No. of Sweep = %3.0f \n',Nrelax)
                  fprintf('Total W differences = %12.5e \n',SumOfDs)
                  Nrelax=Nrelax+1;
          end
        end
```

Notice that the variable Nrelax keeps track how many sweeps have been performed during the relaxation procedure, it is an output argument like W which can be printed if desired. The exit flag, FlagExit, is initially set equal to zero and changed to a value of unity when the total error, SumOfDs, is less than and allowing the relaxation to be terminated.

Figure 7 is the result for $\epsilon = 100$ when a = 30 and b = 20 which means that each of the 600 gridpoints is allowed to have, on average, a difference equal to 1/6 during two consecutive relaxations. The W values are found to be in the range of –154 and 24.8 for = 100. When is set equal to 1 which means that each gridpoint is allowed to have, on the average, a difference of 1/600, the mesh plot of W is shown in Figure 8. The W values are now all equal to or less than zero.

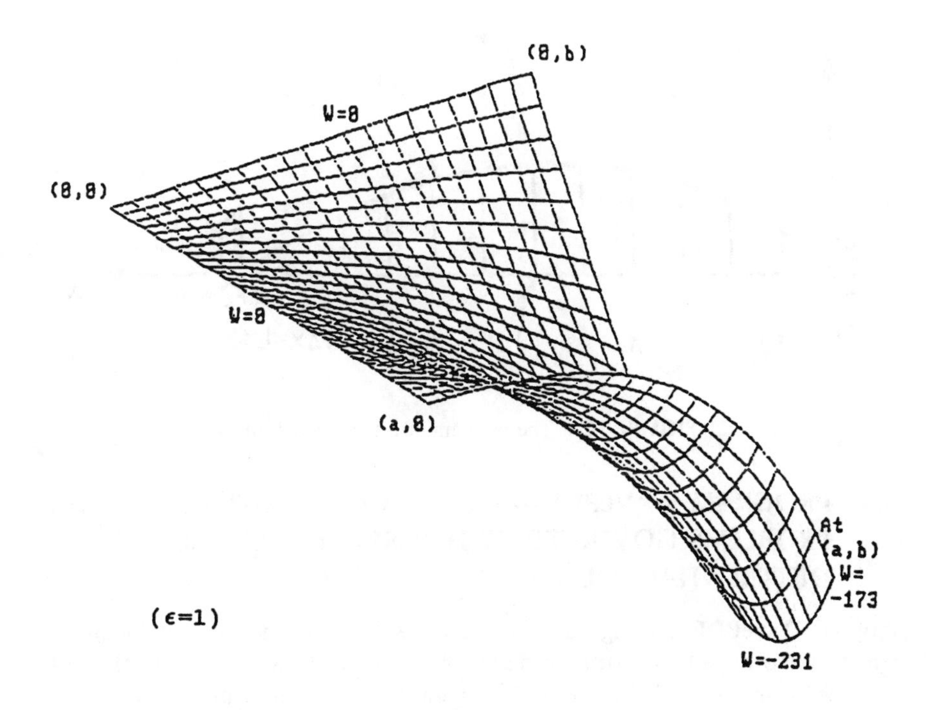

FIGURE 8.

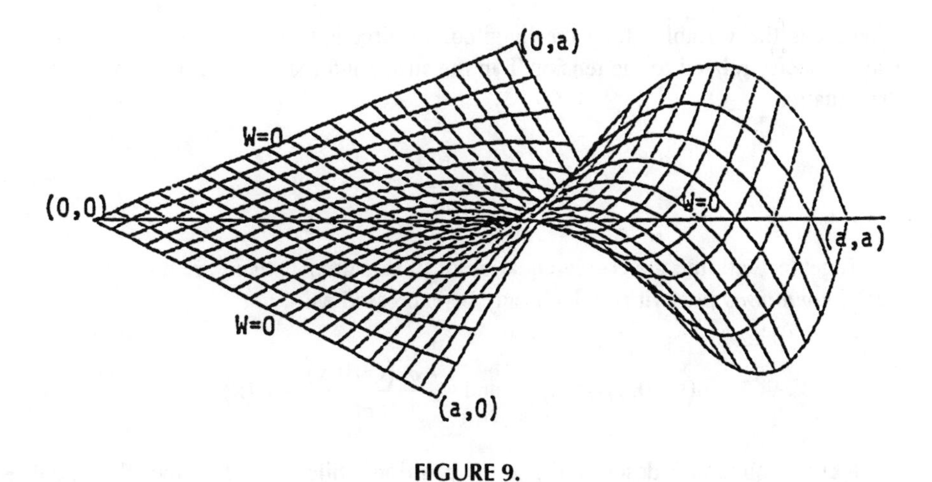

FIGURE 9.

It is noteworthy that if the cross section of the twisted rod is square, that is when a = b, there is no warping along the x = y line as illustrated by the mesh plot shown in Figure 9. This case is left as a homework problem for the readers to work out the details.

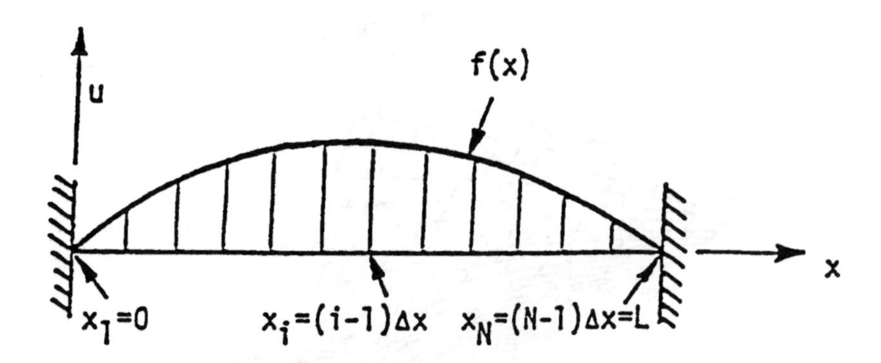

FIGURE 10. The problem of a tightened string.

8.4 PROGRAM WAVEPDE — NUMERICAL SOLUTION OF WAVE PROBLEMS GOVERNED BY HYPERBOLIC PARTIAL DIFFERENTIAL EQUATIONS

Program **WavePDE** is designed for numerical solution of the wave problems governed by the hyperbolic partial differential equation.[1] Consider the problem of a tightened string shown in Figure 10. The lateral displacement u satisfies the equation:

$$\frac{\partial^2 u}{\partial t^2} = a^2 \frac{\partial^2 u}{\partial x^2} \tag{1}$$

where x is the variable along the longitudinal direction of the string and a is the wave velocity related to the tension T in the string and the mass m of the string by the equation:

$$a^2 = \frac{T}{m} \tag{2}$$

Together with the governing equation of u, Equation 1, there are also so-called *initial conditions* prescribed which may be expressed as:

$$u(t=0,x) = f(x) \quad \text{and} \quad \frac{\partial u(t=0,x)}{\partial t} = v_0(x) \tag{3,4}$$

f(x) in Equation 3 describe the initial position while $v_0(x)$ describes the initial velocity of the tightened string. And, also there are so-called *boundary conditions* which in the case of a string at both ends are:

$$u(t, x = x_1 = 0) = 0 \quad \text{and} \quad u(t, x = x_N = L) = 0 \tag{5,6}$$

If the string is made of a single material, T/m would then be equal to a constant. Analytical solution can be found for this simple case. For the general case that the string may be composed of a number of different materials and the mass m is then a function of the spatial variable x. The more complicated the variation of these properties in x and t, the more likely no analytical solution possible and the problem can only be solved numerically.

The finite-difference approximation of Equation 1 can be achieved by applying the central difference for the second derivatives for both with respect to the space variable x and the time variable t. If we consider the displacement u only a finite number of stations, say N, defined with a spatial increment Δx and using a time increment of Δt, then specially, for t at t_i and x at x_j, we can have:

$$\frac{\partial^2 u}{\partial t^2} = \frac{u_{i-1,j} - 2u_{i,j} - u_{i+1,j}}{(\Delta t)^2}$$

and

$$\frac{\partial^2 u}{\partial x^2} = \frac{u_{i,j-1} - 2u_{i,j} + u_{i,j+1}}{(\Delta x)^2}$$

Substituting the above expressions into Equation 1, we obtain:

$$u_{i+1,j} = -u_{i=1,j} + 2u_{i,j} + \left(\frac{a\Delta t}{\Delta x}\right)^2 \left(u_{i,j-1} - 2u_{i,j} + u_{i,j+1}\right) \tag{7}$$

Equation 7 is to be applied for j = 2 through j = N–1. Initially for $t = t_1 = 0$, Equation 7 can be applied for i = 2, then $u_{i-1,j} \equiv u(t_1,x_j)$ term is simply $f(x_j)$ and can be calculated using Equation 3, and the $u_{i,j-1}$, $u_{i,j}$, and $u_{i,j+1}$ terms can be calculated using the forward-difference approximations of Equation 4 which are, for k = 2,3,…,N–1

$$u(t_2,x_k) = u(t_1,x_k) + v_0(x_k)\Delta t = f(x_k) + v_0(x_k)\Delta t \tag{8}$$

Since both f(x) and $v_0(x)$ are prescribed, use of Equation 7 enables all u to be calculated at $t = t_3$ and at all in-between stations x_j, for j = 2,3,…,N–1. When u values at $t = t_2$ and $t = t_3$ are completely known, Equation 7 can again be utilized to compute u at $t = t_4$ for i = 4 and so forth. Because the string is tightened, we expect the magnitudes of u values to continuously decrease in time. That is, for a specified tolerance, we may demand that the computation be terminated when

$$\max_{j=2\sim N-1}. \left|u_{i,j}\right| < \varepsilon \tag{9}$$

It should be noted that the obtained u values are only for the selected increments Δt and Δx. Whether or not the results will change if either increment is reduced, need be tested.

Equation 7 relates the displacement at x_j at $t = t_{i+1}$ to the displacements at the same location x_j at two *previous* instants t_i and t_{i-1}, and also those of its left and right neighboring points at one time increment earlier, $t = t_i$. This is an approximation which ignores the influences of the displacements of its left and right neighboring points at the present time $t = t_{i+1}$, namely $u_{i+1,j-1}$ and $u_{i+1,j+1}$. These influences can be taken into consideration if the central-difference approximation for the curvature term in the x domain is applied for $t = t_{i+1}$ instead of at $t = t_i$ in derivation of Equation 7. Reader is urged to work Problem 5 to find the effect of this change.

A computer program called **WavePDE** has been developed for generating the deflection of the string at N stations for any specified time increment Δt until the deflection are almost all equal to zero throughout. The program allows interactive specification of the values of a, L, Δt, and Δx, and requires the user to define FUNCTION Subprograms F(X) and V0(X). Both FORTRAN and QuickBASIC versions are listed below along with a sample application.

QuickBASIC Version

```
function [SumOfDs,T]=Relaxatn(T)
        SumOfDs=0;
        for i=2:6
            for j=2:19
                Tsave=T(i,j);
                T(i,j)=.25*(T(i-1,j)+T(i+1,j)+T(i,j-1)+T(i,j+1));
                SumOfDs=abs(Tsave-T(i,j));
            end
        end
        for i=7:9
            Tsave=T(i,1);
            T(i,1)=.25*(T(i-1,1)+T(i+1,1)+2*T(i,2));
            SumOfDs=abs(Tsave-T(i,1));
            for j=2:19
                Tsave=T(i,j);
                T(i,j)=.25*(T(i-1,j)+T(i+1,j)+T(i,j-1)+T(i,j+1));
                SumOfDs=abs(Tsave-T(i,j));
            end
            if i>7 Tsave=T(i,20);
                   T(i,20)=.25*(T(i-1,20)+T(i+1,20)+2*T(i,19));
                   SumOfDs=abs(Tsave-T(i,20));
            end
        end
        Tsave=T(10,1);
        T(10,1)=.5*(T(9,1)+T(10,2));
        SumOfDs=abs(Tsave-T(10,1));
        for j=2:19
            Tsave=T(10,j);
            T(10,j)=.25*(2*T(9,j)+T(10,j-1)+T(10,j+1));
            SumOfDs=abs(Tsave-T(10,j));
        end
```

```
end
Tsave=T(10,20);
T(10,20)=.5*(T(9,20)+T(10,19));
SumOfDs=abs(Tsave-T(10,20));
```

Sample Application

The two function subprograms F and V0 listed in the program WavePDE are prepared for a string that has a length equal to 32 cm and is fastened at its two ends. At time t = 0, the string is lifted at x = 24 cm upward up 2 cm and then released from rest. That is, for Equations 3 and 4, we have a particular case of f(x) = x/12 in cm for 0≤x≤24 cm and f(x) = 8(x/4) in cm for 24<x≤32, and $v_0(x) = 0$. Suppose that the wave velocity, a in Equation 1, is equal to 90 cm/sec^2. We may be interested in knowing the lateral displacements at 15 stations between its two ends equally spaced at an increment of $\Delta x = 2$ cm for t>0. To perform the step-by-step calculation according to Equation 7, let us compute these displacements using a time increment of $\Delta t = 0.001$ second and proceed until t reaches 2 second or when the maximum displacement u_{max} is less than or equal to 0.001. In view of the small Δt, the results are to be printed after t has been increased by 0.1 second. An interactive run of the program WavePDE is presented below:

```
Program WavePDE - Wave Motion governed by Parital Differential Equation.
Have you created the supporting FUNCTION Subprograms F(X) & V0(X)?
Enter Y if they are done; otherwise press <Ctrl Break> : Y
Enter the value of a**2 : 8100
Enter the values of Delta t and Delta X : 0.001,2
Enter the number of in-between stations, N : 15
Enter the value of Epsilon : 1.e-1
Enter an ending time : 2
Enter a time increment for printing : 0.1

Time =      0.00000
     0.00000E+00      1.66667E-01      3.33333E-01      5.00000E-01      6.66667E-01
     8.33330E-01      1.00000E+00      1.16667E+00      1.33333E+00      1.50000E+00
     1.66667E+00      1.83333E+00      2.00000E+00      1.50000E+00      1.00000E+00
     5.00000E-01      0.00000E+00
Time =      0.10000
     0.00000E+00      1.66667E-01      3.33327E-01      4.99972E-01      6.66622E-01
     8.33034E-01      9.96975E-01      1.14341E+00      1.20900E+00      1.07281E+00
     7.94835E-01      6.92063E-01      4.98757E-01      3.59021E-01      1.31218E-01
     9.69535E-02      0.00000E+00
Time =      0.20000
     0.00000E+00      1.63165E-01      3.14672E-01      4.22497E-01      4.18258E-01
     2.40004E-01     -2.97670E-02     -1.84360E-01     -2.85290E-01     -5.15269E-01
    -6.31962E-01     -7.46900E-01     -7.49567E-01     -5.64395E-01     -2.87552E-01
    -1.67763E-01      0.00000E+00
Time =      0.30000
     0.00000E+00     -4.49038E-01     -8.17692E-01     -1.01222E+00     -1.11154E+00
    -1.30979E+00     -1.47038E+00     -1.47346E+00     -1452023E+00     -1.21779E+00
    -9.50311E-01     -8.11794E-01     -6.90822E-01     -5.14203E-01     -2.97612E-01
    -1.93379E-01      0.00000E+00
Time =      0.40000
     0.00000E+00     -4.20338E-01     -8.80266E-01     -1.25924E+00     -1.54300E+00
    -1.63233E+00     -1.48975E+00     -1.44211E+00     -1.36224E+00     -1.16807E+00
    -9.80121E-01     -8.28685E-01     -6.99694E-01     -4.55598E-01     -3.70615E-01
    -1.46329E-01      0.00000E+00
Time =      0.50000
     0.00000E+00      2.41942E-01      4.10856E-01      2.85292E-01      3.48808E-02
     1.04799E-02     -8.50767E-02     -3.54692E-01     -6.20998E-01     -7.73858E-01
    -8.03896E-01     -8.02862E-01     -6.00058E-01     -5.28044E-01     -3.08007E-01
    -1.77244E-01      0.00000E+00
```

```
Time =      0.60000
     0.00000E+00      9.71458E-02      3.40795E-01      4.81283E-01      6.95600E-01
     9.75958E-01      1.09782E+00      1.01151E+00      8.96584E-01      8.55300E-01
     6.80964E-01      5.46788E-01      2.18061E-01      5.21467E-02     -7.60440E-02
    -5.90514E-02      0.00000E+00
Time =      0.69999
     0.00000E+00      1.82785E-01      2.88539E-01      5.23221E-01      7.05613E-01
     8.19889E-01      9.58409E-01      1.11394E+00      1.38274E+00      1.59416E+00
     1.82468E+00      1.73587E+00      1.62435E+00      1.43918E+00      1.15483E+00
     6.62980E-01      0.00000E+00
Time =      0.79999
     0.00000E+00      2.06798E-01      2.97231E-01      4.80511E-01      7.23746E-01
     8.07089E-01      9.12777E-01      1.17106E+00      1.20114E+00      1.23721E+00
     1.09641E+00      8.92447E-01      6.88054E-01      5.65833E-01      4.80817E-01
     2.72947E-01      0.00000E+00
Time =      0.89999
     0.00000E+00      1.63375E-01      2.95364E-01      3.78608E-01      4.93753E-01
     4.38827E-01      2.94639E-01      7.43454E-02     -1.05731E-01     -4.12107E-01
    -5.38549E-01     -5.10644E-01     -5.64354E-01     -5.70825E-01     -5.37805E-01
    -3.22995E-01      0.00000E+00

Time =      0.99999
     0.00000E+00     -2.20420E-01     -4.59750E-01     -7.26144E-01     -1.05391E+00
    -1.15153E+00     -1.30936E+00     -1.33672E+00     -1.28539E+00     -1.31363E+00
    -1.19968E+00     -8.88411E-01     -5.39097E-01     -4.63126E-01     -3.50094E-01
    -1.97370E-01      0.00000E+00
Time =      1.10000
     0.00000E+00     -5.64074E-01     -9.91795E-01     -1.41255E+00     -1.64226E+00
    -1.80542E+00     -1.69046E+00     -1.47052E+00     -1.24728E+00     -1.03373E+00
    -1.06005E-00     -8.77848E-01     -6.51027E-01     -4.84975E-01     -3.61382E-01
    -1.27226E-01      0.00000E+00
Time =      1.20000
     0.00000E+00     -9.72051E-03      1.85691E-02      1.95219E-01      1.34458E-01
    -1.24663E-01     -4.45530E-01     -7.12247E-01     -8.02365E-01     -8.67517E-01
    -7.87895E-01     -7.98698E-01     -6.37735E-01     -4.78853E-01     -3.23825E-01
    -1.64963E-01      0.00000E+00
Time =      1.30000
     0.00000E+00      1.34638E-01      3.56986E-01      6.11524E-01      8.14188E-01
     9.11534E-01      8.98481E-01      9.45475E-01      8.16344E-01      6.76114E-01
     4.66450E-01      2.24011E-01     -6.47397E-02     -7.60462E-02     -7.50079E-02
    -1.34715E-01      0.00000E+00
Time =      1.40000
     0.00000E+00      2.27415E-01      4.29710E-01      4.54726E-01      5.69692E-01
     7.20758E-01      1.06781E+00      1.30335E+00      1.47015E+00      1.55998E+00
     1.66749E+00      1.64235E+00      1.53506E+00      1.37629E+00      1.02935E+00
     4.84091E-01      0.00000E+00
Time =      1.50001
     0.00000E+00      2.07191E-01      3.19376E-01      5.14515E-01      5.60746E-01
     8.68705E-01      1.07440E+00      1.12295E+00      1.08696E+00      1.18324E+00
     1.26952E+00      1.23464E+00      1.16266E+00      8.80416E-01      5.14480E-01
     2.67528E-01      0.00000E+00

Time =      1.60002
     0.00000E+00      1.21418E-01      3.38192E-01      3.67370E-01      4.97966E-01
     5.29223E-01      5.31940E-01      2.78686E-01      6.51729E-02     -4.83741E-02
    -2.87173E-01     -9.95912E-01     -8.37233E-01     -5.53924E-01     -1.56820E-01
    -2.05631E-01      0.00000E+00
Time =      1.70002
     0.00000E+00     -1.06744E-01     -2.69642E-01     -4.68614E-01     -7.33737E-01
    -9.10616E-01     -1.19498E+00     -1.31302E+00     -1.25113E+00     -1.20241E+00
    -1.08212E+00     -9.95912E-01     -8.37233E-01     -5.53924E-01     -1.56820E-01
    -1.65612E-01      0.00000E+00
Time =      1.80003
     0.00000E+00     -5.84820E-01     -1.20072E+00     -1.54521E+00     -1.65254E+00
    -1.72072E+00     -1.73798E+00     -1.57696E+00     -1.37549E+00     -1.15779E+00
    -9.02859E-01     -7.17306E-01     -7.00641E-01     -5.41795E-01     -3.65590E-01
    -1.65612E-01      0.00000E+00
Time =      1.90003
     0.00000E+00      7.24233E-02     -3.24723E-02     -1.85636E-01     -1.95257E-01
    -4.46355E-01     -5.88320E-01     -7.38593E-01     -9.34458E-01     -1.01370E+00
    -9.17116E-01     -7.40747E-01     -5.44435E-01     -4.80090E-01     -3.61236E-01
    -2.02691E-01      0.00000E+00
```

```
Time =      2.00004
     0.00000E+00      3.56727E-01      5.43352E-01      7.00064E-01      6.67818E-01
     8.18378E-01      8.30461E-01      8.55482E-01      7.11126E-01      4.01830E-01
     9.60796E-02     -4.12570E-02     -8.85658E-02     -1.59569E-01     -1.91057E-01
    -9.68752E-02      0.00000E+00
```

MATLAB APPLICATIONS

A **MATLAB** version of **WavePDE.m** can be developed to run the sample problem as follows:

```
>>W=zeros(31,21); [W,n]=warping(30,20,31,21,100,W), mesh(W)

function [W,Nrelax]=Warping(a,b,M,N,Epsilon,W)
        dx=a/(M-1); dy=b/(N-1); Nrelax=1; ExitFlag=0;
        while ExitFlag==0;
          SumOfDs=0; Wsave=W(M,N);
          %  Eq. (26)
             W(M,N)=.5*(W(M,N-1)+W(M-1,N)+b*dx-a*dy);
          SumOfDs=SumOfDs+abs(Wsave-W(M,N));
          for jr=2:N-1; j=N+1-jr;
             Wsave=W(M,j);
          %  Eq. (25)
             W(M,j)=W(M-1,j)+(j-1)*dy*dx;
             SumOfDs=SumOfDs+abs(Wsave-W(M,j));
          end
          for ir=2:M-1; i=M+1-ir;
             Wsave=W(i,N);
          %  Eq. (24)
             W(i,N)=W(i,N-1)-(i-1)*dx*dy;
             SumOfDs=SumOfDs+abs(Wsave-W(i,N));
             for jr=2:N-1; j=N+1-jr;
                Wsave=W(i,j);
             %  Eq. (8)
                W(i,j)=.25*(W(i-1,j)+W(i+1,j)+W(i,j-1)+W(i,j+1));
                SumOfDs=SumOfDs+abs(Wsave-W(i,j));
             end
          end
          if SumOfDs<Epsilon, ExitFlag=1; break
             else fprintf('No. of Sweep = %3.0f \n',Nrelax)
```

When this function is applied for generating displacements using a time increment of 0.001 and a storing increment of 0.1, the **MATLAB** commands, the data for plotting, and the resulting graph (Figure 2) having 21 curves each with 17 points are as follows:

```
>> format compact, UPlot=feval('a:WavePDE',8100,0,2,0.001,0.1,15,2,1.e-3)

 UPlot =
   Columns 1 through 7
        0          0          0          0          0          0          0
   0.1667     0.1667     0.1632    -0.4490    -0.4204     0.2418     0.0971
```

```
    0.3333    0.3333    0.3148   -0.8175   -0.8804    0.4105    0.3406
    0.5000    0.5000    0.4227   -1.0120   -1.2594    0.2849    0.4811
    0.6667    0.6666    0.4185   -1.1152   -1.5432    0.0344    0.6953
    0.8333    0.8331    0.2402   -1.3096   -0.6326    0.0099    0.9756
    1.0000    0.9970   -0.0295   -1.4702   -1.4900   -0.0857    1.0973
    1.1667    1.1435   -0.1841   -1.4733   -1.4423   -0.3553    1.0109
    1.3333    1.2091   -0.2850   -1.4519   -1.3624   -0.6216    0.8959
    1.5000    1.0729   -0.5150   -1.2177   -1.1682   -0.7745    0.8546
    1.6667    0.7949   -0.6318   -0.9502   -0.9803   -0.8044    0.6803
    1.8333    0.6922   -0.7467   -0.8117   -0.8288   -0.8033    0.5461
    2.0000    0.4989   -0.7494   -0.6908   -0.6998   -0.6004    0.2175
    1.5000    0.3591   -0.5643   -0.5142   -0.4557   -0.5283    0.0517
    1.0000    0.1313   -0.2875   -0.2976   -0.3707   -0.3082   -0.0764
    0.5000    0.0961   -0.1677   -0.1934   -0.1463   -0.1773   -0.0592
         0         0         0         0         0         0         0
  Columns 8 through 14
         0         0         0         0         0         0         0
    0.1828    0.2069    0.1636   -0.2201   -0.5642   -0.0102    0.1345
    0.2885    0.2974    0.2958   -0.4592   -0.9921    0.0177    0.3566
    0.5232    0.4808    0.3792   -0.7254   -1.4129    0.1941    0.6109
    0.7056    0.7241    0.4946   -1.0531   -1.6427    0.1331    0.8133
    0.8198    0.8075    0.4398   -1.1506   -1.8059   -0.1262    0.9104
    0.9584    0.9133    0.2958   -1.3084   -1.6910   -0.4472    0.8971
    1.1139    1.1717    0.0756   -1.3359   -1.4710   -0.7139    0.9439
    1.3827    1.2018   -0.1045   -1.2846   -1.2477   -0.8039    0.8147
    1.5941    1.2380   -0.4109   -1.3130   -1.0340   -0.8689    0.6744
    1.8246    1.0972   -0.5374   -1.1992   -1.0603   -0.7891    0.4647
    1.7357    0.8933   -0.5069   -0.8880   -0.8781   -0.7997    0.2224
    1.6242    0.6888   -0.5635   -0.5388   -0.6512   -0.6385    0.0661
    1.4390    0.5665   -0.5702   -0.4629   -0.4851   -0.4795   -0.0771
    1.1547    0.4813   -0.5374   -0.3499   -0.3615   -0.3242   -0.0758
    0.6629    0.2732   -0.3299   -0.1973   -0.1273   -0.1652   -0.1351
         0         0         0         0         0         0         0
  Columns 15 through 21
         0         0         0         0         0         0         0
    0.2274    0.2074    0.1218   -0.1063   -0.5849    0.0718    0.3564
    0.4297    0.3197    0.3389   -0.2687   -1.2009   -0.0337    0.5427
    0.4546    0.5150    0.3684   -0.4673   -1.5454   -0.1875    0.6989
    0.5696    0.5614    0.4994   -0.7321   -1.6529   -0.1975    0.6663
    0.7206    0.8695    0.5310   -0.9089   -1.7211   -0.4488    0.8165
    1.0677    1.0753    0.5339   -1.1932   -1.7384   -0.5908    0.8282
    1.3031    1.1240    0.2809   -1.3114   -1.5774   -0.7410    0.8529
    1.4698    1.0881    0.0675   -1.2496   -1.3760   -0.9367    0.7084
    1.5595    1.1846   -0.0461   -1.2011   -1.1582   -1.0517    0.3991
    1.6669    1.2710   -0.2850   -1.0810   -0.9032   -0.9189    0.0934
    1.6417    1.2361   -0.4508   -0.9951   -0.7176   -0.7422   -0.0436
    1.5345    1.1640   -0.5788   -0.8366   -0.7009   -0.5456   -0.0906
    1.3757    0.8816   -0.6134   -0.5535   -0.5419   -0.4810   -0.1611
    1.0289    0.5153   -0.4800   -0.1565   -0.3658   -0.3619   -0.1921
    0.4838    0.2679   -0.2052    0.0041   -0.1657   -0.2030   -0.0974
         0         0         0         0         0         0         0
>> plot(Uplot), text(13.5,1.7,'(1)'), text(15,0.1,'(2)'), text(4.8,0.2,'(3)')
>> text(3.2,-0.9,'(4)')
```

Notice that solid, broken, dotted, and dot center lines and in that order are used repeatedly for plotting the 21 curves in Figure 11. The first four curves have been marked using the text command. Form the 21st column listed above, we observe that maximum initial displacement of 2 cm has only been reduced to 0.8529 cm after two seconds. It should be pointed out that in **WavePDE.m** the computation should also be ended when the sum of the absolute values of displacements is less than the specified tolerance.

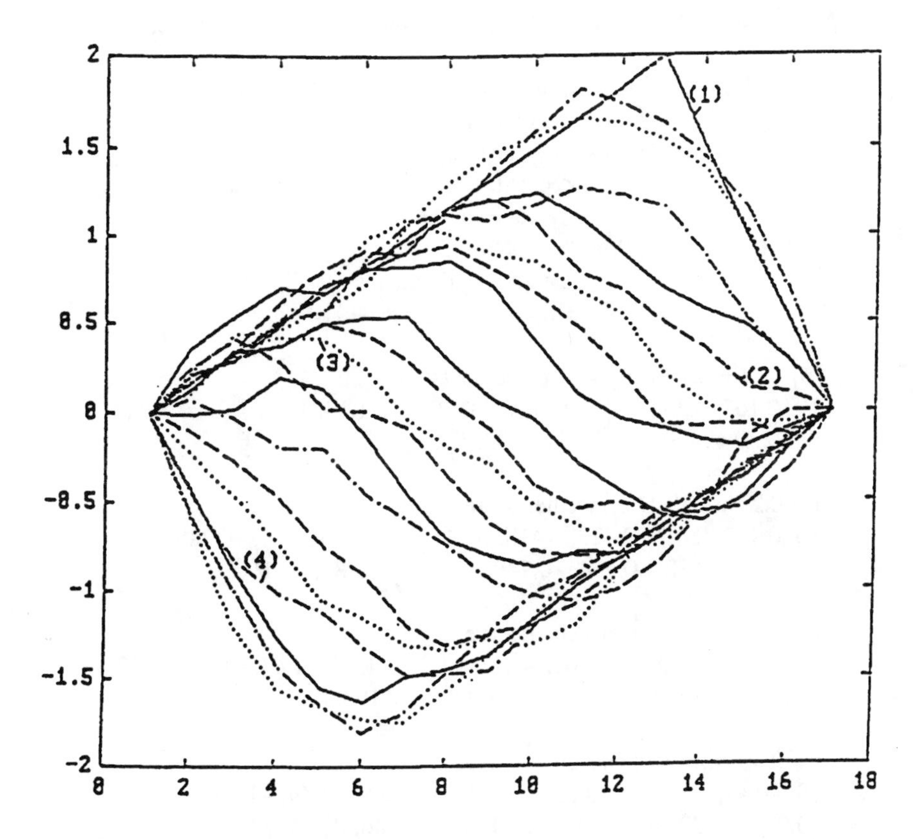

FIGURE 11. Solid, broken, dotted, and dot center lines are used repeatedly for plotting the 21 curves in this figure.

MATHEMATICA APPLICATION

Mathematica can be applied to investigate the string vibration problem by the following interactive operations:

In[1]: = f[fx_]: = If[x>24, 8x/4, x/12]

In[2]: = (asq = 81000; dt = 0.001; dx = 2: n = 15; eps = .1; tend = 2; dtp = 0.1; v0 = 0; np = 100; c = (dt/dx)^2*asq; exitflagg = 0; t = 0)

In[3]: = u0 = Table[0.{j,n + 2}]; Do[x = (j–1)*dx; u0[[j]] = f[x];,{j,n + 2}];

In[4]: = (Print["t = ",N[t,3]," String's Displacements are: "]; Print[N[u03]])

Out[4]: = t = 0 String's Displacements are:
{0, 0.167, 0.333, 0.5, 0.667, 0.833, 1., 1.17, 1.33, 1.5, 1.67, 1.83, 2., 1.5, 1., 0.5, 0}

```
In[5]: = t = t + dt; u = u0 + dt*v0; un = Table[0.{j,n + 2}]; nc = 2;

In[6]: = (While[exitflag == 0, usum = 0; Do[un[[j + 1]] =-u0[[j + 1]] + c*u[[j]]
                    + 2*(1–c)*u[[j + 1]] + c*u[[j + 2]];
            usum = usum + Abs[un[[j + 1]]];,{j,n}];
      t = t + dt; If[(usum<eps)||(t>tend), exitflag = 1; Break,
                    u0 = u; u = un;];
      If[nc = = np, Print["t = ",N[t,3],
                "   String's Displacements are: "];
            Print[N[u,3]]; nc = 1;, nc = nc + 1]])

Out[6]: = t = 0.1 String's Displacements are:
  {0. 0.167, 0.333, 0.5, 0.667, 0.833, 0.997, 1.14, 1.21, 1.07, 0.795,
  0.692, 0.499, 0.359, 0.131, 0.0961, 0}

  t = 0.2 String's Displacements are:
  {0. 0.163, 0.315, 0.423, 0.418, 0.24, –0.0295, –0.184, –0.285,
  –0.515, –0.632, –0.474, –0.749, –0.564, –0.287, –0.168, 0}

  t = 0.3 String's Displacements are:
  {0. –0.449, –0.818, –1.01, –1.12, –1.31, –1.47, –1.47, –1.45, –1.22,
  –0.95, –0.812,–0.691, –0.514, –0.298, –0.193, 0}

  t = 0.4 String's Displacements are:
  {0. –0.42, –0.88, –1.26, –1.54, –1.63, –1.49, –1.44, –1.36, –1.17,
  –0.98, –0.829, –0.7, –0.456, –0.371, –0.146, 0}

  t = 0.5 String's Displacements are:
  {0. 0.242, 0.411, 0.285, 0.0344, 0.00989, –0.0857, –0.355, –0.0622,
  –0.774, –0.804, –0.803, –0.6, –0.528, –0.308, –.177, 0}

  t = 0.6 String's Displacements are:
  {0. 0.0971, 0.341, 0.481, 0.695, 0.976, 1.1, 1.01, 0.896, 0.855,
  0.68, 0.546, 0.217, 0.0517, –0.0764, –0.0592, 0}

  t = 0.7 String's Displacements are:
  {0. 0.183, 0.289, 0.523, 0.706, 0.82, 0.958, 1.11, 1.38, 1.59, 1.82,
  1.74, 1.62, 1.44, 1.15, 0.663, 0}

  t = 0.8 String's Displacements are:
  {0. 0.207, 0.297, 0.481, 0.724, 0.808, 0.913, 1.17, 1.2, 1.24, 1.1,
  0.893, 0.689, 0.567, 0.481, 0.273, 0}

  t = 0.9 String's Displacements are:
  {0. 0.164, 0.296, 0.379, 0.495, 0.44, 0.296, 0.0756, –0.104, –0.411,
  –0.537, –0.51, –0.563, –0.57, –0.537, –0.323, 0}
```

t = 1.0 String's Displacements are:
{0. –0.22, –0.459, –0.725, –1.05, –1.15, –1.31, –1.34, –1.28, –1.31
–1.2, –0.888, –0.539, –0.463, –0.53, –0.197, 0}

t = 1.1 String's Displacements are:
{0. –0.564, –0.992, –1.41, –1.64, –1.81, –1.69, –1.47, –1.25, –1.03,
–1.06, –0.878, –0.651, –0.485, –0.361, –0.127, 0}

t = 1.2 String's Displacements are:
{0. –0.0102, 0.0177, 0.194, 0.133, –0.126, –0.447, –0.714, –0.804,
–0.869, –0.789, –0.8, –0.639, –0.479, –0.324, –0.165, 0}

t = 1.3 String's Displacements are:
{0. 0.134, 0.357, 0.611, 0.813, 0.91, 0.897, 0.944, 0.815, 0.674,
0.465, 0.222, –0.0661, –0.0771, –0.0758, –0.135, 0}

t = 1.4 String's Displacements are:
{0. 0.227, 0.43, 0.455, 0.57, 0.721, 1.07, 1.3, 1.47, 1.56, 1.67,
1.64, 1.53, 1.38, 1.03, 0.484, 0}

t = 1.5 String's Displacements are:
{0. 0.207, 0.32, 0.515, 0.561, 0.869, 1.08, 1.12, 1.09, 1.18, 1.27,
1.24, 1.16, 0.882, 0.515, 0.268, 0}

t = 1.6 String's Displacements are:
{0. 0.122, 0.339, 0.368, 0.499, 0.531, 0.534, 0.281, 0.0675, –0.0461,
–0.285, –0.451, –0.579, –0.613, –0.48, –0.205, 0}

t = 1.7 String's Displacements are:
{0. –0.106, –0.269, –0.467, –0.732, –0.909, –1.19, –1.31, –1.25,
–1.2, –1.08, –0.995, –0.837, –0.533, –0.156, 0.00405, 0}

t = 1.8 String's Displacements are:
{0. –0.585, –1.2, –1.55, –1.65, –1.72, –1.74, –1.58, –1.38, –1.66,
–0.903, –0.718, –0.701, –0.542, –0.366, –0.166, 0}

t = 1.9 String's Displacements are:
{0. 0.0718, –0.0337, –0.188, –0.197, –0.4499, –0.591, –0.741, –0.937,
–1.02, –0.919, –0.742, –0.546, –0.481, –0.362, –0.203, 0}

t = 2.0 String's Displacements are:
{0. 0.0356, 0.543, 0.699, 0.666, 0.816, 0.828, 0.853, 0.708, 0.399,
0.0934, –0.0436, –0.0906, –0.161, –0.192, –0.0974, 0}

The results are in complete agreement with those obtained previously.

8.5 PROBLEMS

PARABPDE

1. Expand the program **ParabPDE** to print out the computed temperature distribution along the entire rod only when the temperature at an interactively specified location reaches an interactively entered value. The time required to reach this condition should be printed as shown below. Call this new program **ParaPDE1**.

   Enter the x value at which a desired temperature is to be specified: 6.0
   Enter the temperature to be reached, in °F: 50
   It takes X.XXXXXE + 00 seconds.

   Notice that a format of E13.5 is to be used to print out the time.
2. Consider the transient heat-conduction problem where $k/c\rho = 0.00104$ ft^2/sec, $\Delta t = 1$ sec, $\Delta x = 1"$, and at the left end of the rod, $x = 0$, the temperature u is equal to 50°F initially but equal to 0°F for $t>0$. Compute the temperatures at $x = 1"$, 2", and 3" when $t = 1$, 2, and 3 seconds.
3. Use the same data as in the program **ParabPDE**, but modify the program for the case when the right end is not insulated but is maintained at 0°F until $t = 100$ seconds and then is heated at 100°F afterwards. Print out the times required for the station at the mid-length reaches to 10°F, 20°F, at 10°F increments until 100°F.
4. Study the effect of changing the value of DT in the program **ParabPDE** on t_{final}, the time required for reaching the uniform distribution of 100°F throughout the rod. Tabulate DT vs. t_{final}.
5. Change the steady-state temperature from 100°F to 75°F and generate a plot similar to that shown in Figure 2.
6. For the rod shown in Figure 1, one-half of the surface insulation, for $x/L = 0.4$ to $x/L = 0.7$, is to be removed and the temperature for that portion of the rod is to be maintained at 70°F. Modify the program **ParabPDE** is accommodate for the computation needed to determine how long it takes to reach a stable temperature distribution.
7. Apply **MATLAB** and **Mathematica** (using **ParaPDE.m**) to solve the transient heat-conduction problem by decreasing Δx from 1 to 0.5.

RELAXATN

1. For the problem treated by the program **Relaxatn** except that the temperature is equal to 100°F along the top boundary, perform the relaxation by upgrading T starting from the top boundary instead of from the bottom boundary to expedite the convergence. Modify the program **Relaxatn** to perform this new sweeping process.
2. Round the right lower corner of the square plate with a radius equal to 5 and consider this corner as an insulated boundary. Modify the program

Relaxatn and rerun the problem to print out the converged temperature distribution.

3. Initially, the temperatures are assumed to be equal to zero everywhere in the plate shown in Figure 12 except those on the boundary. Carry out one complete relaxation (starting from the top row and from left to right, and then down to the next row and so on) cycle to upgrade the unknown temperatures.

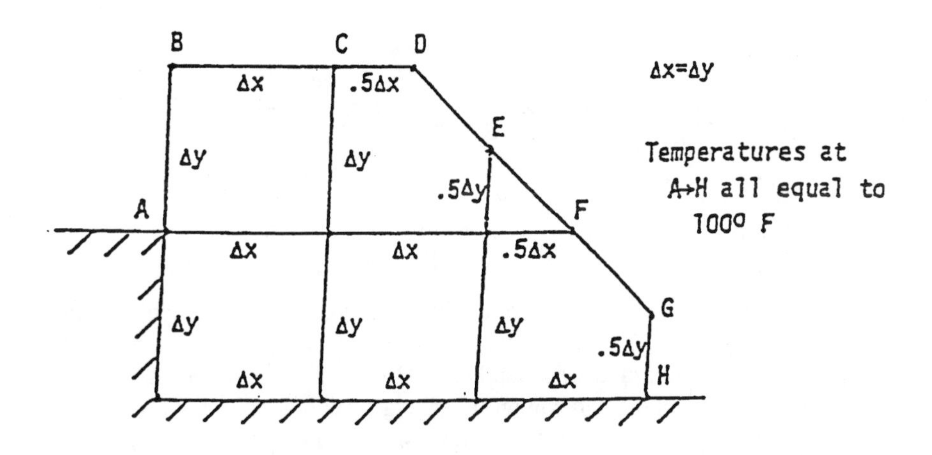

FIGURE 12. Problem 3.

4. Complete the derivation of relaxation equations for the temperatures at the points B and C shown in Figure 2 by use of Equation 18 and by considering the cord BC only. Derive the equation for the point B by averaging the effects of both cords AB and BC.
5. For the problem described in Problem 3, derive a matrix equation of order 5 for direct solution of the five unknown temperatures (two between points A and F, and three along the bottom insulted boundary) based on the finite-difference formulas, Equations 3, 6, 8, 10, 11, or, 12. Compare the resulting temperature distribution with that obtained in Problem 3.
6. Rework Problem 3 if the boundary DEFG is insulated but T_D and T_G remain equal to 100°F.
7. Initially, the temperatures are assumed to be equal to zero everywhere in the shown in Figure 13 except those on the boundary. Carry out one complete relaxation (starting from the top row and from left to right, and then down to the next row and so on) cycle to upgrade the unknown temperatures. The points A, B, and D are on a straight line.
8. Rework Problem 7 if the boundary ABCD is insulated but T_D remains equal to 100°F.
9. Following the same process as in Problem 8, but obtain the direct solution of the temperature distribution for Problem 7.

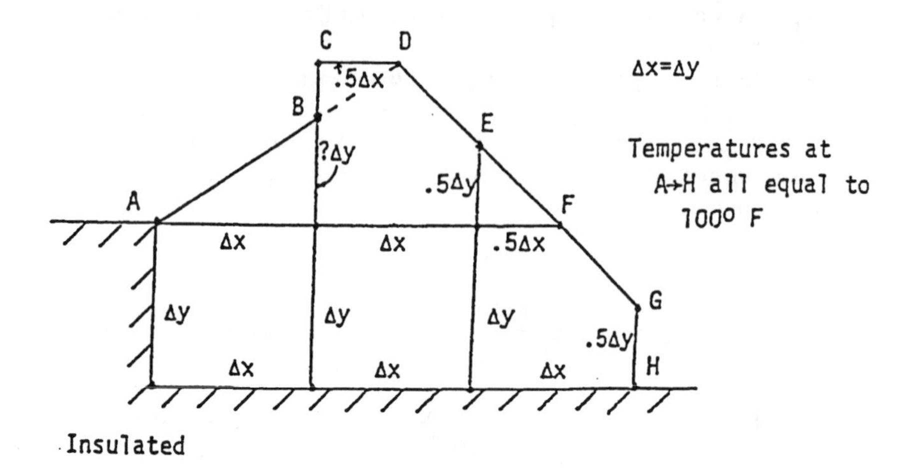

FIGURE 13. Prtoblem 7.

10. The warping of a twisted bar of uniform rectangular cross section already depicted by the mesh plot shown in Figures 5 and 6 also can be observed using the contour plotting capability of **MATLAB**. Apply program **Warping.m** and the command contour to generate a contour plot for a = 30 and b = 20 using = 100 and = 1 (Figures 14A and 14B, respectively).

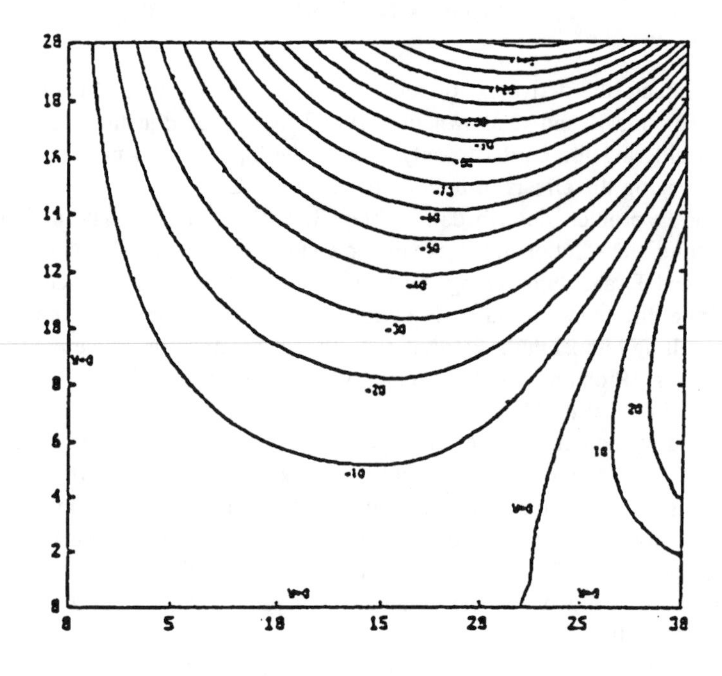

FIGURE 14A. Problem 10.

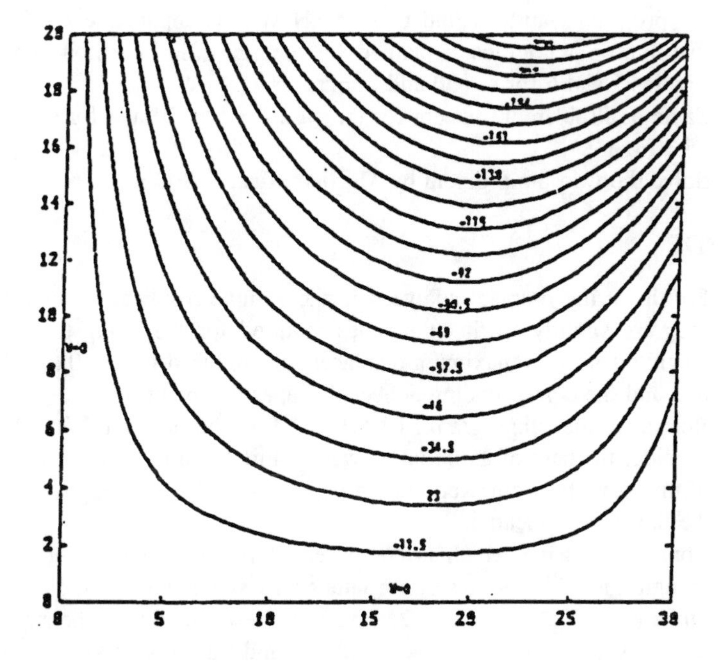

FIGURE 14B. Problem 10.

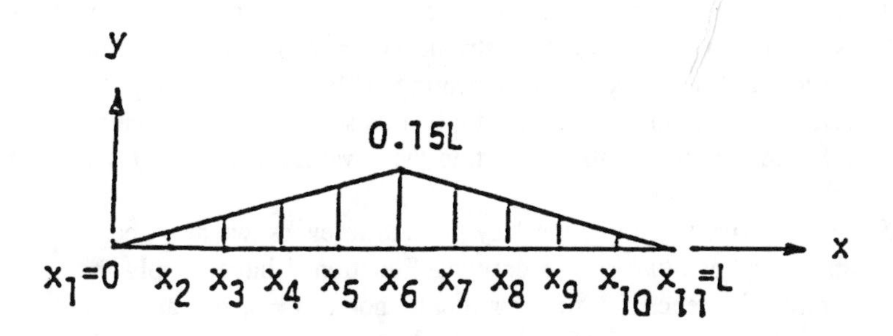

FIGURE 15. Problem 4.

11. Direct solution of the warping function W(X,Y) can also be obtained following the procedure described in Problem 8. For a rectangular cross section ($-a \le X \le a$ and $-b \le Y \le b$) of a twisted rod, the warping W(X,Y) needs to be found only for the upper right quadrant $0 \le X \le a$ and $0 \le Y \le b$ which in general can be divided into a gridwork of (M + 1)x(N + 1). The antisymmetry conditions W(X = 0,Y) = W(X,Y = 0) = 0 reduces to only (M + 1)x(N + 1)-(M + N + 1) = MxN unknowns, i.e., only solving those W's for X>0. That means we have to derive a system of MxN linear algebraic equations: along the boundaries X = a and Y = b, Equations 23 to 25 are to be used and for the interior grid points ($0<X<a$ and $0<Y<b$), Equation 3 is to be used. Generate such a matrix equation and then apply

the program Gauss to find these MxN W's. Compare the resulting W distribution with those obtained by the relaxation method shown in Figures 5 and 6 for a = 30 and b = 20.

12. Same as Problem 11 except for the case a = b = 20 and for comparing with Figure 7.
13. Solve the warping problem by **Mathematica**.

WavePDE

1. For the string problem analyzed in the Sample Application, modify the program slightly so that the times required for the string to have the magnitudes of its maximum displacements reduced to 0.8, 0.6, 0.4, and 0.2, and the corresponding deflected shapes can be printed.
2. Rearrange the subprogram FUNCTION **F** in the program **WavePDE** to consider the case of an initial, upward lifting the mid-third ($8 \leq x \leq 16$ cm) of the string by 1 cm. Rerun the program using the same input data as in the Sample Application.
3. Consider a string which is composed of two different materials even though it is subjected to a uniform tension T so that the left and right one-thirds (i.e., , $0 \leq x \leq 8$ cm and $24 \leq x \leq 32$ cm, respectively) of the string has a wave velocity a = 80 cm/sec while its mid-third (i.e., $8 \leq x \leq 16$ cm) has a wave velocity a = 90 cm/sec. Modify and then rerun program **WavePDE** using the other input same as in the Sample Application.
4. A tightened string of length L equal to 1 ft is lifted as shown in 15 and is released with a velocity distribution $v = \partial y(t = 0,x)/\partial t = 2\sin\pi x/L$ in ft/sec. If the constant T/m appearing in Equation 2 is equal to 8,100 ft^2/sec^2, use a time increment $\Delta t = 0.0005$ sec and a space increment $\Delta x = 0.1L$ and apply Equation 7 to find the y values at t = 0.001 sec and for the stations x_2 and x_3.
5. In approximating Equation 1 by finite differences, we may keep the same approach for $\partial^2 u/\partial t^2$ as in deriving Equation 7 but to apply the second central-difference formula for $\partial^2 u/\partial t^2$ not at $t = t_i$ but at $t = t_{i+1}$. The resulting equation, for $C = (a\Delta t/\Delta x)^2$, is:

$$-Cu_{i+1,j-1} + (2 + C)u_{i+1,j} - Cu_{i+1,j+1} = -u_{i-1,j} + 2u_{i,j}$$

Derive a matrix equation for solving the unknowns u_j for j = 1,2,…,N–1, at $t = t_{i+1}$. Note that the boundary conditions are $u_{i+1,0} = u_{i+1,N} = 0$. Write a program **WavePDE.G** which uses the Gaussian Elimination method to solve this matrix equation and run it for the Sample problem to compare the results.

6. Change the **MATLAB** m file **WavePDE** to solve Problem 4.
7. Apply **Mathematica** to solve Problem 4.

8.6 REFERENCES

1. C. R. Wylie, Jr., *Advanced Engineering Mathematics*, Chapter 9, Second Edition, McGraw-Hill, New York, 1960.
2. J. P. Holman, *Heat Transfer*, McGraw-Hill, New York, 1963.
3. S. Timoshenko and J. N. Goodier, *Theory of Elasticity*, Chapter 11, Second Edition, McGraw-Hill, New York, 1951.

General Index

Note: Programs and commands are indicated in boldface type.

G

H

I

J

K

L

M

N

O

P

Q

R

S

T

U

V

FORTRAN Commands and Programs

QuickBASIC Commands and Programs

MATLAB Commands and Programs

Mathematica Commands and Programs

N

O

P

Q

R

S

T

V

W